Ecological Research Monographs

Series editor
Yoh Iwasa

More information about this series at http://www.springer.com/series/8852

Makoto Kato • Atsushi Kawakita
Editors

Obligate Pollination Mutualism

Springer

Editors
Makoto Kato
Graduate School of Human and Environmental Studies
Kyoto University
Kyoto, Japan

Atsushi Kawakita
Center for Ecological Research
Kyoto University
Otsu, Shiga
Japan

ISSN 2191-0707 ISSN 2191-0715 (electronic)
Ecological Research Monographs
ISBN 978-4-431-56530-7 ISBN 978-4-431-56532-1 (eBook)
DOI 10.1007/978-4-431-56532-1

Library of Congress Control Number: 2017938280

Front cover: A female *Epicephala anthophilia* moth actively depositing pollen on the pistil of a *Glochidion acuminatum* flower in a subtropical forest of Amami Island, Japan. Note that the proboscis, which the moth uses to pollinate, is coated with numerous pollen grains that the moth collected previously on a male flower. Photo by Atsushi Kawakita.

Printed on acid-free paper

This Springer imprint is published by Springer Nature
The registered company is Springer Japan KK
The registered company address is: Chiyoda First Bldg. East, 3-8-1 Nishi-Kanda, Chiyoda-ku, Tokyo 101-0065, Japan

Preface

Organisms cannot live without interacting with other species of organisms, because each organism sits anywhere in a local food web, and because organisms utilizing similar food resource inevitably compete with each other. Although most interactions between a random pair of species are antagonistic, two intimately interacting species occasionally evolve to reduce the cost of the antagonistic interaction and occasionally become cooperative. Accordingly, we are stunned by the fact that the ecosystem is a network of innumerable mutual interactions.

Life on the earth is believed to have originated somewhere in the sea, and the land has been colonized by diverse lineages in the Paleozoic. All the land plants are descendants of the green plants that first colonized the land in the Ordovician, and the land plants at each geological era have shaped the forest ecosystems since the Carboniferous. In modern terrestrial ecosystems, angiosperms are the predominant large-size sessile primary producers and are involved in four mutualisms, i.e., pollination mutualism, seed-dispersal mutualism, mycorrhizal mutualism, and protection mutualism. The conspicuous diversification of angiosperms is hypothesized to have been driven by these mutualisms.

In contrast with terrestrial ecosystem, the medium of the marine ecosystem is sea water, where diverse microorganisms can live as plankton. Accordingly, planktonic photosynthetic microorganisms are primary producers in the ocean, whereas macrophytes are primary producers only in coastal sea. Because biomineralization especially via calcium carbonate is easy in sea water, the hard substrate in the coastal sea is often colonized heavily by diverse reef-building sessile organisms, which flourish as filter-feeders of plankton. The reef-building sessile organisms are furthermore colonized externally and internally by diverse encrusting or boring organisms, thus these symbiotic organisms form cohabitation mutualism. The exceedingly high biodiversity of coral reefs is associated with the chain of cohabitation and has been created at least partly by the complicated interactions between these symbiotic/parasitic organisms. A coral reef is also characterized by the predominance of photosynthetic mutualisms in which sessile organisms host photosynthetic microorganisms and receive their assimilates.

The origin of eukaryotic cells is endoparasitism by intracellular symbiosis of prokaryotes, which evolved into intracellular mutualism. Moreover, although the interaction between prey and predator is a typical antagonism, even the prey and the predator happened to evolve to be mutual, like crop cultivation or livestock farming by humans.

Among these diverse mutualisms, obligate mutualisms have attracted the attention of evolutionary biologists and naturalists alike because they are models for exploring the evolutionary dynamics between interacting organisms and because humans were originally involved in admiration of altruism. Excluding intracellular symbiosis, obligate mutualisms are rare, and the rare examples are the fig–fig wasp and yucca–yucca moth pollination mutualisms, both of which were discovered more than 100 years ago. In both systems, pollen-bringing adult females actively pollinate and oviposit into the ovary, and the larvae of the pollinators grow only by infesting the growing seeds. These plant–pollinator mutualisms are obligate and highly host-specific and must have accelerated reciprocal diversification of plants and pollinator insects. So far, several hundred papers on these pollination mutualisms have been published, but many hypotheses on the evolution of obligate mutualism have not been fully tested because there are only two known systems. The unanswered hypotheses are (1) obligate pollination mutualism evolved from antagonism between seed and seed-infesting insects, (2) high host specificity is reinforced as the result of coevolution, (3) obligate pollination mutualism is maintained by plant sanctions against uncooperative pollinators, and (4) obligate pollination mutualism causes reciprocal diversification of both the plants and the pollinators.

In this book, we provide the third case of obligate pollination mutualism, i.e., the leafflower–leafflower moth mutualism discovered in the plant genus *Glochidion* (Phyllanthaceae). The flowers of *Glochidion* are minute and far from showy, and their reproductive systems have not been explored until recently. The obviously rewardless inconspicuous flowers, low frequency of anthophilous insects, and the high rate of infested seeds caused us to explore the reproductive system of the plants. By rearing seed-infesting larvae, the seed-parasites proved to be gracillariid moths. However, irrespective of extensive observations of insect visitors on *Glochidion* plants, we could not detect the pollinator for several years. By sweeping insects around the inflorescence, the gracillariid moths were collected. By examining the collected female moths, the proboscises were found to be covered by pollen of the *Glochidion* flower. At last, 8 years after the start of observation, the actual pollination behavior of the moth was observed on a tree of *Glochidion acuminatum* one midnight in May 2001 in a subtropical forest in Amami-Oshima Island in the Ryukyu Archipelago, Japan.

Like figs and yuccas, the adult females of the seed-parasitic insect actively pollinate the female flowers and oviposit to the flowers. Unlike figs and yuccas, however, the plant and insect taxa involved in the mutualism are Phyllanthaceae and Gracillariidae. The mutualism of Phyllanthaceae is widespread throughout tropical regions of the world, and the number of Phyllanthaceae species involved in the obligate pollination mutualism is estimated to exceed 500 species. Since the first report of the mutualism in 2003, much data and information have accumulated, e.g., morphological adaptation in plants and insects, pollination behavior, mechanism maintaining the mutualism, phylogenetics of both the plants and insects, and the origin and diversification process of the mutualism. By exploring the third system following figs and yuccas, now we can examine the abovementioned hypotheses on the evolutionary process of obligate pollination mutualism.

Kyoto, Japan Makoto Kato
Shiga, Japan Atsushi Kawakita

Acknowledgments

We wish to thank Yoh Iwasa who suggested that we write this volume. To Eiji Kuno, Masafumi Matsui, Sadao Sakamoto, the late Tamiji Inoue, and the late Mitsuru Hotta, MK is deeply grateful for their guidance of ecology, zoology, and botany and vital encouragements. We thank Ryohei Terauchi, Toshimitsu Fukiharu, Takayoshi Nishida, and Reiichi Miura for their constructive advice and discussion on wide academic disciplines. We are grateful to Toshio Kumata for his kind guidance on taxonomy and biology of Gracillariidae. Toby Kiers came to our laboratory during the postdoctoral summer program of JSPS in 2004 and gave us several substantive help and comments on our studies since then. Glenn Svensson also visited our laboratory in 2006 and 2008 and made great contribution to our understandings on the moth's olfactory sense and the function of floral odor in *Breynia*. We express our thanks to them. During our field trips in various countries, many researchers helped our study in various ways: Yasuhiro Kosaka in Laos, Takao Itioka in Malaysia, Shi-Xiao Luo in China, Elisett in Madagascar, Akira Wong Sato and Juana Rosa Llacsahuanga Salazar in Peru, and Hervé Vandrot in New Caledonia. We thank them for their kind help in our field works. AK is deeply obliged to Issei Ohshima for his kind guidance to the methodology in lepidopteran taxonomy and for discussion on broad topics in Gracillariidae biology.

We thank Teiji Sota and Hidetoshi Nagamasu for reading and providing many comments on our manuscript. The students, postdocs, and staffs associated with our laboratories often served as a critical sounding board in seminars and field excursions. They contributed in many ways to the development of our research, and we would like to thank Takao Itioka, Shoko Sakai, Atsushi Takimura, Hiroki Hata, Shinji Sugiura, Hiromitsu Samejima, Hirokazu Toju, Chisato Kobayashi, Yudai Okuyama, Yuichi Kameda, Remi Tsubaki, Yuta Nakase, Kenji Suetsugu, Eri Yamasaki, Yoshiko Yamane, Yume Imada, Saori Furukawa, Takayuki Ohgue, Akira Wong Sato, Ko Mochizuki, Momoko Igawa, Luna Yamamori, and Sho Sakurai.

Sincere thanks are also due to John N. Thompson, J. Nathaniel Holland, Laurence Després, Anne-Marie Labouche, and Bo Song who have generously provided their original photographs to be used in this book.

Finally, we thank the contributors of this volume, Ryutaro Goto, Tomoko Okamoto, and David Hembry, who have been wonderful collaborators since the early years of our study of Phyllanthaceae. Their studies have brought about major advances in our understanding of obligate pollination mutualism.

Our research was partly supported by the Ministry of Education, Science, Sports and Culture, Grant-in-Aid for Scientific Research: (B) 2000–2002 (12440217, Makoto Kato), (B) 2003–2006 (15370012, Makoto Kato), (A) 2006–2011 (18207002, Makoto Kato), (A) 2010–2014 (25241024, Makoto Kato), (B) 2010–2014 (22247003, Makoto Kato), (A) 2015–2020 (15H02420, Makoto Kato), (B for young scientists) 2010–2011 (22770017, Atsushi Kawakita), (B for young scientists) 2012–2013 (24770018, Atsushi Kawakita), and (B) 2015–2019 (15H04421, Atsushi Kawakita).

Kyoto, Japan Makoto Kato
Shiga, Japan Atsushi Kawakita

Contents

Part I
Natural History

Chapter 1
Introduction

Makoto Kato

Keywords Forest ecosystem • Pollination mutualism • Seed-dispersal mutualism • Protection mutualism • Mycorrhizal mutualism • Biodiversity

1.1 Four Mutualisms in Forest Ecosystems

In cases of moderate temperature and sufficient rainfall, Earth's land areas are green due to coverage by forests. A forest is composed of varied plant species and provides diverse microhabitats for various animals and fungi. Each plant species harbors a specific insect fauna on its leaves and a specific fungal flora on its roots. Moreover, the biodiversity of a forest is determined by the diversity of associations in which plants participate. A rare mycoheterotrophic plant growing in a sacred place in a pristine forest, for example, is simultaneously a symbol of the plant diversity therein and of the complexity of the underground network of plant roots and mycorrhizal fungi. When two organisms intimately interact, the interaction occasionally becomes mutual through evolution even if the interaction is initially antagonistic.

Because a forest ecosystem harbors numerous interactions among various organisms in diverse microhabitats, numerous mutual interactions must occur. Currently, most forests are dominated by flowering plants. A flowering plant comprises four structures: the flower, seed/fruit, leaf/shoot, and root; these structures participate in pollination mutualism, seed-dispersal mutualism, defense mutualism, and mycorrhizal mutualism, respectively (Fig. 1.1). The present biodiversity of the terrestrial ecosystem is the highest that it has been in the last 38 million years, and the history of colonization of land by green plants and codiversification of terrestrial plants and their associates is an important topic in ecology and evolutionary biology.

M. Kato (✉)
Graduate School of Human and Environmental Studies, Kyoto University, Yoshida-Nihonmatsu-cho, Sakyo, Kyoto 606-8501, Japan
e-mail: kato@zoo.zool.kyoto-u.ac.jp

M. Kato, A. Kawakita (eds.), *Obligate Pollination Mutualism*, Ecological Research Monographs, DOI 10.1007/978-4-431-56532-1_1

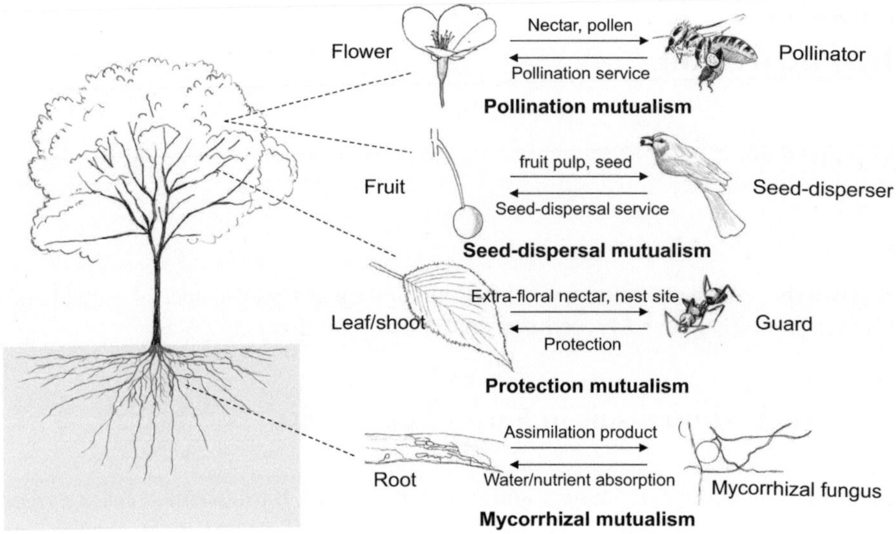

Fig. 1.1 Four mutualisms in the forest ecosystems: pollination, seed dispersal, protection, and mycorrhizal mutualism. For example, a cherry tree (*Cerasus speciosa*) is associated with insect pollinators, bird seed dispersers, ant guards, and mycorrhizal fungi by producing nectariferous flowers, fleshy fruits, extrafloral nectar-producing leaves, and assimilate-rewarding roots, respectively

1.1.1 Pollination Mutualism

Seed plants reproduce in a special manner (i.e., pollination) in which pollen is transported from the anther to the stigma of flowers by wind or migratory animals. Pollination mutualism is a mutual interaction between zoophilous flowers and anthophilous animals, in which the flower provides a floral reward to the pollinator, which in return pollinates the flower. This mutualism accelerates the evolution and diversification of both the plants and the pollinators. The Earth is currently dominated by exceedingly diverse angiosperms, which adopted zoophily at the earliest stage and have developed diverse pollination systems since the Cretaceous. One of the most tightly connected plant–pollinator interactions is obligate pollination mutualism, in which a plant species is pollinated by only one species of insect that exclusively depends on the seeds of the plant to feed its larvae. This book focuses on this mutualism, and in doing so, contributes to our understanding of the dynamics of coevolution and codiversification of intimately interacting plants and animals.

1.1.2 Seed-Dispersal Mutualism

In addition to pollination, seed plants are unique because zygotes develop into seeds through the provision of nourishment by maternal plants. Seeds are covered by a hard seed coat, and accordingly, tolerate unfavorable conditions such as desiccation and low temperature and can be dispersed to appropriate habitats by wind, water, and migratory animals. Seed-dispersal mutualism is the mutual interaction between zoochorous seed plants and seed-dispersing animals, in which the plant provides fruit pulp or a seed to the seed-dispersers as a reward, and the migratory animals provide a seed-dispersal service to the plant. By adopting this mutualism, zoochorous plants acquired the ability to colonize directionally across considerable distances.

1.1.3 Protection Mutualism

Because plants are immobile and rich in products of photosynthesis, they have suffered severe exploitation by herbivorous animals and pathogenic microorganisms. To avoid herbivory and parasitism, plants have developed physical and chemical antiherbivore/antipathogen defenses. In contrast to animals and fungi, plants have developed chemical defense mechanisms because plant cells have vacuoles in which mutation-derived poisonous secondary metabolites can be sequestered, plants have sufficient energy reserves for chemical reactions due to photosynthesis, and poisonous secondary metabolites play an active role in defense against herbivores or pathogens (Wink 1997). In addition, some plant species employ "security guards" to protect them against herbivores and pathogens. For example, a plant provides extrafloral nectar and/or nest sites to ants nesting nearby, and the ants patrol and attack herbivores. In contrast, fungal endophytes infecting the leaves/shoots contribute to defense against virulent pathogens. These associations between plants and antagonists of herbivores/pathogens represent protection mutualisms.

1.1.4 Mycorrhizal Mutualism

The plant root is the underground structure that mechanically supports the aboveground plant body and absorbs water and nutrients from the soil via root hairs. The roots of terrestrial plants, however, sometimes lose root hairs, and can become infected by mycorrhizal fungi. In the mycorrhizal association, the plant provides assimilates to the fungi, which in return provide nutrients and defense against pathogens to the plant. The mycorrhizal mutualism between plants and mycorrhizal fungi occurs in diverse plant taxa and in almost all of the climatic regions. In the rhizosphere, diverse fungi and microorganisms are involved in complex interactions with plant roots. Thus, the forest ecosystem is a complex web of intricate mutual and antagonistic interactions between plants and their animal/fungal associates, in the presence of which the diversification of terrestrial plants and animals must have occurred. The primary purpose of this book is to review the natural history and evolutionary dynamics of a recently discovered obligate pollination mutualism in a complex forest ecosystem. Before focusing on the obligate pollination mutualism, we review the history and life history of the terrestrial plants and their associates, with a focus on the above-mentioned four mutualisms.

Chapter 2
History and Natural History of Plants and Their Associates

Makoto Kato

Keywords Terrestrialization • Tracheophyte • Mycorrhiza • Origin of pollination • Angiosperm diversification • Bee • Seed dispersal • Bird • Mammal • Cultivation mutualism

2.1 Origin of Land Plants

Life on Earth originated in the sea; thus, land is a frontier for aquatic organisms. Although colonization of land by plants occurred in the Ordovician about 450 million years ago (ma; Field et al. 2015), aquatic microorganisms colonized land as early as 3500 mya (Beraldi-Campesi 2013). When aquatic microorganisms (including cyanobacteria) headed to land, they encountered adverse terrestrial conditions such as drought, extreme diurnal and seasonal temperature changes, low nutrient supply, and strong sunlight. These microorganisms are considered to have achieved terrestrialization by developing a tough, pigmented cell wall, an agglutinated colony structure, and a symbiotic lifestyle. Although modern lichens are associations of internal algae with external advanced fungi belonging to Ascomycetes and Basidiomycetes, colonylike associations of microorganisms, including basal algae and fungi, are thought to have colonized land before the Phanerozoic. Colonization of terrestrial habitats by photosynthesizing multicellular organisms occurred in a clade of green plants containing chlorophyll a and b (Delaux et al. 2012). The order Charales in the division Charophyte adapted to life in ephemeral freshwater pools, which frequently dry up. The body of the Charales is a monoploid gametophyte and comprises a main axis and lateral branchlets occurring in whorls on each node of the axis (Fig. 2.1). The reproductive organs consist of the nucleus (i.e., archegonia-producing ova) and globules (i.e., antheridia-producing flagellate sperm; Fig. 2.2). The sperm swim to the archegonium and fertilize the ovum. The zygote develops into an oospore, which is protected by its tough cell wall containing an unusually stable polymer, sporopollenin. The oospore can resist

M. Kato (✉)
Graduate School of Human and Environmental Studies, Kyoto University, Yoshida-Nihonmatsu-cho, Sakyo, Kyoto, 606-8501, Japan
e-mail: kato@zoo.zool.kyoto-u.ac.jp

M. Kato, A. Kawakita (eds.), *Obligate Pollination Mutualism*, Ecological Research Monographs, DOI 10.1007/978-4-431-56532-1_2

Fig. 2.1 Two derived orders of Charophyta: Charales (**a**–**b**) and Coleochaetales (**c**–**d**). (**a**) *Chara australis*; (**b**) *Nitella flexilis*; (**c**, **d**) *Coleochaete* sp. growing on reed shoots in a pond near Biwa Lake, Japan

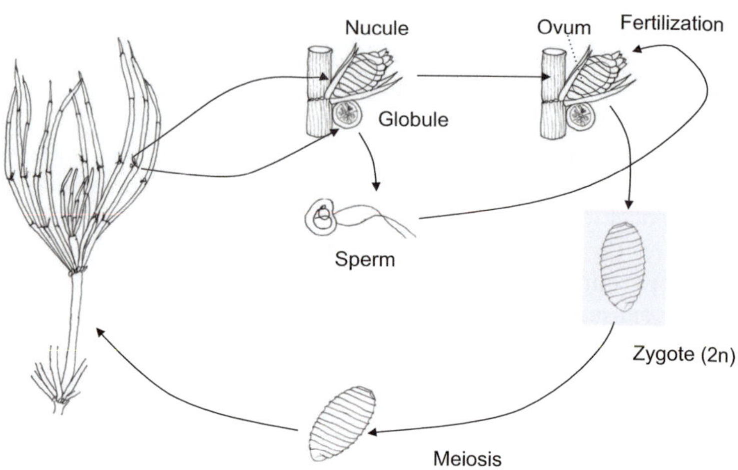

Fig. 2.2 Life cycle of a charophyete, *Chara braunii*

desiccation by remaining in a dormant state, and undergoes meiosis before germination of protonema. Another charophyte order, Coleochaetales, is a parenchymatous disclike alga that grows on substrata in waterfront habitats (Fig. 2.1). In the *Coleochaete*, plural meiospores are produced from a zygote, suggesting that embryos are protected and nourished in maternal tissue (Graham and Wilcox 2000).

The earliest plant that accomplished terrestrialization was the bryophyte (Fig. 2.3), which has acquired a thin extracellular waxy cuticle. Bryophytes possess a precursor of lignan, but lack actual lignin (Weng and Chapple 2010). Sexual reproduction of bryophytes only occurs in wet habitats on rainy days because flagellate sperm released from the antheridium must reach the archegonia by swimming within the water film (Fig. 2.4). The most important adaptation to terrestrial life in bryophytes is protection of zygotes from desiccation and shortage of assimilates. For this purpose, zygotes are protected in maternal gametophyte tissue as embryos, and growth of the sporophyte embryo is facilitated by nutrients provided by the host gametophyte. The parasitic sporophytes finally bear spores, which are coated with a sporopollenin-coated cell wall and are generally dispersed by the wind. Biotic spore dispersal, however, occurs in arctic mosses of the family Splachnaceae. These mosses grow on the dung of herbivorous mammals, and the colored sporophytes emit stinking volatiles to attract flies, which disperse the spores to other dung deposits (Koponen 1990). Bryophytes comprise three distinct clades: Hepaticae (liverwort), Musci (moss), and Anthocerotophyta (hornwort; Fig. 2.5); however, the phylogenetic relationships among these groups are unclear (Cox et al. 2014). Liverworts are thalloid or foliose, mosses are foliose, and hornworts are thalloid. In contrast to mosses and hornworts, liverworts have intracellular oil bodies, which contain various terpenoids and lipophilic aromatic compounds and might contribute to antiherbivore defense. Similar to vascular plants, sporophytes of mosses and hornworts have stomates, which cannot be closed.

2.2 Plants Standing Up from the Land

In the Silurian, a lineage of plants stood up from the land by acquiring upright or clambering shoots with vascular bundles and ramified roots. The former contributed to efficient transportation of water within the plant, and the latter contributed to anchoring and supporting the plant and absorbing water and nutrients from the soil. These vascular land plants are known as tracheophytes, and comprise three extant monophyletic groups: lycophytes (division Lycophyta), ferns (division Polypodiopsida including Sphenophyta), and seed plants (division Spermatophyta; Fig. 2.6). The phylogeny of the three groups is Lycophyta, Polypodiopsida, and Spermatophyta, whereas lycophytes and ferns have been treated as pteridophytes. Vascular bundles are complex structures comprising two organic monomers, cellulose and lignin. Cellulose is a linear chain of β-linked D-glucose, and lignin is synthesized via lignan from phenylpropanoid, which is synthesized from the amino

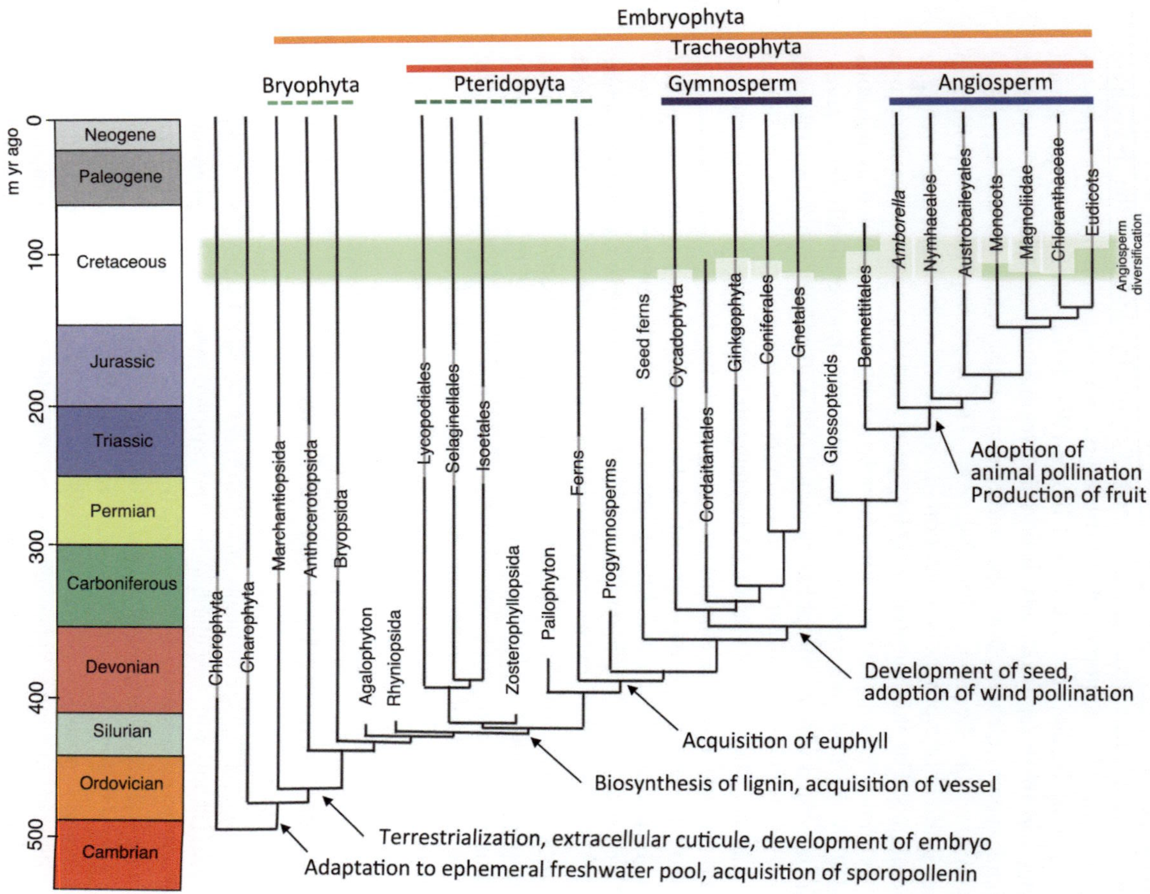

Fig. 2.3 History of diversification of the land plants. The groups shown in broken lines are not monophyletic. Phylogeny and divergence times are based on Kenrick and Crane (1997), McCourt et al. (2004), and Zeng et al. (2014)

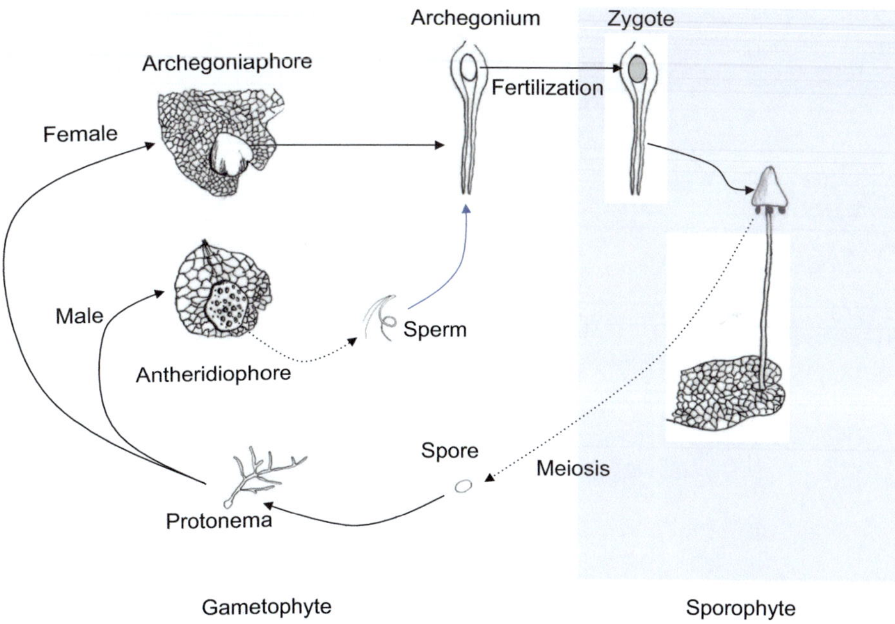

Fig. 2.4 Life cycle of a bryophyte, *Conocephalum conicum*. In bryophytes, small sporophytes parasitize gametophytes. Broken lines show airborne transportation and a *blue* line shows migration in water film

acid phenylalanine. Acquisition of tough vascular bundles enabled tracheophytes to rise up from the land and overpower bryophytes through competition for light. Cellulose and lignin accumulated in the lateral meristem in shoots and formed the cambium, which contributed to secondary growth of shoots and formation of woody shoots and trunks. In pteridophytes, gametophytes (i.e., prothallium) in a monoploid generation and sporophytes in a diploid generation grow separately and independently, although gametophytes are small and lack vascular bundles (Fig. 2.7). Gametophytes are dioecious or monoecious, and have an antheridium and archegonium. As in bryophytes, flagellate sperm released from the antheridium must reach archegonia by swimming within a water film, suggesting that pteridophytes cannot live outside damp habitats. Fertilization occurs in the archegonia, and growth of the zygotes is nourished by the gametophytes. The diploid sporophytes bear spores in the sporangium on fronds (Fig. 2.6). Spores of modern ferns are frequently infested by various insects such as stathmopodid and tineid moths and mirid bugs (Sawamura et al. 2009). Spores released from sporangia are dispersed by the wind, and germinate to form prothallia. Spore dispersal mutualism has not been reported in pteridophytes. The first shrublike forest on Earth formed in damp habitats during the Devonian. The woody plants became greater in height and in trunk diameter, and forests in the Carboniferous were composed of large pteridophyte trees (e.g., *Lepidodendron*, *Sigillaria* [Lycophyta], and *Calamites* [Pteridophyta]). Although arboreal lycophytes became extinct during the Paleozoic,

Fig. 2.5 Three clades of bryophytes: liverworts (**a–c**), moss (**d–e**) and hornworts (**f**). (**a–c**) *Conocephalum conicum*. (**d**) *Polytrichum commune*. (**e**) *Pogonatum minflexum*. (**f**) *Megaceros flagellaris*. On gametophytes, antheridia (**a**), archegonia (**b**, **d**), and sporophytes (**c**, **e**, **f**) are observed

herbaceous lycophytes have survived to the present (Fig. 2.6). The woody trunks of these Paleozoic arboreal plants were embedded underground and turned to coal without decomposing, in part because these forests were formed in swamps where submerged woody deposits tend not to decompose, and because efficient wood-decomposing fungi had not emerged.

2.3 Formation of Mycorrhiza

The roots of tracheophytes are multicellular organs ramifying in the soil. Each root is covered by monocellular root hairs, which mediate absorption of water and nutrients. The root hairs are morphologically and ecologically similar to rhizoids of bryophytes. The absorption function of tracheophyte roots, however, is often mediated by symbiotic mycorrhizal fungi. Cooperative associations between algae and fungi have been widespread since the early stages of terrestrialization (Field et al. 2015) because they could compensate for their functional defects through symbiosis (i.e., algal photosynthesis and fungal absorption of water and nutrients).

Fig. 2.6 Growing habits of Lycopodiophyta (**a–c**) and Pteridophyta (**d–f**): Isoetopsida (**a**) Lycopsida (**b–c**) Sphenopsida (**d**) and Polypodiopsida (**e–f**). (**a**) *Isoetes japonica*. (**b**) *Lycopodium serratum* the leaf of which is mined by a pallopterid fly larva. (**c**) *Lycopodium annotinum* with strobili. (**d**) *Equisetum arvense* with strobili. (**e**) Sori of *Stegnogramma pozoi* with sori, which are infested by stathmopodid moth larvae (shown by *arrows*). (**f**) Prothallia of *Stegnogramma pozoi*

The origin of mycorrhizal symbiosis may be traced back to the Precambrian symbiosis between algae and fungi. The most widespread mycorrhizal symbiosis in the present era is arbuscular mycorrhizal symbiosis. In this symbiosis, the minute hyphae of the basal fungi of the phylum Glomeromycota invade a root cell of land plants and form a highly ramified structure (i.e., an arbuscule; Fig. 2.8), where exchange of assimilates and nutrients between the plant cell and the fungi occurs. Although there is the potential for conflicts of interest between the plants and their mycorrhizal fungi, mycorrhizal mutualism is maintained by the sanction of the plant and fungus against noncooperative behavior by their partners (Kiers et al. 2011). In arbuscular mycorrhiza, the hyphae extending from host roots produce minute underground sporangia, which release spores in the soil. Mycorrhizal symbiosis is believed to be ancient (Simon et al. 1993) because Glomeromycota

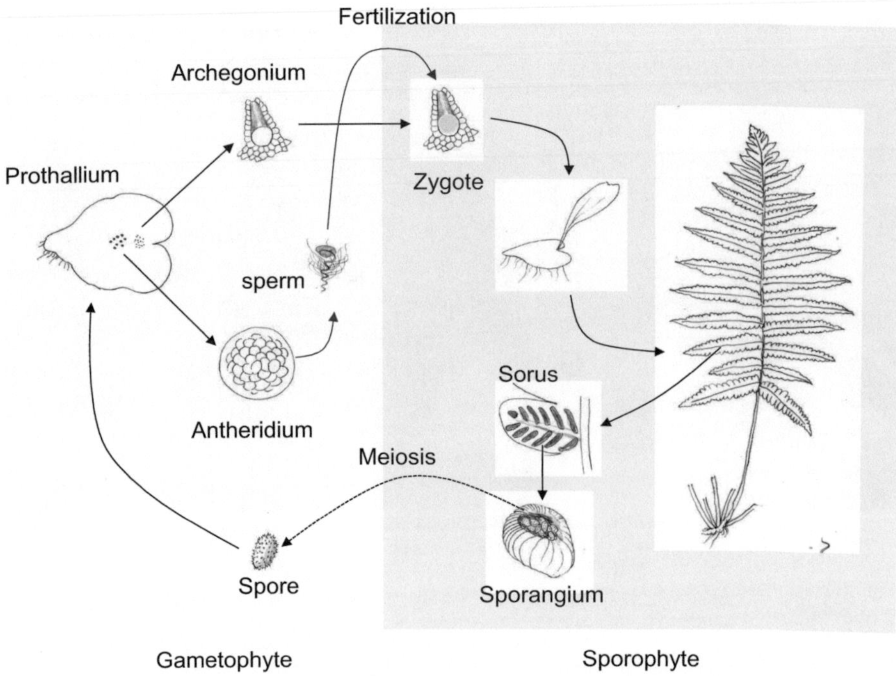

Fig. 2.7 Life cycle of a pteridophyte, *Stegnogramma pozoi*. In pteridophytes, gametophytes and sporophytes grow separately and independently. A broken line shows airborne transportation and *blue* lines show migration in water film

is a basal lineage in the fungal phylogenetic tree, because almost all land plants (including bryophytes) engage in the symbiosis, and because fossilized hyphae of Glomeromycota are found from the Ordovician when only bryophytes grew on land (Redecker et al. 2000). Symbiosis with Glomeromycota is known even in liverworts, which do not have roots. In basal thalloid liverworts, hyphae of the mycorrhizal fungi invade cells of the thallus, and enhance plant fitness by promoting phosphorus and nitrogen uptake from soil (Humphreys et al. 2010). These facts suggest that the initial colonization of land by liverworts was assisted by these arbuscular mycorrhizal fungi. Liverworts and hornworts are symbiotic with arbuscular mycorrhizal fungi of the Glomeromycota and Mucoromycotina, whereas most mosses with multicellular rhizoids are free from mycorrhizal fungi (Bidartondo et al. 2011; Desirò et al. 2013). The sporophytes and gametophytes of Pteridophytes are symbiotic with arbuscular mycorrhizal fungi (Zubek et al. 2010; Ogura-Tsujita et al. 2016). Furthermore, the roots of almost all seed plants, with the exception of several plant families (e.g., Brassicaceae) and aquatic plants, are also symbiotic with Glomeromycota (Table 2.1; Heijden et al. 2015).

The most diverse fungal lineage is the subkingdom Dikarya, which is well-adapted to terrestrial life by losing flagella, acquiring tolerance to desiccation, and developing aerial spore dispersal (James et al. 2006). Dikarya are characterized by a

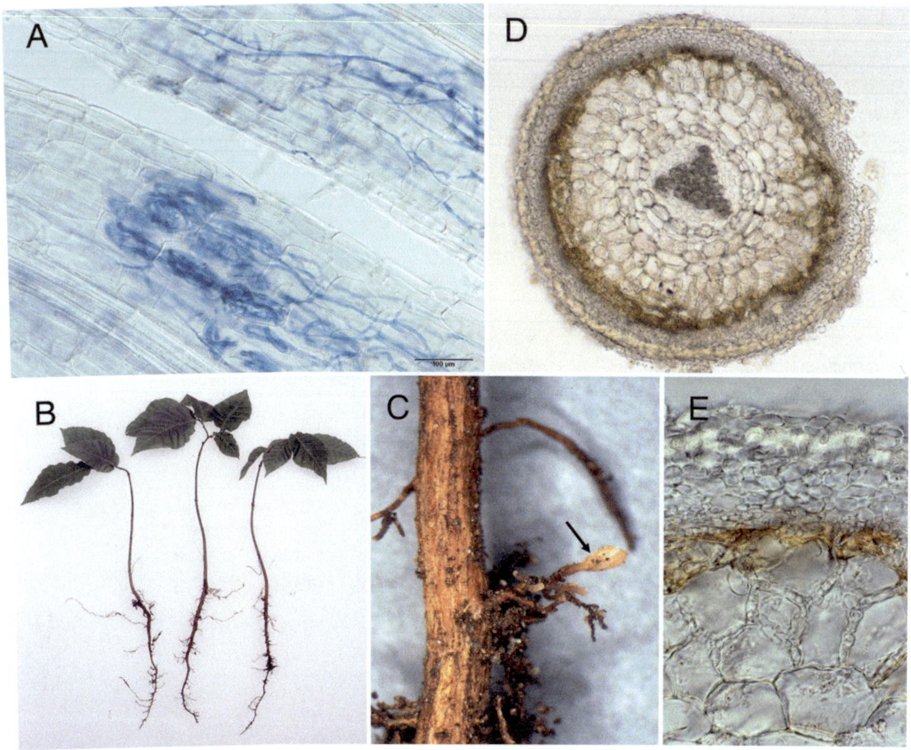

Fig. 2.8 Arbuscular mycorrhiza of *Dioscorea japonica* (**a**) and ectomycorrhiza of *Fagus crenata* (**b–e**). (**a**) Roots invaded by hyphae of Glomeromycotal fungi. (**b**) Seedlings. (**c**) A root system with nonmycorrhizal and mycorrhizal (shown by an *arrow*) roots. (**d**) A cross-section of the mycorrhizal root. (**e**) External fungal layer and fungal hyphae invading interspace of root cells

dikaryon (i.e., each cell contains two unfused nuclei) and comprises two phyla, Ascomycota and Basidiomycota. Dikarya grow in a filamentous (as hyphae) or unicellular (as yeasts) manner and have diverse habits including consumption of organic materials and fluids, decomposition of plant litter or wood, and parasitization of various organisms and mycorrhizal symbionts. Dikarya typically form ectomycorrhiza, in which a root is thickly enveloped by a hyphal sheath and its highly branched hyphae penetrate deep into the intercellular space of inner root cells but not into the root cells (Fig. 2.8). Ectomycorrhiza are formed on the roots of a few plant families (e.g., Pinaceae, Fagaceae, Betulaceae, Salicaceae, Myrtaceae, and Dipterocarpaceae). Ectomycorrhizal fungi produce large fruit bodies (i.e., mushrooms) from the ground in forests comprising pines, larches, firs, hemlocks, oaks, beeches, birches, poplars, myrtles, and dipterocarps. The diversity and abundance of mushrooms of these mycorrhizal fungi in these forests contribute to the diversity and abundance of mycophagous insects and fungivorous vertebrates. Photosynthetic organisms relinquish their surplus carbohydrate assimilates in various ways. Aquatic algae secrete surplus polysaccharides on the surface of their

Table 2.1 Mycorrhizal types

Mycorrhizal type	Fungi	Plants	Morphology of Mycorrhiza	Host specificity
Arbuscular mycorrhiza	Mucoromycotina	Basal liverworts	Hypha invades thallus cells and occupies mucilage filled space	Low
Arbuscular mycorrhiza	Glomeromycota	Liverworts and hornworts	Hypha invades thallus cells and forms arbuscules in the cells	Low
Arbuscular mycorrhiza	Glomeromycota	Vascular plants except aquatic plants, Brassicaceae, Crassulaceae, Orobanchaceae, and Proteaceae	Hypha invades plant's root cell and forms arbuscules in the cells	low
Ectomycorrhiza	Basidiomycota (Agaricales, Boletales, Russulales) and Ascomycota	Pinaceae, Fagaceae, Betulaceae, Salicaceae, Dipterocarpaceae; a part of Myrtaceae, Rosaceae, and Fabaceae	Hypha surrounds roots and penetrates in interspace of root cells	High
Ericoid mycorrhiza	Ascomycota (Helotiales) and some Basidiomycota (Sebacinales)	Ericaceae	Hypha surrounding fine hair root invades epidermal cells and forms intracellular coil	High?
Orchid mycorrhiza	Basidiomycota (*Ceratobasidium*, *Sebacina*, *Tulasnella* and *Russula*)	Orchids	Hypha invades root cells and forms coils (peloton), which are digested by the orchid cell	High

Heijden et al. (2015)

algal bodies to prevent fouling by organisms such as macrophytes, and to deter predation by zooplankton. In contrast, land plants provide large quantities of assimilates to symbiotic mycorrhizal fungi, which in turn contribute to translocation of water and inorganic nutrients from soil to plant roots and to protection of roots against soil pathogenic fungi and bacteria. Moreover, the mycorrhizal system is parasitized by diverse mycoheterotrophic plants such as Orchidaceae, Triuridaceae, and Burmanniaceae (Bidartondo 2005). The intricate underground network of mutual and antagonistic plants and fungi makes the interactions and dynamics complex. In addition to roots, the internal tissues of aerial parts of plants are also inhabited by diverse fungi, bacteria, and other microorganisms (Table 2.1; Hardoim et al. 2015). These microbial endophytes are either pathogenic or beneficial to plants, either obligate or opportunistic, and are transmitted either vertically or horizontally. The most common obligate mutual endophytic fungi are ascomycetes of the genera *Balansia*, *Epichloe*, and *Neotyphodium* (Clavicipitaceae), which promote plant growth and protect the host plant against biotic and abiotic stresses by producing antibiotics or other secondary metabolites, stimulating plant growth, or activating the expression of particular host genes. Plant roots are also inhabited by unique endophytes, which have dark septate hyphae (Mandyam and Jumpponen 2005). Infection by dark septate endophytes (DSEs) can have beneficial or pathogenic effects on plants depending upon the environmental factors and interactions with root mycorrhizal fungi. Because mutualistic endophytes may be pathogenic, interactions between plants and endophytes must be evolutionarily dynamic. It is beneficial to a plant to host an endophyte with low pathogenicity at a cost, because the endophyte may protect against invasion by more virulent pathogens. Thus, the high infection rate and high diversity of obligate mutualistic endophytes in plants suggest the prevalence of protection mutualism involving plant pathogens.

2.4 Appearance of Insects

Following the terrestrialization of green plants, arthropods also colonized land. The subphylum Hexapoda of the phylum Arthropoda is a genuinely terrestrial clade, but its phylogenetic linkages are unknown. Recent molecular phylogenetic analysis suggested that the sister group of Hexapoda is Remipedia (Regier et al. 2010), which is a group of small swimming crustaceans with numerous pairs of swimming appendages that live in brackish groundwater in anchialine caves. Hexapoda and Remipedia are morphologically distinct: Remipadea have up to 42 pairs of swimming appendages, in contrast to three pairs of legs in Hexapoda. However, an autosynapomorphy between Hexapoda and Remipedia has been found (i.e., the presence of a mandibular venom gland; Regier et al. 2010), which suggests that the route of terrestrialization of the hexapod class Entognatha (the most basal wingless hexapods, exemplified by springtails) was through coastal aquifers. Having colonized land, a Hexapoda lineage developed an external mouthpart structure in the Silurian, which is the origin of insects (i.e., the class Insecta; Fig. 2.9). The most important innovation of insects was the acquisition of wings. Wings are thought to have originated in aquatic insects because the most basal winged insect is Ephemeroptera, whose larvae live in aquatic habitats, and because the morphological and developmental characteristics of ephemeropteran gills resemble those of wings. Insects with wings (i.e., Pterygota) appeared in the Devonian and underwent adaptive radiation in diverse niches, such as herbivores, grazers, fungivores, detritivores, predators, scavengers, parasitoids, and parasites, in aquatic and terrestrial ecosystems (Fig. 2.9). The great diversity of insects is considered to have resulted from adoption of herbivory, because the five orders that feed on seed plants (i.e., Hemiptera, Coleoptera, Diptera, Hymenoptera, and Lepidoptera) achieved megadiversity. The great diversity of herbivorous insects is closely related to their high host specificity (Ehrlich and Raven 1964).

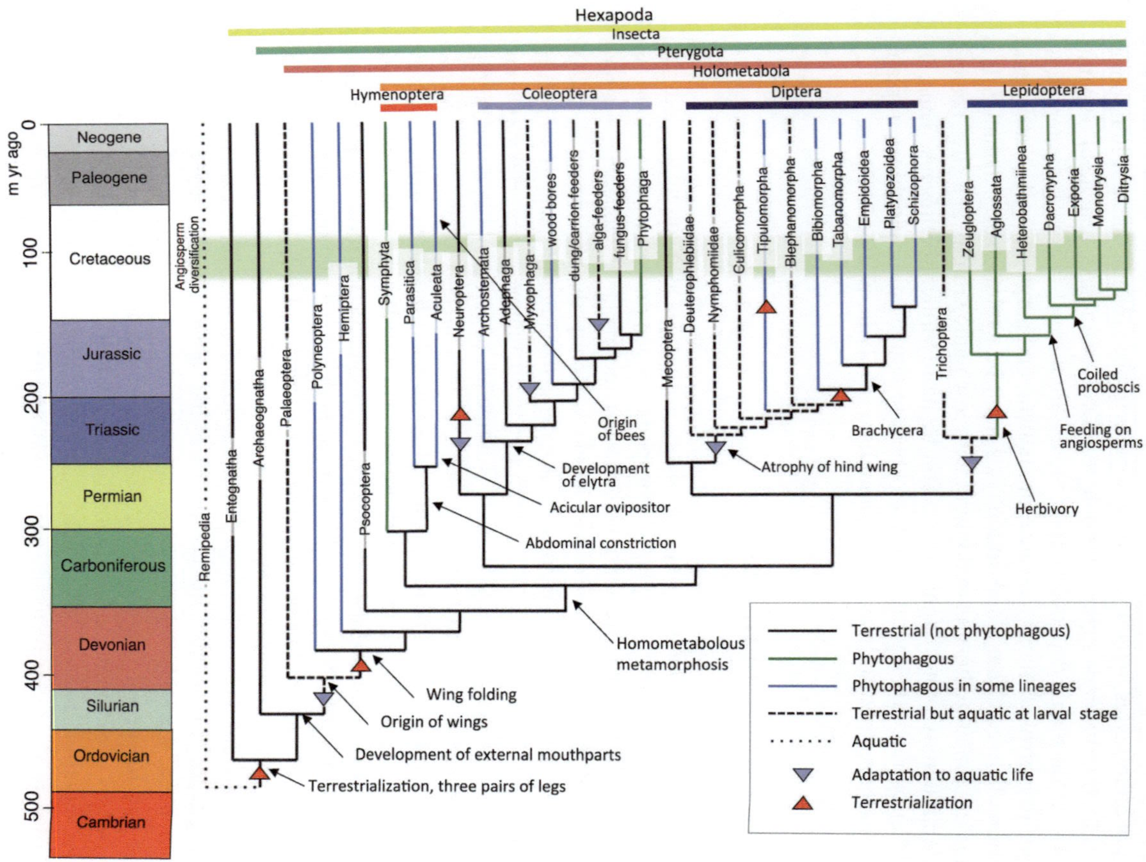

Fig. 2.9 History of diversification of insects focusing on crossing the border between terrestrial and aquatic ecosystems and adoption of herbivorous life. Dominant pollinators evolved from the herbivorous lineages. Phylogeny and divergence times are based on Wiegmann et al. (2011), Imada et al. (2011), Ronquist et al. (2012), Misof et al. (2014), and Regier et al. (2015)

In general, plants defend against herbivore predation by producing toxic secondary metabolites, and herbivorous insects have evolved mechanisms to detoxify or tolerate these agents. Accordingly, coevolutionary arms races between plants and herbivorous insects have reinforced the host specificity of herbivorous insects. Herbivory originated at least several times in various insect orders from algal grazers, detritus feeders, and fungal feeders, whereas the origins of herbivory are unclear. Bryophytes are infested by limited groups of insects, such as bugs (Peroidae and Tingidae), crane flies (Cylindrotomidae), flies (Rhagionidae and Agromyzidae), and moths (e.g., Micropterigidae, Pyralidae). Ferns are infested by various insect herbivores, including leaf miners (agromyzid and anthomyiid flies, stathmopodid moths, and chrysomelid beetles) and defoliators (diverse lepidopterans, hemipterans, and sawflies). The diversity of fern-feeding insects is considerably lower than that of angiosperm-feeding insects, likely due to the low morphological complexity of ferns (Hendrix 1980). Herbivores of lyophytes are only known in lepidopterans and pallopterid flies (Fig. 2.10; Kato 2002). Seed plants (particularly angiosperms) are infested by exceedingly diverse lineages of the above-mentioned five orders of insects, the diversity of which accounts for one-quarter of global biodiversity.

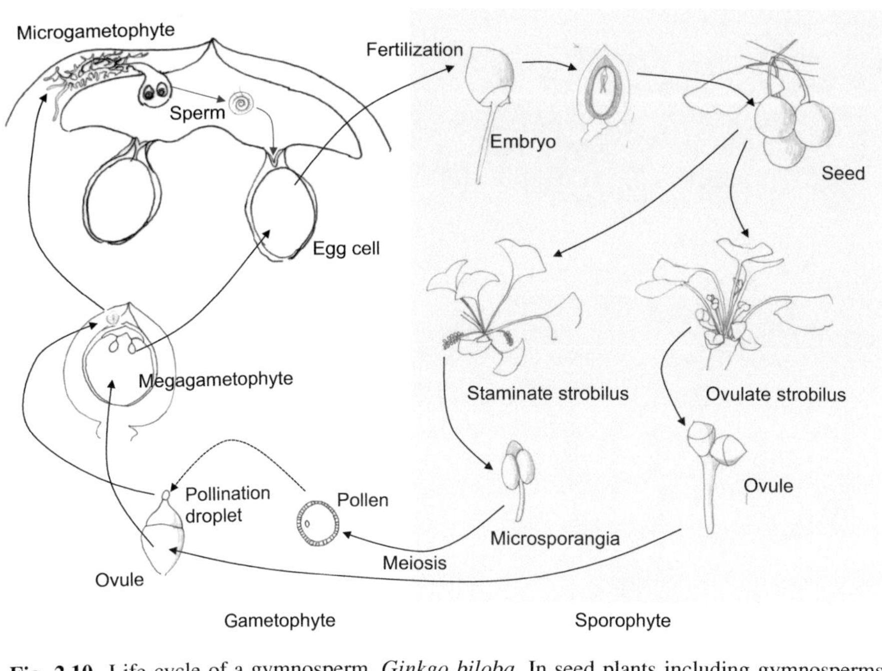

Fig. 2.10 Life cycle of a gymnosperm, *Ginkgo biloba*. In seed plants including gymnosperms, male and female gametophytes are parasitic to sporophytes. In *Ginkgo*, male gametophytes growing in an ovule produce sperm, which fertilize egg cells of female gametophytes. A broken line shows airborne transportation, and *blue* lines show migration in fluid

2.5 Origin of Pollination and Invention of Seeds

Evolution from pteridophytes to seed plants occurred through radical morphological and life history modification. In the first step, spores of a tracheophyte lineage become dimorphic (i.e., megaspores and microspores). Next, the megaspores and microspores germinate without being released and, respectively, become megagametophytes and microgametophytes on maternal plants. The megagametophyte is parasitic to maternal sporophytes (the ovule harbors megagametophytes on megasprorophylls) and produces archegonia. Using nutrients supplied by sporophytes, the archegonia produce egg cells that are fertilized and grow as seeds on the sporophyte. The microgametophytes become enclosed in a tough coat with a sporopollenin shell (pollen harbors microgametophytes on microsporophylls) and then are released and transported to megagametophytes by the wind (Fig. 2.11). Thus, in contrast to free-sporing pteridophytes, in seed plants, the whole microgametophyte rather than sperm moves to the archegonia by the wind rather than by swimming in a water film. This new process of pollen transfer to ovules is pollination.

Fig. 2.11 Extant conifers. (**a**) *Araucaria humboldtensis* (Araucariaceae). (**b**, **c**) *Parasitaxus usta*, parasitic to root of another conifer, *Falcatifolium taxoides* (both host and parasite are Podocarpaceae). (**d**) *Sciadopitys verticillata* (a monotypic family Sciadopityaceae endemic to Japan). (**e**) *Chamaecyparis obtuse* (Cupressaceae) with male cones. (**f**) *Cephalotaxus harringtonia* with female flowers secreting pollination droplets from ovules. (**g**) *Pinus thunbergii* (Pinaceae) with male and female cones. (**a**–**c**) Endemic to New Caledonia; (**d**–**f**) endemic to Japan

The successfully transported microgametophyte produces antheridia. In ginkgo, the antheridia release flagellate sperms in the ovule and the sperm fertilizes an egg in an archegonium. In conifers, however, the successfully transported microgametophyte produces an unflagellated sperm cell, which is guided to the archegonia via a pollen tube. The zygote is nourished by the maternal sporophyte, develops within ovules, and matures as a seed. This modification of life history liberated the seed plants from dependence on aquatic habitats for reproduction. Based on the protective structure of ovules, seed plants are classified into gymnosperms and angiosperms, and the gymnosperms are paraphyletic. The extant members of gymnosperms are the cycads (Cycadophyta), ginkgo (Ginkgophyta), conifers (Pinophyta), and gnetophytes (Gnetophyta; Fig. 2.11). The oldest gymnosperm fossil is from the Carboniferous, and there are many extinct gymnosperm groups such as Bennetiales, Caytoniales, and Glossopteridales.

In most gymnosperms, pollen is dispersed by wind and caught by pollination droplets secreted from ovules. However, some cycads and gnetophytes are pollinated by insects. In *Gnetum gnemon* var. *tenerum* in tropical regions of Southeast Asia, both the megasporophyll and microsporophyll emit an unpleasant odor after sunset, secrete pollination droplets from ovules of the megasporophyll and residual ovules of the microsporophyll, and are visited by nocturnal nectar-seeking moths (Fig. 2.12). The pollination droplet contains sugar and is harvested by moths as a reward for

Fig. 2.12 Entomophily of a gnetophyte, *Gnetum gnemon tenellum*. (**a**) Male strobili visited by a pyralid moth. (**b**) Female strobili visited by a pyralid moth. (**c**, **d**) Male and female strobili secreting pollination droplets

pollination (Kato and Inoue 1994). The insect pollination system of this *Gnetum* is different from that of zoophilous angiosperms in that pollinators are attracted by an odor emitted from the sporophyll and the pollen is received by pollination droplets secreted from ovules. Recent fossil evidence indicates that insect feeding on pollen, pollination droplets, and reproductive tissues of extinct gymnosperm lineages date back to the Permian (Labandeira 2010). The Mesozoic anthophilous insects comprise mandibulate insects (e.g., Coleoptera and Hymenoptera), which feed on pollen and pollination drops, and long proboscid insects (e.g., Neuroptera, Mecoptera, and Diptera), which utilize fluid secreted from plants such as pollination droplets (Ren et al. 2009). Whereas some of these insects are considered to contribute to pollination of extinct gymnosperms, these associations were extinguished by the late Cretaceous and were succeeded by new types of associations with angiosperms.

In addition to adoption of pollination, seed plants developed a novel system to disperse their offspring. A seed contains an embryo and is covered by the seed coat, which consists of the inner tegmen and outer testa. Thus, the embryo is protected from environmental stresses such as desiccation and can be dispersed from mother plants by wind or other agents. Although the seeds of gymnosperms are naked, in some gymnosperm genera, such as *Ephedra* (Gnetales) and *Taxus* (Coniferes), inner cone scales are modified to enclose the seed and become red and fleshy (Fig. 2.13). Consequently, the cone has an appearance similar to a red fruit and is

Fig. 2.13 Reproductive organs of gymnosperms. (**a–c**) *Cycas revoluta*. (**d**) *Torreya nucifera*. (**e**) *Taxus cuspidata*. (**f**) *Gnetum cuspidatum*. (**g**) *Ephedra gerardiana*. (**a**, **d–g**) seeds. (**b–c**) megasporophyll mined by a cerambycid beetle larva. In the seeds of *Taxus* and *Ephedra*, a red fleshy aril is developed

dispersed by birds. All of the extant gymnosperms are woody plants, including an archaic tree of the monotypic genus *Sciadopitys* (Sciadopityaceae) that is found only in Japan, and a plant-parasitic achlorophyllous treelet, *Parasitaxus usta* (Podocarpaceae), which is endemic to New Caledonia (Fig. 2.11). Most gymnosperms have arbuscular mycorrhiza symbiotic with fungi of Glomerales. In contrast, the conifer family Pinaceae is unique in being symbiotic with ectomycorrhizal fungi of the basidiomycetes. Gymnosperms are infested by various insect herbivores such as leaf miners (gelechid, gracillariid, coleophorid, and tortricid lepidopterans), defoliators (various families of lepidopterans, and sawflies), and wood bores (siricid sawflies, buprestid, cerambycid, curculionid, scolytid, and platypodid beetles). Because gymnosperms flourished in the Mesozoic, relict plant–herbivore associations have remained; seeds of a basal conifer *Agathis* (Araucaliaceae) are infested by a basal lepidopteran, *Agathiphaga* (Agathiphagidae), buds of *Abies* fir are bored by xyelid sawflies, microsporophylls of pines are infested by xyelid sawflies, and the rotten wood of pines is infested by basal cupedidid beetles (suborder Archostemata).

2.6 Invention of Flowers and Fruits

Megagametophytes and microgametophytes of a lineage of gymnosperms became reduced and protected in multilayered, delicately enfolded structures of the megasporophyll and microsporophyll. The structure harboring the compactly assembled reproductive organs is called the flower, and was invented by the angiosperm clade (Fig. 2.14). The megasporophyll and the microsporophyll in gymnosperms, respectively, became the pistil and stamen in angiosperms. The pistils and stamens are subtended by tepals. When the inner and outer tepals differentiate, they are termed petals and sepals, respectively. The flower structure is considered to have evolved to protect gametophytes against biotic and abiotic environmental stresses (e.g., attack by herbivores and pathogens, rain, and desiccation), and so that tepals could contribute to attracting pollinators. The pistil comprises the basal ovary, intermediate style, and apical stigma; these parts function to house ovules, support/project stigma, and receive pollen, respectively. Pollen attached to the stigma germinates and extends as a pollen tube in the style, and passes into the ovule. Thus, the long style facilitates pollen tube elongation. Microgametophytes in angiosperms became reduced in size and cell number (i.e., a pollen tube is a single multinuclear cell with one pollen tube nucleus and two sperm nuclei). The reduction of pollen size reduced time from pollination to fertilization. Full-grown megagametophytes (i.e., embryo sac) only have seven cells with eight nuclei (i.e., an egg cell, two synergids, three antipodals, and two polar nuclei). The reduction in megagametophyte size also reduced the time from fertilization to seed set. In contrast to the long time from pollination to seed set (~1 year in conifers), the rapid fertilization and seed set in angiosperms contributed to adaptation to an annual herbaceous life cycle. After entering ovules, the pollen tubes fertilize the egg and the polar nuclei,

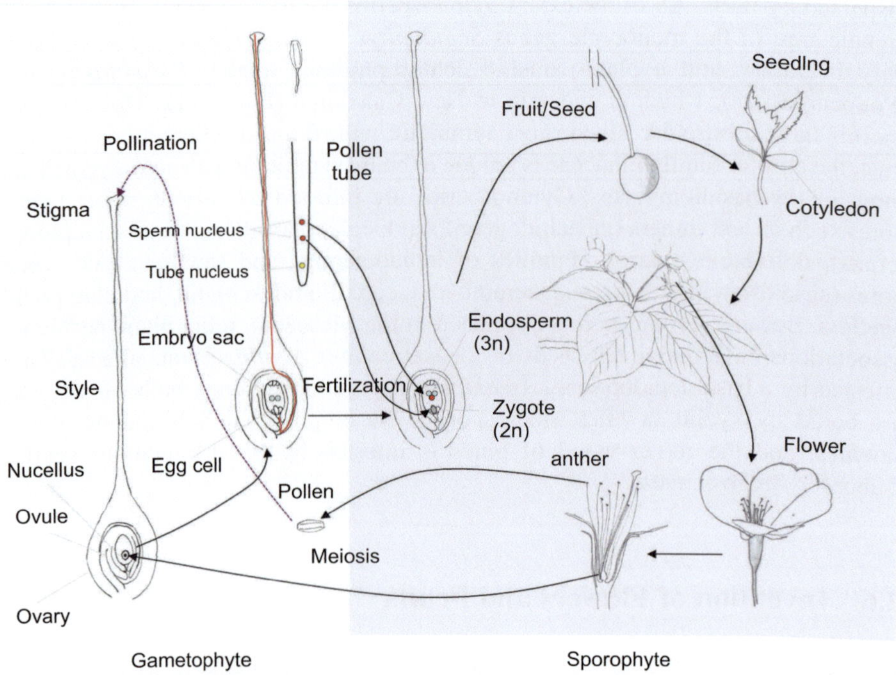

Fig. 2.14 Life cycle of an angiosperm, *Prunus speciosa*. A purple *broken* line shows transportation by animals. A pollen tube extends in the tissue of the pistil and reaches an egg cell of an embryo sac through a micropyle

which develop into the seed embryo and endosperm, respectively. This double fertilization is unique to angiosperms.

The earliest known angiosperm fossil is the aquatic herbaceous plant *Archaefructus liaoningensis* in the early Cretaceous (Sun et al. 2002), whereas angiosperms are thought to date back to the Jurassic. The pollen fossil data suggest that angiosperms became dominant over gymnosperms in the middle Cretaceous. Compared to leaves and shoots, pollen and spores are small but are protein-rich resources for herbivores, and accordingly, some herbivorous insects became specific feeders on pollen or spores. Although their nutritional quality is similar, pollen and spores evolved differently. A pollen-feeding insect transporting pollen from a flower to another conspecific flower facilitates pollination by accident, whereas a spore-feeding insect cannot serve a fern. Thus, the appearance of pollen-feeding insects enabled the innovative evolution of seed plants. Because pollen is costly, nectar became a popular reward for pollinators. Unlike the naked ovule in gymnosperms, the ovule of angiosperms is enveloped by several fused carpels; this structure is the ovary, which develops into the fruit. The fruit morphology is remarkably diverse among angiosperm species, and adapted for protection against abiotic and biotic stresses and for seed dispersal (Table 2.2). In general, dehiscent fruits release seeds, which are the dispersal units, but seed-containing indehiscent

Table 2.2 Seed dispersal syndromes

Syndrome	Agent of seed dispersal	Reward	Plant's adaptation	Plant examples
Nonbiotic				
Anemochory	Wind	Absent	Dust seed	Orchidaceae
	Wind	Absent	Seed or fruit with developed wings	Pinaceae, Ulmaceae, Dioscoreaceae, Aceraceae, Dipterocarpacea, Oleaceae, *Alsomitra* (Cucurbitaceae)
	Wind	Absent	Seed or fruit with developed plumes	Salicaceae, Aceraceae, *Clematis, Pulsatilla* (Ranunculaceae), Ascrepiadaceae, Asteraceae
	Wind	Absent	Air bladderlike fruit	*Cardiospermum* (Sapindaceae), *Physalis* (Solanaceae)
Hydrochory	Water	Absent	Floating fruit with developed cork layer or spongy tissue	*Cocos* (Arecaceae), *Barringtonia* (Lecythidaceae), Rhizophoraceae, *Terminalia* (Combretaceae), *Paliurus* (Rhamnaceae), Ipomoea (Convolvulaceae), Trapaceae, Scaevola (Goodeniaceae)
Autochory	Spring-operation	Absent	Fruit with ballistics	*Impatiens* (Balsaminaceae), *Geranium* (Geraniaceae), *Oxalis* (Oxalidaceae)
Barochory	Gravity	Absent	Heavy seed?	
Biotic, but not mutual				
Epizoochory	Migratory animals with fur and feathers	Absent	Fruit with hooked spines	Herbacious plants such as *Geum* (Rosaceae), *Achyranthes* (Amaranthaceae), *Xanthium, Bidens* (Asteraceae)
	Migratory animals with fur and feathers	Absent	Adhesive appendages on fruit	Herbacious plants such as *Oplismenus* (Poaceae), *Desmodium* (Fabaceae), *Torilis* (Apiaceae), *Adenocaulon* (Asteraceae)

(continued)

Table 2.2 (continued)

Syndrome	Agent of seed dispersal	Reward	Plant's adaptation	Plant examples
Biotic and mutual				
Endozoochory (Ornithochory)	Frugivorous bird (e.g., passeriformes)	Juicy pulp or other tissue of fruit/seed	Red/black fleshy fruit with hard seed coat	Zingiberaceae, Myricaceae, *Malus, Rosa, Rubus, Prunus, Pyrus* (Rosaceae), *Ribes* (Grossulariaceae), *Viscum* (Santalaceae), Ficus (Moraceae), Ericaceae, Symplocaceae, Vitaceae, *Breynia, Glochidion* (Phyllanthaceae), *Vibur-num* (Adoxaceae), *Lonicera* (Caprifoliaceae), Rubiaceae etc.
Endozoochory (Chiropterochory)	Frugivorous bat	Sweet pulp or other tis-sue of fruit	Aromatic fleshy fruit with hard seed coat	Musaceae, *Ficus* (Moraceae)
Endozoochory	Frugivorous nonflying mammal	Sweet pulp or other tis-sue of fruit	Aromatic fleshy fruit with hard seed coat	Myricaceae, Actinidiaceae, *Durio* (Malvaceae), *Nephelium, Dimocarpus* (Sapindaceae), *Diospyrus* (Ebenaceae), *Mangifera* (Anacardiaceae), *Lansium* (Meliaceae), *Garcinia* (Clusiaceae), *Ficus, Artocarpus* (Moraceae)
Endozoochory (Piscichory)	Omnivorous fish	Creamy pulp or other tis-sue of fruit	Fleshy fruit with hard seed coat	trees in flood plain in Amasonia, Annonaceae, *Virola* (Myristicaceae), *Lucuma, Pouteri*a (Sapotaceae), Elaeocarpaceae, *Ficus* (Moraceae) etc.
Synzoochory	Scatter-hoard-ing mammal (e.g., rodent)	Seed itself	Nut (fruit with hard shell) or hard shelled seed	*Quercus, Litocarpus, Castanopsis* (Fagaceae), *Corylus* (Betulaceae), Juglandaceae

(continued)

Table 2.2 (continued)

Syndrome	Agent of seed dispersal	Reward	Plant's adaptation	Plant examples
	Scatter-hoarding bird (e.g., raven, nut-cracker, jay)	Seed itself	Nut (fruit with hard shell) or hard shelled seed	*Pinus, Picea* (Pinaceae), *Quercus* (Fagaceae), *Corylus* (Betulaceae)
Myrmecochory	Ant	Elaiosome	Seed with elaiosome	*Anemone, Corydalys, Hepatica* (Ranunculaceae), Primulaceae, Violaceae, *Erythronium* (Liliaceae), *Asarum* (Aristolochiaceae)

Howe and Smallwood (1982), Sorensen (1986), and van Oudtshoornand and van Rooyen (2013)

fruits are themselves dispersed. Although fruits are larger and heavier than spores and pollen, the fruits of some plants can be dispersed by wind if they have developed wings or plumes (anemochory). In addition, although most fruits are heavier than water, the fruits of some riparian plants have developed corky or spongy tissues to enable them to float and thus be dispersed by water (hydrochory). Some herbaceous plants developed automatic seed dispersal mechanisms, in which seeds are flicked off by the spring function of dehiscent fruits (autochory). Some herbaceous plants developed fruits with an attachment apparatus (e.g., hooked spines and adhesive mucilage), which attach to the fur of mammals and are dispersed by them (epizoochory). The predominant means of seed dispersal in extant angiosperms, however, is mutualistic zoochory, in which plants delegate seed dispersal to specific migrant animals by offering them rewards.

2.7 Diversification of Angiosperms

The estimated number of angiosperm species is 352,000, which outnumbers that of gymnosperms (1000), pteridophytes (13,000), and bryophytes (20,000; The Plant List 2013). The exceeding diversity of angiosperms is closely related to zoophily for many reasons: first, zoophilous plants can achieve pollination even on the forest floor where wind is rarely available; second, because even small, rare, isolated plants can achieve pollination if reliable pollinators move among the same species; and third, because the pollinators evolve floral fidelity to maximize foraging efficiency and the pollinator's behavior canalizes gene flow, which contributed to speciation and diversification. Zoophilous plants are estimated to account for 87.5% of the total number of angiosperm species (Ollerton et al. 2011). Molecular phylogenetic studies (Angiosperm Phylogeny Group 2003) have revealed the

Fig. 2.15 Flowers of basal angiosperms. (**a**, **b**) Male flowers of *Amborella trichopoda* (Amborellaceae). (**c**) A flower of *Zygogynum baillonii* (Winteraceae) visited by a micropterigid moth. (**d**) A flower of *Nuphar japonicum* (Nymphaeaceae) visited by ephydrid flies. (**e**) Flowers of *Illicium anisatum* (Schisandraceae)

phylogeny of angiosperms: (Amborellaceae (Nymphaeace (Austrobaileyales (Chloranthaceae (Magnoliid (monocots (Ceratophyllales, Eudicots))))))). This shows that the most basal extant angiosperm is *Amborella* (the monotypic family Amborellaceae, endemic to New Caledonia). *Amborella* is a dioecious evergreen shrub. Its female flower has 7–8 white tepals, several staminodes, and a spiral of several free carpels; in contrast, the male flower has 6–15 white tepals and 10–21 spirally arranged stamens (Fig. 2.15). Pollen is readily released from anthers by wind, whereas various unspecialized insects visit and pollinate male and female flowers, suggesting ambophily (adoption of zoophily and anemophily simultaneously; Thien et al. 2003). The *Amborella* pollination system suggests that the earliest angiosperm adopted insect pollination by developing attractive petals in addition to anemophily.

Petals of zoophilous flowers likely played an important role in angiosperm diversification, because most pollinators are diurnal sight-sensed animals that are attracted by the nongreen color of petals (Table 2.3). Absorption of extraviolet light by a part of the petal can contribute to guiding pollinators to the nectary. Floral odor is also important for attracting pollinators, particularly in nocturnally opening flowers. Most zoophilous flowers secrete nectar, the sugar concentration of which

Table 2.3 Pollination syndrome of diverse seed plants

Pollination type	Floral property								
Pollination agent	Color[1]	Symmetry[2]	Shape[3]	Anthesis[4]	Reward[5]	Odor[6]	Distribution[7]	Characteristics of pollination system	Plant exampls
Wind pollination									
Wind	g	n	a	d	–	–	W	Enormous number of light pollen grains are shed	Diverse seed plants
Water pollination									
Water current	g	n	a	d	–	–	W	Thread like pollen istransported by water	Aquatic monocots
Water current and wind	g	n	a	d	–	–	TR	buoyant anthur is also blown by wind	*Enhalus*
Ambophily									
Insect and wind	w	n	o	d	p	–	NC	Pollinated by wind and unspecialized insect	*Amborella*
Bee pollination									
Bumblebee	w, y, p	z, a	b, c, o, s, t	d	n, p	+	HA, SA	Homeothermic bees with high floral constancy	Diverse angiosperms
Honeybee	w, y, p	z, a	b, c, o, s, t	d	n, p	+	PA, AF, OR	Scout–recruit foraging using letter 8 dance	Diverse angiosperms
Stingless bee	w, y, p	z, a	b, c, o, s, t	d	n, p	+	TR	Scout–recruit foraging using pheromone	Diverse angiosperms
Euglossine bee (female)	w, y, p, r	z, a	c, s, t	d	n, p	+	SA	Long-proboscid bees trapline scattered flowers	Diverse angiosperms
Euglossine bee (male)	y	z	c	d	fr	+	SA	Floral fragrace is harvested and reused	*Coryanthes*
Small solitary bee	w, y, p	z, a	b, c, o, s, t	d	n, p	+	W	Flowers often having nectar guids	Diverse angiosperms

(continued)

Table 2.3 (continued)

Pollination type	Floral property								
Pollination agent	Color[1]	Symmetry[2]	Shape[3]	Anthesis[4]	Reward[5]	Odor[6]	Distribution[7]	Characteristics of pollination system	Plant exampls
Anthophorine bee	w, y, p, r	z, a	c, s, t	d	n, p	+	W	Long-proboscid bees trapline scattered flowers	Diverse angiosperms
Leafcutter bee	w, y, p, r	z, a	c, s, t	d	n, p	+	W	Middle-sized bees adopting buzz-pollination	Fabaceae etc.
Carpenter bee	w, y, p	z, a	b, c, o, s, t	d	n, p	+	W	Large-sized bese adopting buzz-pollination	Fabaceae etc.
Oil-collecting bee	w, y, p	z, a	c, t	d	o, p	+	W	Bees collect and use oil in sterad of nectar	Several families
Wasp pollination									
Hunting wasp	w, y	a	o	d	n	+	W	Nectar accessible to wasps	Diverse angiosperms
Social wasp	g	a	o	d	–	+	AU	Green-leaf volatiles attract prey-seeking wasps	*Epipactis*
Sphecid wasp	bl	z	c	d	–	+	AU	Deceptive pollination with pseudocopulation	*Cryptostylis*
Ant	g	a	o	d	n	+	W	Subaerial small flowers accesible by ants	Subaerial plants
Fig wasp	g	n	a	d	s	+	TR	Obligate mutualism	*Ficus*
Lepidopteran pollination									
Butterfly	w, y, p, r	a	b, c, o, s, t	d	n	+	W	Style and pistils protruded from corolla	Diverse angiosperms
Hawkmoth	w	a	s, t	n, d	n	+	W	Long-distance transportation of pollen	Diverse angiosperms
Nocturnal moth	w	a	b, o, s, t	n	n	+	W	Flowers emit odor in night	Diverse angiosperms

Nocturnal moth	g	a	a	n	d	+	W	Pollination drop is harvested by moths	*Gnetum*
Yucca moth	w	a	o	n	s	+	NA	Obligate mutualism	Diverse angiosperms
Leafflower moth	g	a	a	n	s	+	TR	Obligate mutualism	Diverse angiosperms
Dipteran pollination									
Hoverfly	w, y	a	b, c, o	d	o, p	+	W	Hoverflies harvest nectar and pollen	Diverse angiosperms
Blow fly	br	a	o	d	–	+	W	Deceptive pollination, flowers mimic rotting Flesh	*Rafflesia*
Drosophilid fly	w	a	a, o	d	n	+	TR	Brood-site pollination	*Colocasia*
Fungus gnat	g	a	o	d	n	+	HA	Highly specific mutualism	*Mitella*
Fungus gnat	bl, g	a	o	d	–	+	HA	Deceptive pollination	*Arisaema*
Phorid fly	g	n	a	d	fl	+	W	Brood-site pollination	*Aristolochia*
Mosquito	w, p	a	o	d	n	+	W	Utilizing nectar instead of blood	*Persicaria* etc.
Gall midge	g	n	a	d	fl	+	OR	Brood-site pollination	*Artocarpus*
Coleopteran pollination									
Scarabaeid beetle	w	a	a, o	d	p	+	TR	Flowers emit strong odor	Diverse angiosperms
Carrion beetle	w, bl	n	a	n	c	+	TR	Flowers emit strong aminoid odor	*Amorphophallus* etc.
Dung beetle	bl	z	o	n	–	+	OR	Deceptive pollination via floral fetid odor	*Orchidantha*
Cerambycid beetle	w	a	b, o	d	p	+	W	Pollen is main reward	Diverse angiosperms
Small beetle	w	a	b, o	d	p	+	W	Pollen is main reward	Diverse angiosperms

(continued)

Table 2.3 (continued)

Pollination type	Floral property								
Pollination agent	Color[1]	Symmetry[2]	Shape[3]	Anthesis[4]	Reward[5]	Odor[6]	Distribution[7]	Characteristics of pollination system	Plant exampls
Hemipteran pollination									
Mirid bug	w	a	o	d, n	fl	+	OR	Brood-site pollination	*Macaranga*
Thrips pollination									
Thrips	w, g	a	o	d, n	p, fl	+	W	Brood-site pollination	*Dioscorea* etc.
Amphipod pollination									
Amphipod	bl	a	c	d, n	p	+	OR	Cryptic ground flower visited by soil organisms	*Aspidistra*
Bird pollination									
Hummingbird	r, y	a, z	c, t	d	n	–	NA, SA	Deep flowers secrete ample nectar	Diverse angiosperms
Sunbird	r, y	a, z	c, t	d	n	–	AU, OR, AF	Deep flowers secrete ample nectar	Diverse angiosperms
Spiderhunter	r, y	a, z	c, t	d	n	–	OR	Deepest flowers secrete ample nectar	Diverse angiosperms
Honeycreeper	r, y	a, z	c, t	d	n	–	SA	Deep flowers secrete ample nectar	Diverse angiosperms
Honeyeater	r, y	a, z	c, t	d	n	–	AU, PC	Deep flowers secrete ample nectar	Diverse angiosperms
White-eye	r	a, z	c, t	d	n	–	AF, OR, AU	Deep flowers secrete ample nectar	Diverse angiosperms
Bulbul	r	a, z	c, t	d	n	–	AF, OR	Deep flowers secrete ample nectar	Diverse angiosperms

Mammal pollination									
Echo-locating bat	w	a	b, c, t	n	n	–	SA	Echo-acoustic signals to lure flying bats	Diverse angiosperms
Flying fox (megabat)	w	a	b, c, t	n	n	+	TR	Flowers emit strong odor in night	Diverse angiosperms
Arboreal mammals	w	a	c, o	n	n	+	TR	Large flowers secrete large amount of nectar	Diverse angiosperms
Terrestrial mammals	w	a	b,o	n	n	+	AU, AF	Strong musty odor and large amount of nectar	*Protea* etc.
Monkey	p	a	o	d	n	–	TR	Red-colored, abundant nectar	Fabaceae

[1] *bl* black, *br* brown, *g* green, *p* pink, *r* red, *w* white, *y* yellow
[2] *a* actinomorphic, *n* nonsymmetric, *z* zeigomorphic
[3] *a* apetalous, *b* brush, *c* campanulate, *o* open, *s* long-spurred, *t* tubular
[4] *d* diurnal, *n* nocturnal
[5] *c* copulation site, *fl* floral tissue, *fr* fragrance, *n* nectar, *o* oil, *p* pollen, *s* seed
[6] + present, – absent
[7] *AF* Africa, *AU* Australian, *HA* Holoarctic, *NA* North America, *NC* New Caledonia, *OR* Oriental, *PA* Palaearctic, *PC* Pacific, *SA* South America, *TR* Pan-tropics, *W* worldwide

Fig. 2.16 Diversity of floral rewards of angiosperms. (**a**) A flower of *Dalechampia tiliifolia* (Euphorbiaceae) secreting resin for euglossine bees. (**b**) Female inflorescence of *Anthurium* sp. (Araceae) attracting male euglossine bees by floral odor. (**c**) A flower of *Thladiantha nudiflora* (Cucurbitaceae) secreting floral oil and visited by an apid bee, *Ctenoplectra cornuta*. (**d–g**) Inflorescence and flowers of *Arisaema negishii* lacking floral reward (Araceae). (**d**) Male spathe. (**e**) Cross-section of a spathe. (**f**) Male spadix. (**g**) Female spadix. In a female spathe, there are dead fungus gnats attracted by the floral odor

is 10–50%, which is lower in bird- and moth-pollinated flowers. The association between flowers and pollinators is generally mutual because pollinators offer flowers pollination services, and in turn, flowers provide pollinators with rewards such as nectar and pollen. More rare floral rewards include floral oil (several genera in Orchidaceae, Primulaceae, Malpighiaceae, Cucurbitaceae, and Scrophulariaceae) for oil-collecting bees (Fig. 2.16), floral resin (*Dalechampia* in Euphorbiaceae) for euglossine bees (Fig. 2.16), and a part of ovules or seeds (Agavaceae, Ranunculaceae, Saxifragaceae, Cactaceae, Moraceae, and Phyllanthaceae), which are detailed below. In contrast to mutual relationships, some plants achieve pollination by deception without offering a reward. For example, flowers of jack-in-the-pulpit (*Arisaema* spp., Araceae; Fig. 2.16) attract fungus gnats (Mycetophilidae) by emitting a mushroomlike odor (Barriault et al. 2010). The perennial monocot is dioecious and has an erect inflorescence known as a spadix, which is covered by a spathe. Male and female spathes are similar, although male spathes have an aperture at the basal seam, whereas female spathes do not. Both male and female spadices attract fungus gnats by emitting a floral odor, and the fungus gnats attracted to a male spadix escape from the spathe aperture with their bodies dusted with pollen. Orchids, which exhibit a diverse floral morphology, use various deceptive pollination mechanisms, including generalized food deception, food-deceptive floral mimicry, brood-site imitation, pseudoantagonism, and sexual deception (Jersáková et al. 2006). For example, flowers of some orchid genera attract nectar-seeking bees by visually mimicking other nectariferous plants. In contrast, some orchids attract mate-seeking males of sphecid wasps or sciarid fungus gnats by chemically mimicking the sexual pheromone of pollinator females, and are pollinated by the male pollinators during pseudocopulation (Schiestl et al. 1999, Blanco and Barboza 2005, Gaskett 2011). The insect pollinators of basal angiosperms in the early to mid-Cretaceous are considered to have been Mesozoic insects such as beetles, mecopterans, dipterans, sawflies, and micropterigid moths (Fig. 2.9). A relic plant–pollinator interaction remains evident in New Caledonia, where *Zygogynum*, a genus of the basal plant family Winteraceae, is pollinated by micropterigid moths (Fig. 2.15; Thien et al. 1985). However, the principal pollinators of modern diverse angiosperms are mammals (particularly bats and nonflying arboreal mammals), birds, and insects belonging to four of the five above-mentioned megadiverse orders: Coleoptera (particularly Staphylidae, Nitidulidae, Scarabaeidae, Cerambycidae, Chrysomelidae, and Curculionidae), Diptera (particularly Syrphidae, Bombyliidae, Drosophilidae, and Calliphoridae), Hymenoptera (particularly bee families; i.e., Apoidea), and Lepidoptera (particularly various families of long-tongued moths and butterflies). All of these diverged simultaneously with angiosperms after the late Cretaceous.

2.8 Evolutionary History of Hymenopterans

Bees have played the most prominent role as pollinators of angiosperms. Before focusing on their role, we review the prehistory of bees. Hymenoptera is an order of insects with hyaline wings, and comprises two suborders, Symphyta and Apocrita. Symphyta, although nonmonophyletic, are the sawflies, characterized by lack of a body constriction between the thorax and abdomen. Larvae of most sawflies are external or internal herbivores (Fig. 2.17). The most basal clade of Symphyta, Xyelidae, appeared in the Triassic, and the larvae of extant xyelids are bud borers of firs, male cone borers of pines, and defoliators of walnut leaves. In contrast to Symphyta, Apocrita comprises parasitic or hunting wasps (the ant is one of the hunting wasp clades) and is characterized by a body constriction between the thorax and abdomen; specifically, the constriction exists between the first and second segments of the abdomen. This body constriction is thought to have evolved to handle their ovipositors efficiently and powerfully (Iwata 1971). In parasitic wasps (Parasitica), ovipositors are utilized to insert eggs into the insect hosts (in parasitoid wasp superfamilies, e.g., Chalchidoidea, Ichneumonoidea, and Proctotrupoidea) or rarely into plants (in several genera in Chalchidoidea and Ichneumonoidea, and the gall wasp superfamily, Cynipoidea; Fig. 2.17). A great number of parasitoid wasp

Fig. 2.17 Diverse hymenopterans associated with plants: (**a–c**) Suborder Symphyta; (**d–f**) Suborder Apocrita. (**a**) A sawfly *Tenthredo nigropicta* (Tenthrenididae) visiting flowers of *Angelica polymorpha* (Apiaceae). (**b**) Aggregation of larvae of a sawfly, *Diprion nipponicus* (Diprionidae) on pine tree *Pinus densiflora*. (**c**) A larva of a sawfly *Cimbex conatus* (Cimbicidae) on Japanese alder *Alnus japonica*, mimicking a land snail *Euhadra amaliae*. (**d**) A braconid wasp inserting her ovipositor into a fruit of *Phyllanthus pulcher* in Laos. (**e**) A vespid wasp *Vespula shidai amamiana* visiting flowers of *Psychotria rubra* (Rubiaceae). (**f**) A sphecid wasp visiting flowers of *Euphorbia jolkinii* (Euphorbiaceae)

species are associated with eggs, larvae, pupae, and even adults of various insects in diverse habitats and constitute important natural enemies particularly of phytophagous insects. The parasitoid community is diverse, particularly in endophytic insects such as leaf miners (Askew 1980), which contributes to regulation of host population dynamics (Kato 1996a). In parasitic wasps, Parasitica, a shift from insect-parasitic life to plant-parasitic life has occurred in a few clades (e.g., Braconidae, Torymidae, and Agaonidae). The last family is unique because all of the members are parasites of fig seeds or parasitize fig wasps (Fig. 2.27). Fig wasps are obligate, exclusive pollinators of *Ficus* (Moraceae), and are estimated to have originated in the late Cretaceous (Machado et al. 2001). Hunting wasps (Aculeata) evolved from parasitic wasps by developing a series of behaviors (e.g., hunting prey, constructing nests, transporting prey to their nests, laying eggs in the host, and sometimes resupplying prey). During this evolutionary process, ovipositors lost their ability to insert eggs and were only utilized to inject paralyzing venom into insect prey (Iwata 1971). The nesting habit likely evolved to escape from attack by predators and parasitoid wasps and from inclement weather conditions. To maintain a clean nest interior, hunting wasps developed sophisticated behavior such as grooming eggs, applying antimicrobial substances to their nest, and exterminating parasites. Aculeata comprises two superfamilies, Vespoidea and Apoidea. The former includes spider wasps (Pompilidae), potter wasps and hornets (Vespidae), and ants (Formicidae). The hunting wasps of Vespoidea are important predators of herbivorous insects, and contribute to regulation of herbivore populations. In Vespoidea, eusociality has evolved twice in Vespidae and Formicidae, and a shift from carnivory to pollenivory has occurred once in Masarinae in Vespidae.

Ants are apterous eusocial wasps and flourish in various ecosystems, particularly in the canopy of tropical rainforests (Hölldobler and Wilson 1990). Because numerous workers forage for insects by walking away from the nest, herbivorous insects around the nest are heavily predated. Thus, some plants have evolved to attract ants by providing extrafloral nectar or domatia for nest sites. Extrafloral nectaries are located on leaf laminae, petioles, rachids, bracts, stipules, pedicels, fruits, and so forth of the following families: Rosaceae, Fabaceae, Passifloraceae, Euphorbiaceae, Malvaceae, and Bignoniaceae (Bentley 1977). The myrmecophytic plants that offer ants cavities for nesting are known as "ant plants." Most ant plants have hollows as domatia in their shoots, trunks, or rhizomes (Table 2.4; Beattie and Hughes 2002). The high diversity of ant plant taxa in tropical regions suggests that protection mutualism with ants has been a successful strategy against herbivory. Borneo is renowned for its high diversity of the ant tree genus *Macaranga*, and most *Macaranga* species are associated with ants, which nest in shoot cavities of the host tree (Fig. 2.18). The association is highly host-specific and obligatory, because only the seedlings colonized by the ant colony can grow thereafter (Itioka 2005). If the symbiotic ant colony is artificially killed, the host tree would blast due to heavy damage by herbivores. The aggressiveness of symbiotic ants (i.e., the intensity of the antiherbivore defense) is negatively correlated with the chemical defense capacity of the host plant (Itioka et al. 2000).

The epiphytic myrmecophytic ferns, *Lecanopteris* and *Platycerium* (Fig. 2.19), are also inhabited by symbiotic ants, which not only patrol the fern and host tree to hunt herbivores but also kill other epiphytes and lianas climbing the host tree (Tanaka and Itioka 2011). Accordingly, dipterocarp trees harboring epiphytic ant ferns are almost free of other epiphytes and climbing lianas. Ants are generally not efficient pollinators, because apterous ants cannot transport pollen directly between arboreal flowers and because antimicrobial substances secreted from the ant integument are detrimental to pollen viability (Beattie et al. 1984). Weedy short plants, however, sometimes depend on ants for pollination, as described in subsequent chapters. Some ants have shifted their food from insects to seeds or seed appendages, and seed-harvesting ants sometimes contribute to seed dispersal (Fig. 2.20). Seeds of some angiosperm species have fleshy appendages (i.e., elaiosomes) that contain lipid and protein (Beattie and Hughes 2002). Because the elaiosome is firmly attached to the seed, ants cannot free themselves. Seeds with elaiosomes are

Table 2.4 Protection mutualism between plants and its guards

Mutualism type	Guard	Plants	Plant's reward	Guard service	Host specificity	Colonization/ transmission	Region
Extrafloral nectar-involved mutualism	Ant	Bignoniaceae, Euphorbiaceae, Fabaceae, Malvaceae, Passifloraceae, Rosaceae etc.	Extrafloral nectar	Attack on herbivores around nectaries	Low	Migration	Worldwide
Rhizome domatia-involved mutualism	Ant	epiphytic plants such as *Lecanopteris, Platycerium* (Polypodiaceae), Myrmecodia, Hydnophytum (Rubiaceae)	Domatia in rhizome hollow	Attack on herbivores and on other epiphyte and climbers	High	Colonization by a queen	Southeast asia
Stem domatia-involved mutualism	Ant	*Korthalsia* (Arecaceae), *Piper* (Piperaceae), *Dischidia* (Asclepiadaceae), *Macaranga* (Euphorbiaceae), *Medinilla* (Melastomataceae), *Crypteronia* (Crypteroniaceae), *Myrmecorandia* (Rubiaceae)	Domatia in stem hollow sometimes with food body	Attack on herbivores on host plants	High	Colonization by a queen	Southeast asia
Stem domatia-involved mutualism	Ant	*Cecropia* (Cecropiaceae), *Clidemia* (Melastomataceae), *Hirtella* (Chrysobalanaceae), *Cordia* (Boraginaceae), *Duroia* (Rubiaceae)	Domatia in stem hollow sometimes with food body	Attack on herbivores on host plants	High	Colonization by a queen	Neotropics
Stem domatia-involved mutualism	Ant	*Leonardoxa* (Fabaceae), *Cola* (Sterculiaceae), *Canthium* (Rubiaceae)	Domatia in stem hollow sometimes with food body	Attack on herbivores on host plants	High	Colonization by a queen	Africa
Stem domatia-involved mutualism	Ant	*Acasia* (Fabaceae)	Domatia in stem hollow sometimes with food body	Attack on herbivores on host plants	High	Colonization by a queen	Australia

Leaf domatia-involved mutualism	Predatory and fungivory mites	*Annona* (Annonaceae), *Cinnamomum* (Lauraceae), *Cupania* (Sapindaceae), *Terminalia* (Combretaceae), *Vitis* (Vitaceae), *Elaeocarpus* (Elaeocarpacea), *Viburnum* (Adoxaceae), *Coffea* (Rubiaceae), *Tabebuia* (Bignoniaceae)	Leaf domatia in vein axil	Attack on herbivorous mites on leaf	Medium	Migration	Worldwide
Fungal endophyte-involved mutualism	*Colletotrichum Curvularia* (Ascomycetes)	diverse plants	Habitat and assimilate	Protect the host from biotic and abiotic stresses	High	Vertically or horisontary transmitted	Worldwide
Fungal endophyte-involved mutualism	*Epichloë*, *Neotyphodium* (Ascomycetes)	Poaceae, Cyperaceae	Habitat and assimilate	Protect the host from biotic and abiotic stresses	High	Vertically transmitted	Temperate zone
Bacterial endophyte-involved mutualism	Bacteria	Diverse plants	Habitat and assimilate	Protect the host from biotic and abiotic stresses	High	Vertically or horisontary transmitted	Worldwide

Bentley (1977), Tsai et al. (1994), and Romero and Benson (2005))

Fig. 2.18 Protection mutualism in Southeast Asia. (**a–e**) Obligate mutualism between a myrmecophyte *Macaranga bancana* and an ant *Crematogaster* sp.: (**a**) A fast-growing juvenile tree in a forest gap. (**b**) Young leaves patrolled by many ants. (**c**) Exit holes of the ant's nest on a shoot with food bodies around stipules. (**d, a**) Founder queen ants living in a hollow of a seedling. (**e, a**) Scale insects kept inside the ant's nest. (**f**) Facultative mutualism of an ant *Oecophylla smaragdina* with a facultative host plant *Strychnos nux-blanda* in Laos. The ant's nest is spun on a branch with a few leaves tied together

harvested by the ants, transported to their nest, and fed to the larvae. Infestation of the seed by adult and larval ants is prohibited by its hardness. Accordingly, the undamaged seed is removed from the nest and is discarded in a waste disposal area, where it germinates. Although the seed is not transported a great distance, the newly colonized site is rich in nutrients, and root pathogens are removed from the seed by antimicrobial substances secreted by the ants. Myrmecochory has evolved independently in at least 11,000 plant species in 77 families, and the elaiosomes derived from various seed or fruit tissues have undergone convergent evolution (Lengyela et al. 2010).

Fig. 2.19 Epiphytic ferns *Platycerium* sp. (**a**) and *Lecanopteris crustacea* (**b**) growing on canopy branches of dipterocarp trees. In the litter within basal fronds of the former fern, myrmecophytic ants and specialized cockroaches live. In the cavity of the hollow rhizomes of the latter fern, myrmecophytic ants nest. These ants are very aggressive and attack almost all herbivores and gnaw off all shoots of climbing liana and epiphytes (**c**) thus the canopy around the fern is free from lianas and epiphytes (Photo by T. Itioka)

Fig. 2.20 Seed dispersal of *Aquilaria* sp. (Thymelaeaceae) by ants in a tropical rain forest in Borneo. A seed with elaiosome is transported by ants to their nest (**a**), and afterward the seed without elaiosome is transported to their waste site (**b**) and discarded

2.9 Diversification of Bees

Bees are members of the superfamily Apoidea, and are derived from their sister family, Sphecidae sensu lato. Sphecids are solitary hymenopterans, which dig burrows belowground using their mandibles and legs and rear their larval offspring in the nest by feeding them hunted insects. When bees changed their larval food from hunted prey to pollen and nectar, wasps became bees (Iwata 1971). Although the basic morphology of bees is similar to that of sphecid wasps, bees have developed novel derived characteristics (i.e., delicately branched hair on their body for collection of pollen and an extended proboscis for collection of nectar; Fig. 2.21). Nectar and pollen are sources of carbohydrate and protein, respectively, for bee larvae. Because bees are dependent on flowers for food for themselves and their offspring, they are the most frequent visitors to angiosperms. To collect nectar

Fig. 2.21 Diverse morphology and behavior of bees: (**a**) A female of *Lasioglossum sibiriacum* (Halictidae) visiting flowers of *Hydrangea macrophylla* (Hydrangeaceae). (**b**–**c**) Exceedingly long-tongued bees *Amegilla elephas* (Apidae) visiting an exceedingly long-tubed flower of *Etlingia* sp. (Zingiberaceae) in Sumatra. (**d**) A queen of *Bombus senex* (Apidae) endemic to mountain forests in Sumatra. (**e**) *Bombus diversus* visiting a zygomorphic flower of *Salvia glabrescens* (Lamiaceae). (**f**) A hovering female of *Euglossa tridentata* (Apidae) in Panama. (**g**, **h**) Workers of the giant honeybee *Apis dorsata* visiting flowers of *Dryobalanops lanceolata* (Dipterocarpaceae) and *Dillenia suffruticosa* (Dilleniaceae). (**i**) Nests of *Apis dorsata* hung below the boughs of an emergent tree in a tropical rain forest in Lambir Hills National Park in Borneo during a mass-flowering event in 1996 (**g**–**i**, Photo by Tamiji Inoue)

and pollen efficiently, bees visit limited types of flowers. Their frequent and selective visits to flowers and thick coats of branched hair have made them the most important pollinators of diverse angiosperms (Table 2.1). The total number of bee species is estimated to exceed 20,000, most of which are solitary species nesting below ground or in plant cavities (Michener 2007). In contrast to many other insects, the center of bee diversity is located not in tropical rain forests but in scrub land in Mediterranean climate regions, which harbor diverse rare oligolectic (depending on a few flower species) solitary bees. Oceanic islands are unique in that the bee faunae are only composed of solitary bees nesting in plant cavities, suggesting that colonization of oceanic islands occurred by drifting of nests in wood. The small size and scanty hair of solitary bees residing on oceanic islands (e.g., *Hylaeus* in Colletidae, which transport pollen not on their bodies but in their gastric contents) are thought to have enhanced the geitonogamy of the insular woody plants, which increases the risk of inbreeding and accordingly the evolution of dioecy by these insular plants (Kato and Nagamasu 1995). As a result of coevolution with bees, melittophilous (bee-pollinated) flowers became white or yellow, fragrant, nectariferous, and often zygomorphic. The proboscises of some bee species (particularly Anthophorini and Euglossini in Apidae) are strongly extended as a result of the arms race with nectariferous plants, which have tubular corollas or spurs. These long-tongued bees forage scattered deep flowers by following defined routes over long distances. This behavior is known as traplining, and has been observed in *Amegilla* (Anthophorini) in tropical regions of Asia (Fig. 2.21; Kato 1996b) and in *Euglossa* (Euglossini) in the Neotropics (Fig. 2.21; Janzen 1971). Euglossine bees are termed orchid bees, because males are attracted to specific orchids and aroids by the floral odor (Fig. 2.16).

In contrast to nectariferous flowers, another type of coevolution occurred in oil-secreting flowers, which offered oil instead of nectar to bee pollinators as a reward. The forelegs of some oil-collecting bees (*Redivia*, Melittidae) in South Africa are remarkably extended as a result of an arms race with plants that have two oil-secreting floral spurs (Scrophulariaceae; Buchmann 1987). The principal reward for pollinators is not nectar but pollen in some plants such as *Caesalpinia* (Fabaceae), *Diospyros* (Ebenaceae), *Melastoma* (Melastomataceae), *Pyrola* (Ericaceae), *Solanum* (Solanaceae), and *Vaccinium* (Ericaceae). On these flowers, leaf-cutter bees (Megachilidae), carpenter bees (Xylopopini, Apidae), and bumblebees (Bombini, Apidae) grasp and vibrate anthers and efficiently collect the dislodged pollen (Knudsen and Olesen 1993). The pollination system involving this technique is known as buzz pollination. To reduce pollen removal, some plants have evolved a dimorphy of anthers. For example, in *Melastoma*, pollen of yellow anthers is harvested by bees as a reward, but pollen of purple anthers contributes to actual pollination without being harvested (Luo et al. 2008). As with hornets (vespid wasps) and ants, eusociality has evolved in at least five clades of bees: *Lasioglossum* (Halictidae), *Ceratina* (Xylocopinae, Apidae), bumblebee (Bombinae, Apidae), honeybee (Apinnae, Apidae), and stingless bee (Meliponinae, Apidae; Sakagami and Maeta 1989, Michener 2007). Retention of workers specialized in foraging nectar and pollen made eusocial bees more competitive and efficient pollinators. The morphological differentiation between queen and worker is weak in *Lasioglossum* and *Ceratina*, but is prominent in highly eusocial bees: bumblebees, stingless bees, and honeybees.

Bumblebees are densely furry bees (Fig. 2.21) that were originally distributed in the Northern Hemisphere and South America. Most bumblebee species are adapted to temperate and boreal climates, however, there are a few tropical species in South America. Bumblebees nest in underground cavities such as abandoned vole nests. A queen bumblebee can found a nest by herself. Bumblebees have coats of long hairs to which pollen readily attaches, and they can maintain their body temperature at about 30–38 °C, even in cold weather. Bumblebee workers exhibit characteristic foraging behavior, in that each individual specializes in a single plant species and continues to visit only flowers of this species (Heinrich 1979). This results in flower constancy by bumblebees. Thus, the bumblebee is a reliable pollinator for plants, because each bumblebee worker transports ample pollen to other flowers of the same plant species. Because bumblebee larvae are brooded in a mass of pollen and nectar, not in separate cells, worker size varies greatly. Workers with different body sizes visit flowers of different sizes, because morphological matching between flower and bee is a critical determinant of their foraging efficiency. A bee colony can utilize diverse flower species because it contains workers of various sizes. In addition, to enhance foraging efficiency, bumblebee workers avoid visiting flowers that have recently been visited by other conspecifics (Kato 1988). The floral cues perceived by the bumblebees involve a repellent scent that they deposit (Stout et al. 1998) and floral electric fields that they charge (Clark et al. 2013).

In contrast to bumblebees, honeybees are only distributed in Eurasia and Africa and their center of diversity is in the tropical region of Southeast Asia, where five honeybee species coexist. Giant honeybees construct large uncovered nests below branches of emergent trees (the giant honeybee *Apis dorsata*; Fig. 2.21) or on cliffs (the Himalayan giant honeybee *Apis laboriosa*), whereas Asian and European honeybees nest in hollows of tree trunks. Honeybees reproduce by swarming of a queen and her workers, because a queen cannot found a nest herself. Because larvae are brooded in separate standardized cells, workers are similar in size. Honeybees forage using a scout–recruit strategy. Scouting workers that have completed scouting of a floral resource return to the nest and transmit information on the location of a promising flowering patch to colony members via figure-eight dances. The workers rush to the flowers and collect the concentrated floral resource. Due to the high efficiency of the scout–recruit foraging strategy, honeybees have a competitive advantage over other solitary bees, particularly in landscapes where flowers are distributed in clumps, as in tropical rain forests. The predominance of social bees (particularly honeybees and stingless bees) explains the inferiority of solitary bees and the low diversity of bee species in tropical forests. In tropical rainforests in Borneo, five species of honeybee are the predominant pollinators, and giant honeybees are important pollinators of canopy trees, including Dipterocarpaceae, Anacardiaceae, and Fabaceae (Momose et al. 1998). The canopy trees mass-flower for about 3 months at an interval of 4–6 years, and the mass flowering is triggered by drought and/or low temperature for several weeks (Sakai et al. 2006). During the nonflowering period, there are no giant honeybees in the forest. When mass flowering begins, the giant honeybees appear from somewhere (possibly nearby swamp forests), and the number of colonies hanging below branches of emergent trees increases rapidly. Because giant honeybees can fly at least several kilometers, they can pollinate flowers of distant, isolated, rare canopy trees. In the Neotropics, where honeybees are not distributed, eusocial stingless bees were the dominant pollinators before the introduction of European honeybees and the subsequent expansion of Africanized honeybees (Roubik 1992). Extensive observations of flower visitors and pollen attachment to the visitors assist determination of the pollination systems of plant species. By compiling these data, plant pollination systems in different climatic zones can be compared. In five types of forest in Asia, bee pollination (melittophily) is predominant, whereas the main pollinator bee groups vary among climatic zones. In temperate regions in Japan, bumblebees and solitary bees are the most important pollinators, whereas long-tongued bees (*Amegilla*), carpenter bees (*Xylocopa*), and leaf-cutting bees (Megachilidae) are also important pollinators in subtropical and tropical monsoon forests (Fig. 2.22; Kato et al. 2008). For more than a century, European honeybees have been introduced for apiculture to regions in which they were absent (e.g., the New World, Australia, and oceanic islands). The native bee faunae of oceanic islands, which were originally dominated by cavity-nesting solitary bees, were affected by artificial introduction of the honeybee, particularly the Ogasawara Islands in Japan (Kato et al. 1999), Mauritius (Hansen et al. 2002), New Caledonia (Kato and Kawakita 2004), and the Canary Islands (Dupont et al. 2004).

2.10 Lepidopterans as Herbivores and Pollinators

Although bees are indeed important pollinators in most regions of the world, butterflies may be more eye-catching on flowers. Lepidoptera, to which butterflies belong, is the only order that has preadapted to herbivory, and almost all of the members have evolved as herbivores. Trichoptera is the sister group of Lepidoptera, whose larvae are aquatic detritivores or algal grazers. Both orders have the ability to spin silk, although their wings are covered by different structures: scales in Lepidoptera and hairs in Trichoptera. The earliest clade of lepidopterans known is Micropterigidae, which originated in the Jurassic. Micropterigid moths have functional mandibles but lack a tubular proboscis and their larvae are grazers of liverworts (Fig. 2.23). In the Japanese Archipelago, there are 25 species of micropterigid moths of which 22 are specialists of the liverwort species of the genus *Conocephalum*. In micropterigid moths with weak flight activity, vicariant speciation without host plant shift has occurred in the Japanese Archipelago (Imada et al. 2011). Other basal moths with functional mandibles include the agathiphagids, whose larvae are seed parasites of the basal conifer *Agathis* (Araucariaceae); and heterobathmiids, whose larvae are leaf miners of the relict angiosperm *Nothofagus* (Nothofagaceae; Fig. 2.23; Kristensen 1984). Other

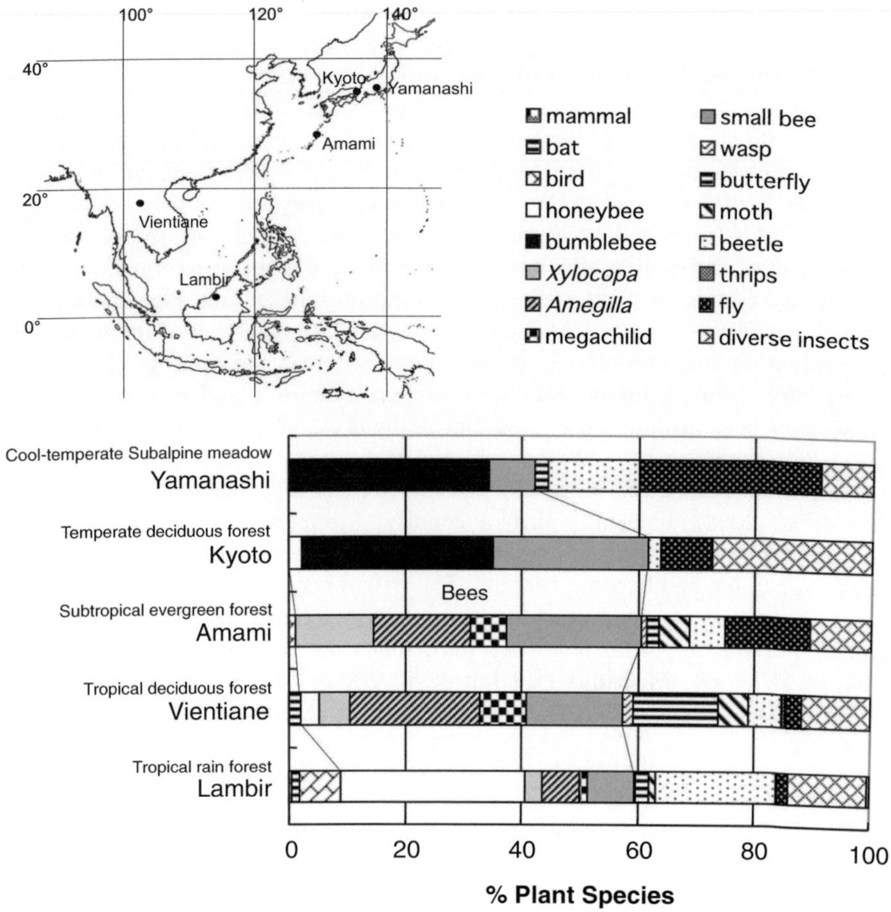

Fig. 2.22 Community-level comparison of pollination systems of plants among various types of forests in Asia: subalpine coniferous forest and meadow in Mt. Kushigata, Yamanashi, cool temperate deciduous forest in Ashiu, Kyoto, subtropical evergreen forest in Amami, tropical monsoon forest in Laos, tropical rainforest in Lambir, Malaysia (Modified from Kato et al. 2008)

members of Lepidoptera have tubular proboscises (suborder Glossata), which suggests that these long-tongued moths evolved after the appearance of nectariferous angiosperm flowers. Although lepidopterans are frequent visitors to flowers, their contribution to pollination is less prominent than that of bees, probably because lepidopterans with exceedingly long proboscises can imbibe nectar without touching stamens and pistils. However, some plants have evolved pollination by lepidopterans. For example, nocturnal flowers with exceedingly long floral tubes or spurs (e.g., Orchidaceae, Caryophyllaceae, Onagraceae, Thymelaeaceae, Rubiaceae, Solanaceae, Campanulaceae, and Apocynaceae) are pollinated by nocturnal long-tongued moths belonging to Pyralidae, Geometridae, Noctuidae, and Sphingidae (Fig. 2.24). Diurnal long-tubed flowers in Liliaceae,

Fig. 2.23 Basal lineages of Lepidoptera. (**a–b**) *Neomicropteryx nipponensis* (Micropterigidae). (**c–d**) *Heterobathmia pseuderiocrania* (Heterobathmiidae). (**e–f**) *Issikiocrania japonicella* (Eriocraniidae). Host plants are a liverwort *Conocephalum conicum* (Conocephalaceae) (**a**), *Nothofagus glauca* (Nothofagaceae) (**c**), and *Fagus crenata* (Fagaceae) (**e**)

Ericaceae, and Lamiaceae are mainly pollinated by butterflies and diurnal hawk-moths. Microlepidopterans have rarely been reported to be pollinators, with the exception of the yucca moth family Prodoxidae, which participates in obligate pollination mutualism with Agavaceae and Saxifragaceae, as detailed below (Pellmyr and Thompson 1992; Pellmyr et al. 1996a, b).

Most members of the proboscid lepidopteran clade, Glossata, are herbivores of angiosperms, whereas some are associated with conifers, pteridophytes, and rarely, with bryophytes. The basal clades of Glossata are small internal herbivores such as leaf miners (Eriocranidae, Nepticulidae, Opostegidae, Heliozeidae, Incurvariidae, Gracillariidae, Tischeridae, Elachistidae, and Cosmopterigidae), leaf gallers (a part of Gracillariidae), shoot borers (Pyralidae and Cossidae), cambium miners (Opostegidae), seed borers (a part of Gracillariidae and Pyralidae), fruit borers (Carposinidae), case bearers (Adelidae, Incurvariidae, Tineidae, Psychidae, and Coleophoridae), leaf rollers (Pyralidae and Tortricidae), and leaf tiers (Gelechidae, Oecophoridae, Pyralidae, and Tortricidae; Fig. 2.25). To break away from the internal life style, the lepidopteran larvae had to adopt novel habits such as draping nettling hair, developing an aposematic or cryptic coloration, mimicking poisonous models, or only being active during the night. By shedding their shelters in this way, lepidopteran larvae could become larger. The scaling-up of lepidopteran size must have made a considerable impact on host plants, and thus had greater selective pressure on plant antiherbivore mechanisms. Thus, the plant–herbivore chemical

Fig. 2.24 Lepidopterans visiting angiosperm flowers for floral nectar: (**a**) A noctuid moth visiting flowers of *Pieris japonica* (Ericaceae). (**b**) A pyralid moth visiting flowers of a pitcher plant *Nepenthes vieillardii* (Nepenthaceae) in New Caledonia. (**c**) A pyralid moth visiting flowers of *Uncaria rhynchophylla* (Rubiaceae) in Laos. (**d**) Pyralid moths *Bradina trigonalis* visiting flowers of *Trachelospermum asiaticum* (Apocynaceae). (**e**) A sphingid moth *Neogurelca himachala* visiting a flower head of *Cirsium sieboldii* (Asteraceae). (**f**) A papilionid butterfly *Papilio bianor* visiting flowers of *Albizia julibrissin* (Fabaceae). (**g**) A papilionid butterfly *Papilio machaon* visiting a flower of *Lilium lancifolium* (Liliaceae)

arms race resulted in the high host specificity of lepidopterans, and the rarity of a shift in their host plant is hypothesized to have facilitated their diversification (Ehrlich and Raven 1964; Fordyce 2010).

2.11 Dipterans also Pollinate Flowers

Diptera is a two-winged insect order with great diversity, several members of which are notorious for their bite, pest damage, dirtiness, nuisance, and insanitation. The most basal clades of extant Dipterans (Deuterophlebiidae, Nymphomyiidae, and Blephariceridae) are aquatic alga-grazers or detritus-feeders, and there are several basal clades of aquatic dipterans, such as Culicomorpha (Brian et al. 2011), suggesting that dipterans originated in an aquatic habitat (Fig. 2.9). Larvae of terrestrial dipterans lack legs and are adapted to mining media rich in organic

Fig. 2.25 Diversity of morphology and feeding habits in lepidopteran larvae: (**a**) A case-bearing larva of *Adela* sp. feeding on fallen leaves (Adelidae). (**b**) Linear mines of *Stigmella betulicola* (Nepticulidae) on a leaf of *Betula grossa* (Betulaceae). (**c**) Larvae of *Cuphodes* sp. (Gracillariidae) mining a leaflet of *Caesalpinia decapetala* (Fabaceae). (**d**) Blotch mine of *Phyllonorycter* sp. (Gracillariidae) on a leaf of *Salix subfragilis* (Salicaceae). (**e**) A larva of *Apatetris elaeagnella* (Gelechidae) mining a leaf of *Elaeagnus pungens* (Elaeagnaceae). (**f**) A larva of *Elachista canis* (Elachistidae) mining a leaf of a dwarf bamboo *Sasa nipponica* (Poaceae). (**g**) Leaf mines of *Cosmopterix zieglerella* (Cosmopterigidae) on a leaf of *Humulus japonicas* (Cannabaceae). (**h**) A larva of *Eudocima tyrannus* (Noctuidae). (**i**) A larva of *Callopistria* sp. (Noctuidae) on a fern *Stegnogramma pozoi* (Thelypteridaceae). (**j**) A larva of *Jocheaera alni* (Noctuidae) on *Acer ginnala* (Sapindaceae). (**k**) A larva of *Papilio machaon* (Papilionidae) feeding on leaves of *Angelica pubescens* (Apiaceae). (**l**) A thorn-covered larva of *Euthalia* sp. (Nymphalidae) on *Cratoxylum* sp. (Hypericaceae)

matter in wet habitats. Herbivory originated in various clades of these terrestrial dipterans. The herbivorous dipterans comprise leaf miners (Sciaridae, Rhagionidae, Tephritidae, Agromyzidae, Ephydridae, and Anthomyiidae), gallers (Cecidomyiidae), shoot borers (Diopsidae and Chloropidae), fruit borers (Tephritidae and Drosophilidae), and so forth. A basal group of the snipe fly family Rhagionidae, Spaniinae is associated with liverworts and mosses, and its association with bryophytes is thought to have been maintained for more than 100 million years since the Jurassic (Imada and Kato 2016a, b). The mouthparts of dipterans are adapted not to bite solid food but to lick liquid or powdery food. Thus, many dipteran adults consume floral nectar, pollen, tree sap, and animal secretions, and some dipterans with aciform mouthparts suck the blood of vertebrates. Among

Fig. 2.26 Diversity of morphology and habits of dipterans in Japan: (**a–b**) An algae-grazing larva (**a**) and adults (**b**) of the basal dipteran *Deuterophlebia nipponica* (Deuterophlebiidae). (**c**) A fungus gnat *Gnoriste mikado* (Mycetophilidae) visiting a flower of *Mitella furusei* var. *subramosa* (Saxifragaceae). (**d**) A fungus gnat *Coelosia* sp. (Mycetophilidae) visiting a flower of *Mitella pauciflora*. (**e**) A male of a mosquito *Tripteroides bambusa* visiting a flower of *Persicaria filiformis* (Polygonaceae). (**f**) Tipulid flies *Elephantomyia* sp. visiting a flower head of *Cirsium nippponicum* var. *yoshinoi* (Asteraceae). (**g**) A syrphid fly *Melanostoma mellinum* visiting male flowers of *Carex foliosissima* (Cyperaceae)

these dipterans, hoverflies (Syrphidae), bee flies (Bombyliidae), blow flies (Calliphoridae), and fruit flies (Drosophilidae and Tephritidae) are frequent visitors to flowers. However, their contribution to pollination is less prominent than that of bees, because their visit frequency, floral constancy, and pollen attachment are inferior to those of bees.

Among anthophilous dipterans, fungus gnats of the family Mycetophagidae are unique exclusive pollinators of the perennial genus *Mitella* (Saxifragaceae; Fig. 2.26; Okuyama et al. 2004). The larvae of pollinator gnats feed on mosses, which grow in the same habitat as *Mitella*. The inconspicuous flowers with branched linear petals secrete species-specific blends of volatiles, and attract specific species of fungus gnats (Okamoto et al. 2015). Accordingly, loss of a floral volatile triggers a change in the pollinator, causing speciation of pollinator-mediated plants.

Some dipterans participate in brood-site pollination mutualisms, in which adult females pollinate and oviposit on flowers and the pollinator larvae develop in

Fig. 2.27 Flowers involved in brood-site pollination by phorid flies (**a**, **b**), drosophilid flies (**c**, **d**) and thrips (**c**). (**a**) *Aristlochia kaempferi* in Japan. (**b**) *Aristlochia* sp. in Peru. (**c**, **d**) *Alocasia odora* in Taiwan. (**e**) *Chloranthus serratus* in Japan

vegetative tissue of inflorescences (Sakai 2002). In this association, host specificity of the pollinator is not high, pollinators sometimes have alternative hosts, and the plants often have copollinators. Brood-site pollination mutualism has been reported between *Alocasia* (Araceae) and drosophilid flies, between *Aristolochia* (Aristolochiaceae) and phorid flies, and between *Artocarpus* (Moraceae) and cecidomyiid flies (Fig. 2.27). In the last case, male inflorescence infected by a few species of fungi is the growth medium for the pollinator larvae (Sakai et al. 2000). The associations between these plants adopting brood-site pollination and the pollinators utilizing floral tissue are less reciprocally obligate than the associations between the plants providing seeds as rewards and the host-specific pollinators whose larvae utilize the seeds, probably because floral tissue is a perishable resource for the pollinator larvae, and because pollination service did not increase fitness of the pollinator.

2.12 Birds with Red Fruits and Red Flowers

In addition to winged insects, birds and bats are essential partners of plants. The bird (class Aves) is an offspring of the dinosaur clade (clade Dinosauria), and the bat (order Chiroptera) is a clade of the class Mammalia. Because birds originated and achieved air supremacy in the late Jurassic, bats, which appeared in the late Cretaceous were forced to become nocturnal. Another flying vertebrate, the pterosaur (clade Ornitodera), was extant in the Mesozoic, but was extinct by the end of the Cretaceous. Due to their acquisition of wings, birds in forest ecosystems have undergone adaptive radiation as insectivores and frugivores. Frugivorous birds play a pivotal role in dispersing the seeds of angiosperms. Fleshy fruits with red, yellow, purple, black, or (rarely) blue colors attract frugivorous birds (Fig. 2.28), which only consume the juicy pulp of the fruits and cast seeds with feces or pellets at

Fig. 2.28 Diverse fleshy fruits of angiosperms in Japan: (**a**) *Lonicera chamissoi* (Caprifoliaceae). (**b**) *Lonicera caerulea*. (**c**) *Symplocos sawafutagi* (Symplocaceae). (**d**) *Diospyros japonica* (Ebenaceae). (**e**) *Glochidion obovatum* (Phyllanthaceae). (**f**) *Viscum album* (Santalaceae). (**g–j**) *Cyrtosia septentrionalis* (Orchidaceae). Pericarps of fruits (**a–c**, **f**, **g**) or outer seed coat (**e**) are brightly colored for seed dispersal by birds

various sites, sometimes far from the mother plant. This directed, long-distance seed dispersal by birds likely enabled angiosperms bearing fleshy fruits to colonize new, isolated, suitable habitats, and even a desert oasis or isolated oceanic island. The fact that a great number of plant species bear fleshy fruits suggests that seed dispersal by birds favored evolution of a juicy pulp and brilliantly colored pericarp. Ornithochory (seed dispersal by birds) has also evolved in orchids, which generally bear numerous dust seeds lacking an endosperm. The red fruits of the mycoheterotrophic orchid *Cyrtosia* growing in the dark on the forest floor are ingested by thrushes, and are dispersed by being discarded in the feces (Suetsugu et al. 2015).

Some insectivorous and frugivorous birds have developed slender bills and long extensible tongues, and imbibe nectar from deep flowers, which are pollinated by the birds. The anthophilous birds comprise hummingbirds (Trochillidae), sunbirds and spiderhunters (Nectariniidae), honeyeaters (Meliphagidae), bulbuls (Picnonotidae), white-eyes (Zosteropidae), Hawaiian honeycreepers (Drepanidini of Fringillidae), and so forth (Table 2.2). Bird-pollinated (ornithophilous) flowers have red, tubular, tough corollas that secrete ample dilute nectar containing ~20% sugar (Cronk and Ojeda 2008; Fig. 2.29). In the Neotropics where hummingbirds

Fig. 2.29 Ornithophilous flowers in the world: (**a**), *Durio lowianus* (Malvaceae) in Borneo. (**b**) *Camellia rusticana* (Theaceae). (**c**) *Brachyotum quinquenerve* (Melastomataceae). (**d**) *Guzmania monostachia* (Bromeliaceae). (**e**) *Bomarea* sp. (Alstroemeriaceae). (**f**) *Syzygium acre* (Myrtaceae). (**g**) *Metrosideros boninensis* (Myrtaceae). (**h**) *Orobanche boninsimae* (Orobanchaceae). (**a**) Sarawak; (**b**) Japan; (**c–e**) Peru; (**f**) New Caledonia; (**g–h**) Ogasawara Islands, Japan. The putative bird pollinators of the last two plant species in the oceanic islands are now extinct

flourish, there are many ornithophilous plants in Heliconiaceae, Zingiberaceae, Bromeliaceae, Gesneriaceae, Apocynaceae, Rubiaceae, and Lamiaceae (Bawa 1990). Phylogenetic studies of monkeyflowers (*Mimulus*, Phrymaceae) suggest that hummingbird pollination in North America has evolved twice from insect pollination, and has resulted in drastic changes in the color and shape of flowers (Beardsley et al. 2003). In Southeast Asian tropical rainforests, in which sunbirds and spiderhunters reside, the ornithophilous plants include Zingiberaceae, Musaceae, Bombacaceae, Sapotaceae, and Loranthaceae (Momose et al. 1998). Sunbirds and spiderhunters must perch on nearby branches to collect floral nectar, whereas hummingbirds can imbibe nectar during hovering. In fact, in Southeast Asian tropical forests, some epiphytes of Loranthaceae have inflorescences whose axis functions as a stout perch for bird pollinators (Yumoto et al. 1997), and the red ground flowers of some tall gingers are pollinated by nectar-seeking spiderhunters on the forest floor. Bird pollination (ornithophily) is also prevalent in Australia, New Zealand, and New Caledonia, where honeyeaters (Meliphagiidae) are frequent visitors to flowers (Fig. 2.29). In Hawaii, there are diverse ornithophilous plants, the flowers of which are pollinated by several Hawaiian honeycreepers with diverse bill morphologies. A recent molecular phylogenetic analysis revealed that these honeyeater-like birds belong to the finch family (Fringillidae) and have undergone adaptive radiation on the islands in the last 5.7 million years (Lerner et al. 2011). In the Ogasawara Islands in Japan there is an endemic nectarivorous bird *Apalopteron familiare*, which was demonstrated to be a member of Zosteropidae. These necatariferous birds on oceanic islands are now threatened by artificial invasions of rats and feral cats, as well as destruction of forests (Kawakami and Higuchi 2013). In the Ogasawara Islands, three of four endemic bird species have become extinct within the last 200 years, which has hampered pollination of some putatively ornithophilous plants (Fig. 2.29). In temperate regions of the Japanese Archipelago, common nectarivorous birds are the brown-eared bulbul and Japanese white-eye, which visit flowers of *Camellia* (Camelliaceae), *Loranthus* (Loranthaceae), and *Myoporum* (Scrophulariaceae). Because the song of the Japanese white-eye has been praised by Japanese people since the Edo period, they were kept as caged birds to perform in song contests. However, they have also been introduced to various oceanic islands, such as Hawaii, where they occupy the niche of the extinct Hawaiian honeycreepers.

2.13 Associations Between Plants and Mammals

Although the megafauna of the current terrestrial ecosystem are dominated by mammals, there is a long suppressed prehistory of mammals. Egg-laying mammals originated in the Triassic, and placental and marsupial mammals have undergone adaptive radiation after the dinosaurs became extinct at the end of the Cretaceous. In the Mesozoic, diverse herbivorous dinosaurs must have depended upon pteridophytes and gymnosperms. During the Paleogene, many clades of mammals adopted herbivory and diversified as angiosperm grazers. Because mammals have fur, they are utilized by some plants as seed dispersers by means of their sticky or prickly fruits, which readily attach to mammalian fur (Table 2.2). The mammals that have adapted to arboreal life are primates and rodents, which can access the fruits on canopy trees. The seeds of these arboreal fleshy fruits are dispersed by mammals through their intestines. In the tropics, many cauliflorous trees that bear flowers directly on the trunk are extant (Fig. 2.30). These cauliflorous habits are thought to be related to seed dispersal (and/or pollination) by arboreal mammals. Southeast Asian rainforests are renowned for their abundance and diversity of native delicious fleshy fruits, such as durian (*Durio*, Malvaceae), mangosteen (*Garcinia*, Clusiaceae), rambutan (*Nephelium*, Sapindaceae), langsat (*Lansium*, Meliaceae), mango (*Mangifera*, Anacardiaceae), jackfruit (*Artocarpus*, Moraceae), starfruit (*Averrhoa*, Oxalidaceae), banana (*Musa*, Musaceae), salak (*Salacca*, Arecaceae), and so forth. These fruits likely evolved to be dispersed by arboreal mammals, particularly primates such as the orangutan, gibbon, and macaque, which have a gustatory sense similar to that of humans. Thus, these fruits are considered delicious, and are sold in local markets (Fig. 2.31).

In contrast to the fleshy fruits, nuts of Fagaceae, Betulaceae, and Juglandaceae are dry indehiscent fruits, each containing one seed. The nut is surrounded by a tough shell (equals the pericarp), and the copious starch stored in the nut guarantees rapid initial seedling growth. The large, nutrient-rich nuts are targeted by various seed-infesting (seminivorous) insects, birds, and mammals, which are thought to be seed predators. Among these seminivorous animals, some rodents (e.g., voles and squirrels) and corvid birds such as ravens and jays conceal surplus nuts in scattered caches for future consumption. The scatter hoarding of food items (i.e., caching) is underpinned by their episodic-like memory, future planning, and mental attribution (Grodzinski and Clayton 2010). A portion of the stored nuts remains intact without being utilized and then germinates at the caches. Thus, in seed dispersal by scatter-hoarding animals, the seed itself is the reward for the seed-dispersing seed

Fig. 2.30 Delicious fruits in tropical rain forests in Borneo. (**a**) *Baccaurea racemosa* (Phyllanthaceae). (**b**) *Artocarpus integer* (Moraceae). (**c**) *Durio kutejensis* (Malvaceae). Cauliflorous habits of these fruits are associated with seed dispersal by mammals

Fig. 2.31 Wild or semidomesticated, native edible seeds and fruits sold in local markets: (**a**) Traditional Asian nuts (from the top clockwise): ginkgo (raw and roasted), shii (*Castanopsis sieboldii,* Fagaceae), kaya (*Torreya nucifera,* Taxaceae), and walnut (*Juglans regia,* Juglandaceae) sold at observance of the close of winter at Yoshida-Jinja temple in Kyoto; (**b**) Native Amazonian fruits sold in a market in Tarapoto, Peru (from the top clockwise): macambo (*Theobroma bicolor*, Malvaceae), cocona (*Solanum sessiliflorum*, Solanaceae), palta (*Persea americana*, Lauraceae), umari (*Poraqueiba sericea* Icacinaceae), camucamu (*Myrciaria dubia*, Myrtaceae), cherimoya (*Annona cherimolina*, Annonaceae), papaya (*Carica papaya*, Caricaceae); (**c–d**) Native Malayan fruits sold in a market in Miri, Sarawak, Malaysia: (**c**) Chempedak (*Artocarpus integer*, Moraceae), rambutan (*Nephelium lappaceum,* Sapindaceae); (**d**) Salak (*Salacca zalacca*, Arecaceae), belimbing merah (*Baccaurea angulate,* Euphorbiaceae)

predators. These nuts were an important source of food for humans until the development of agriculture (Fig. 2.31). Nut-bearing trees are apt to mass flower and mast synchronously at an interval of one or a few years. The mast seeding maximizes seed escape because seed predators consume a decreasing proportion of available seeds with increasing seed production (Fletcher et al. 2010). Thus, episodic and synchronous mast seeding contributes to escape from heavy attack and efficient seed dispersal by these vertebrates in the masting year (Vander Wall 2010). In addition to fruits and seeds, flowers also provide various mammals with nectar as a reward for pollination. Particularly in tropical forests, some nocturnal flowers are visited by nectarivorous mammals such as rodents, primates, and bats, among which bats are the most important pollinators. In Southeast Asia, tubular flowers of *Musa* (Musaceae), *Mucuna* (Fabaceae), and *Fagraea* (Gentianaceae), and brushlike flowers such as *Barringtonia* (Lecythidaceae) are white in color, emit a fermentation odor, secrete copious nectar at night, and are pollinated by fruit-bats (flying foxes or megabats), which search for fragrant flowers using their olfactory sense (Momose et al. 1998). In contrast, the Neotropics harbor echolocating fruit-microbats, which are pollinators of nocturnal flowers of the following families: Agavaceae, Bromeliaceae, Cactaceae, Fabaceae, Malvaceae, Campanulaceae, Bignoniaceae, Gesneriaceae, and Solanaceae (Fleming et al. 2009). A chiropterophilous plant of the genus *Mucuna* (Fabaceae) in the Neotropics has unique petals, which resonate the ultrasonic wave emitted by echolocating microbats and guide the bats to the floral nectar (von Helversen and von Helversen 1999, Simon et al. 2011).

2.14 Cultivation Mutualism

Humans, as primates, have evolved as erect-walking omnivorous hunter–gatherer species who collect fruits and vegetables; hunt mammals, birds, and reptiles; catch insects; and fish aquatic organisms. About 4000–12,000 years ago, several human populations started to cultivate useful plants that yielded nutritious seeds, fruits, leaves, and tubers (Larson et al. 2014). The cultivated plants became domesticated through careful management of their reproduction by humans, and eventually became crops that can only grow with management and protection by humans. Thus, the interaction between human and crops is an obligate cultivation mutualism that originated from an antagonistic plant–herbivore interaction. Although various parts of diverse crop species have been utilized by humans, the most important crops are cereals (Poaceae), which yield a large quantity of nutritious and storable harvest irrespective of their minute seed size. In the process of domestication of cereal crops, nonshattering habit and larger seed size have been selected, and hybridization and polyploidization of wild plant species have also been performed. Ancient human civilizations originated in 10 regions, precisely where domestication and cultivation of crops commenced (Diamond 2002). Human–crop cultivation mutualism has swept the Earth during the past several thousand years, and maize, rice, and wheat production reached 1018, 738, and 711 million metric tons in 2013, respectively (FAOSTAT 2014).

Cultivation mutualism has only evolved between plants and humans. Although the chimpanzee has an episodic memory (Martin-Ordas et al. 2010), even rudimentary plant cultivation has not been observed. In aquatic systems, however, cultivation mutualism has evolved between algae and algivorous damselfish. The damselfish, *Stegastes nigricans*, living in a coral reef ecosystem manages algal farms by defending territory against invading grazers and by weeding of unpalatable algae (Fig. 2.32). As a result, the algal farms are dominated by one filamentous red alga species, *Polysiphonia* sp. (Hata and Kato 2006). The alga is the staple food of the fish, and the alga only grows in territories inhabited by the fish, suggesting an obligate cultivation mutualism. In coral reef ecosystems harboring many algivorous damselfish species, fish species engaged in intensive farming typically utilize

Fig. 2.32 Damselfish–alga cultivation mutualism between *Stegastes nigricans* and a rodophyte *Polysiphonia* sp. observed in a coral reef of the Ryukyu Archipelago, Japan: (**a**) Exclusively distributed algal farms, each of which is occupied by a damselfish. (**b**) A closeup of an algal mat kept by a damselfish. (**c**) A damselfish removing algivorous sea urchin *Echinometra mathaei* out of its territory. (**e**) A damselfish weeding unpalatable algae (Photos by Hiroki Hata)

monoculture, whereas fish species engaged in extensive farming tend mixed-crop farms comprising several species of algae (Hata et al. 2010). In algal cultivation, neither episodic memory nor management of algal reproduction is necessary because the algae can colonize a new patch by releasing spores, and the algal farms are generally colonized by their offspring or conspecifics.

2.15 Obligate Pollination Mutualism

As described above, some seed plants have evolved palatable seeds as a reward in seed-dispersal mutualism and cultivation mutualism. In these mutualisms, seed predators became seed-dispersers and cultivators, respectively. In addition, another mutualism that involves rewarding of seeds between plants and pollinating seed-predators has been established (i.e., obligate pollination mutualism). More than 100 years ago, two popular systems were discovered in *Ficus* (Moraceae) and *Yucca* (Agavaceae), which are pollinated by fig wasps (Agaonidae) and yucca moths (Prodoxidae), respectively (Cunningham 1888, Riley 1873). The intimate

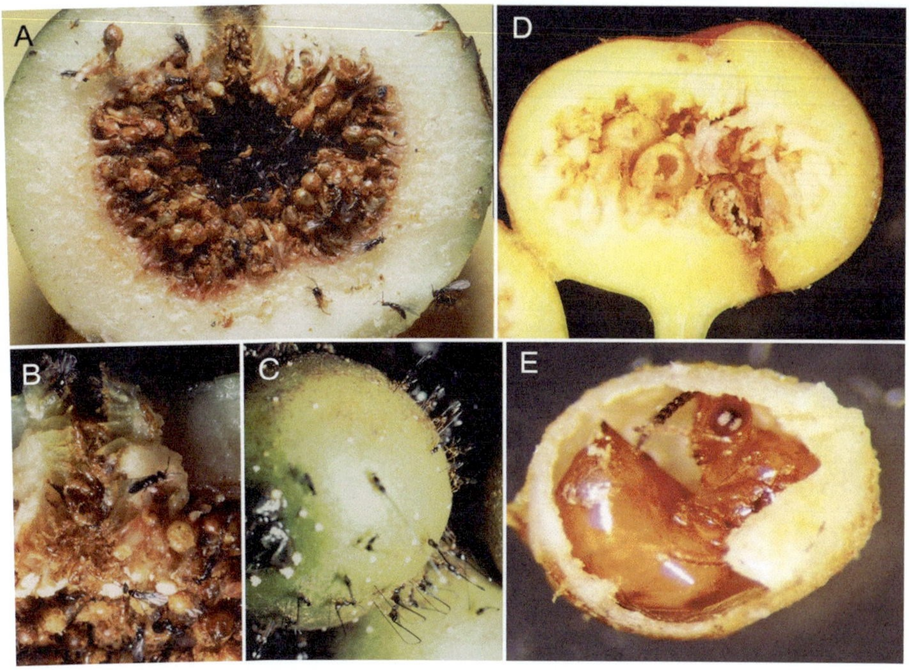

Fig. 2.33 Fig–fig wasp obligate pollination mutualism: (**a–c**) *Ficus* sp. in Borneo (Photo by Tamiji Inoue). (**d–e**) *Ficus septica* in Okinawa, Japan. (**a**) A cross-section of a syconium, showing numrous small galled ovules which produce pollinator fig-wasps. (**b**) The exit passage, through which newly emerged fig-wasps go out of the syconium after collecting pollen from the male flowers just before the exit passage. (**c**) Parasitic fig-wasps trying to insert their long ovipositors into the syconium. (**d**) A cross-section of a syconium showing infested fig seeds and the tunnel that emerged male fig-wasp bored. (**e**) Emerged female fig wasps in an infested seed

pollination mutualisms between figs and fig wasps (Fig. 2.33) and those between yuccas and yucca moths (Fig. 2.34) have long intrigued biologists because they are some of the most sophisticated examples of insect pollination. All of the species of figs and yuccas are pollinated exclusively by females of host-specific fig wasps and yucca moths, respectively, which lay eggs in the pollinated fig/yucca ovules and nourish their larvae with the resulting seeds (Janzen 1979, Powell 1992, Weiblen 2002, Pellmyr 2003, Herre et al. 2008). Some pollinating fig wasps and all of the pollinating yucca moths are morphologically and behaviorally adapted to "actively" pollinate flowers to ensure that larval food (i.e., seeds) is produced for their offspring (Pellmyr 1997, Kjellberg et al. 2001, Pellmyr and Krenn 2002, Jousselin et al. 2003). These mutualistic relationships were recognized more than a century ago; the importance of fig wasps for fig fruit maturation has been known since ancient times (Condit 1947), whereas the basic natural history of the yucca–yucca moth mutualism unfolded with Riley's observations in the late nineteenth century (Riley 1872, 1880, 1881). Detailed accounts of active pollination by fig

Fig. 2.34 Yucca–yucca moth obligate pollination mutualism: (**a–d**) *Yucca filamentosa* and its pollinator *Tegeticula yuccasella* in New York, USA. (**e–f**) *Yucca elata* in Arizona, USA. (**a**) Habitat of the plant; (**b**) A female yucca moth ovipositing in the ovary. (**c**) A female yucca moth actively pollinating the stigma. (**d**) A female yucca moth with a large clump of pollen on the mouthpart, gathered using specialized tentacles. (**e**) Dehisced capsule showing a feeding tunnel made by a yucca moth larva. (**f**) Fruits borne on tall stalk

wasps were provided more recently by Galil and Eisikowitch (1969) and Ramírez (1969).

Although long recognized as classic examples of obligate mutualisms, these associations have become principal models in various fields of ecology and evolutionary biology over the past few decades (Thompson 1994, 2005, Herre et al. 1999, 2008, Weiblen 2002, Cook and Rasplus 2003, Pellmyr 2003). Most prominently, these systems provide ideal models for studying the coevolutionary dynamics of species interactions and the mechanisms that maintain the stability of the mutualisms. The high degrees of species specificity in these associations allow straightforward assessments of the effects of coevolutionary selection acting on each species, and the high species richness (>750 and >40 species of figs and yuccas, respectively) allows a comparative approach for studying the outcomes of reciprocal selection. In addition, the costs and benefits of the mutualism for the plant can easily be measured by counting the number of seeds produced/destroyed, rather than by quantifying nutrients or protection, which most other mutualisms trade (Heil and McKey 2003, Mueller et al. 2005, Kiers and Denison 2008); this facilitates analyses of how mutualisms are maintained in the face of apparent destabilizing potential. In fact, major breakthroughs in our understanding of the coevolutionary process and mutualism stability were brought about by studies of these obligate pollination mutualisms (Pellmyr and Thompson 1992, Pellmyr and Huth 1994, Pellmyr et al. 1996a, b, Molbo et al. 2003). Recently, a remarkable pollination mutualism analogous to the fig–fig wasp and yucca–yucca moth mutualisms was discovered between the tree genus *Glochidion* (Phyllanthaceae, formerly Euphorbiaceae) and the moth genus *Epicephala* (Gracillariidae; Kato et al. 2003). In the long history of plants and their associates, mutualism is one of the most interesting interactions because their host specificity is exceedingly high and because they have attained remarkable codiversification. We focus on this mutualism in the following chapters.

Chapter 3
Biology of the Obligate Pollination Mutualism

Makoto Kato and Atsushi Kawakita

Keywords Active pollination behavior • *Breynia* • *Epicephala* • *Glochidion* • *Gomphidium* • *Phyllanthus* • Species specificity

3.1 Phyllanthaceae with Greenish Inconspicuous Flowers

Most zoophilous plants are adapted to attract reliable pollination agents by providing floral rewards such as nectar and pollen and by advertising their flowers with nongreen color and nonleafy odor, and facilitate pollination by protruding styles and stamens from flowers. The great diversity of angiosperm flowers is believed to be a product of coevolution and codiversification between plants and their pollinators.

However, there are some zoophilous plants whose floral characters do not correspond to these floral characteristics. For example, monoecious plants in several genera of Phyllanthaceae have small, greenish, inconspicuous flowers whose styles and stamens are fused and lodged at basal parts of flowers. Phyllanthaceae (leafflower family) is a diverse, globally distributed plant family, whereas their diversity peaks at tropical regions. Most Phyllanthaceae plants are herbaceous or shrubby plants, including some arboreal genera such as *Glochidion* and *Margaritaria*.

Glochidion, a monoecious tree genus of Phyllanthaceae has minute apetalous female flowers with highly specialized styles (Airy Shaw 1978; Chakrabarty and Gangopadhyay 1995). The genus consists of >300 species ranging from tropical Asia to Australia and Polynesia (Govaerts et al. 2000). Although its pollination system is unknown, *Glochidion* trees usually bear many fruits, most of which are infested by small moth larvae. The exclusively high proportion of the fruits infested

M. Kato (✉) • A. Kawakita
Graduate School of Human and Environmental Studies, Kyoto University, Yoshida-Nihonmatsu-cho, Sakyo, Kyoto 606-8501, Japan

Center for Ecological Research, Kyoto University, 2-509-3 Hirano, Otsu, Shiga 520-2113, Japan
e-mail: kato@zoo.zool.kyoto-u.ac.jp

M. Kato, A. Kawakita (eds.), *Obligate Pollination Mutualism*, Ecological Research Monographs, DOI 10.1007/978-4-431-56532-1_3

by seed-parasitic insects reminds us of the situation where a fraction of seeds in a fruit of figs and yuccas is infested by fig wasps and yucca moths.

The moths associated with *Glochidion* fruits are members of Gracillariidae, most species of which are leaf miners of various angiosperms. The leaf-mining gracillariid larvae are dorsoventrally flattened, endophytic tissue feeders, and pupate in flat cocoons woven by spun silk. Thus far, no gracillariid moths have been known to be pollinators, although they infrequently visit flowers to probe floral nectar with their elongated proboscises. Compilation of circumstantial evidence of plant reproduction and our nocturnal observation on pollination revealed that the plants are certainly pollinated by these female moths whose larvae infest seeds of the flower. The discovery of obligate pollination mutualism in *Glochidion* gave us a strong incentive to study pollination systems of other related genera in Phyllanthaceae.

The genus *Breynia* is a close relative of the genus *Glochidion*, and comprises 35 species of monoecious shrubs, distributed in tropical and subtropical regions of Asia, Australia, and the Pacific Islands (Webster 1994; Govaerts et al. 2000). Unlike *Glochidion* with dehiscent fruits, *Breynia* has fleshy fruits. Observations of pollination of two *Breynia* species in Japan and Laos additionally clarified that this genus also adopts obligate pollination mutualism.

Phyllanthus sensu stricto is one of the largest genera of angiosperms and comprises more than 800 species including herbaceous annuals and perennials, woody shrubs, and small trees (Govaerts et al. 2000). Although at least some herbaceous annual *Phyllanthus* species are pollinated by ants (Kawakita and Kato 2009), the great diversity of *Phyllanthus* species in some regions of the world led us to consider that the diversity might be related to specific animal pollination systems. Among the approximately 15 subgenera currently recognized, *Gomphidium* is a group of small trees comprising about 150 species restricted to Australia, New Guinea, and Polynesia (Holm-Nielsen 1979). Notably, this subgenus has undergone extensive diversification in New Caledonia (115 species) and now constitutes the largest genus on the islands (Schmid 1991). Thus, we made observations of pollination of the diverse *Phyllanthus* species in New Caledonia.

In this chapter, we describe obligate pollination mutualism of the three plant genera *Glochidion*, *Breynia*, and *Phyllanthus*, by focusing on their remarkable floral morphology, pollinator behavior, seed parasitism, host specificity, and the nature of the mutualism.

3.2 Mutualism in *Glochidion*

3.2.1 Flowers of Glochidion

Glochidion flowers are dimorphic, consisting of a pedunculate male flower with unfolded sepals and connate ellipsoid stamens, with a sessile or shortly pedunculate female flower composed of folded sepals and a united columnar style. The style has a narrow pit at its lobed tip. The inner surface of the pit is the stigma. This cryptic stigma is unlikely to be pollinated by wind or by ordinary insect visitors. We therefore made field observations of the pollination of three *Glochidion* species in Japan.

Trees of *G. acuminatum* (Fig. 3.1a) have male and female flowers in separate axillary clusters on each branch (Fig. 3.1b). Male flowers are aggregated at the base, and the female flowers are at the apical part of each branch. Female flowers are generally six-ovuled. Neither male nor female flowers secrete nectar.

3.2.2 Nocturnal Flower Visitor

Observation on insect visits to *G. acuminatum* flowers was made at Amami-Oshima Island, in southwestern Japan. In the daytime, the flowers are rarely visited by insects, although various insects pass by the inflorescence. However, beginning in the evening and continuing until midnight, the flowers are visited frequently by a gracillariid moth, *Epicephala anthophilia*. Female moths visit male flowers to collect pollen by inserting their proboscis into the anthers (Fig. 3.1c). After flight migration among trees or within a tree, female moths visit female flowers. Female *Epicephala* moths netted around the *Glochidion* trees almost always possess numerous *G. acuminatum* pollen grains on their ciliated proboscides (Fig. 3.1f), whereas males are never found to carry pollen grains. The behavior of the female moths on female flowers is remarkable. Visiting a cluster of female flowers, a female uncoils its proboscis, deposits the pollen grains onto the cryptic stigma (Fig. 3.1d), and then bends its abdomen to insert its long ovipositor into the stylar pit (Fig. 3.1e). The series of pollination and oviposition behavior usually lasts up to 1 min. The female walks along the branch, tapping the flowers with its proboscis and visiting each female flower sequentially to repeat the stereotypic pollination/oviposition behavior.

3.2.3 Active Pollination

Pollen attachment and moth eggs in female flowers can be examined by dissecting the styles under a microscope. Normally, a female lays an egg in a flower just above the ovules at the base of each oviposited style (Fig. 3.1h), and an average female flower receives 1.7 eggs (Table 3.1). Oviposited flowers are consistently pollinated (Fig. 3.1g), whereas unoviposited flowers are very rarely pollinated (Fig. 3.2). The average number of pollen grains deposited on a stigma after a single moth visit is less than 10 but is enough to fertilize all six ovules in an ovary.

Active pollination and the oviposition into styles by *Epicephala* moths are also observed on *G. zeylanicum* and *G. obovatum*, which have different style structures

Fig. 3.1 Flowers of *Glochidion acuminatum* and its pollinator *Epicephala anthophilia*. (**a**) General appearance of the tree. (**b**) Branches bearing many male and female flowers. (**c**) A female moth collecting pollen on a male flower with its proboscis (photo courtesy of Bruce Anderson). Note that the proboscis surface is covered with pollen. (**d**) A female moth depositing pollen on stigma. (**e**) An ovipositing *Epicephala* moth. (**f**) Pollen-loaded proboscis of a female moth (*arrow*). Bar = 0.5 mm. (**g**) Apical view of a female flower, showing the pollinated cryptic stigma. Bar = 0.5 mm. (**h**) Cross-section of a female flower with two eggs (*arrows*). Abbreviations are: st, style; te, tepal; ov, ovule. Bar =1 mm. (**i**) Branch with fruits. (**j**) Cross-section of a fruit. One seed (*arrows*) has been destroyed by a moth larva. Bar =1 cm

Table 3.1 Comparison of moth pollination, oviposition, and seed infestation among three *Glochidion* species

Glochidion species	Pollen grains	Laid eggs	Ovules	Intact seeds	Infested seeds	Sterile/ Aborted seeds
acuminatum	7.8 ±4.8 (30)	1.7 ±0.8 (262)	6.1 ±0.4 (364)	3.3 ±1.7 (364)	1.8 ±1.7 (364)	0.9 ±1.2 (364)
zeylanicum	32.4 ±12.3 (51)	2.3 ±1.1 (51)	10.3 ±0.9 (104)	2.1 ±2.7 (104)	7.6 ±2.9 (104)	0.6 ±1.0 (104)
obovatum	26.7 ±11.2 (40)	1.6 ±0.6 (40)	11.9 ±0.4 (56)	3.1 ±2.6 (56)	4.8 ±2.5 (56)	1.3 ±0.6 (56)

Although the typical ovule number is different among the three species, 20–54% of seeds were intact even after multiple oviposition by *Epicephala* moths and infestation by nonpollinating seed-parasitic moths. Means ± SD are shown. Numbers in parentheses represent the number of examined fruits

(Figs. 3.2 and 3.3). *G. zeylanicum* has ovoid, budlike female flowers whose styles are almost completely enclosed by sepals (Fig. 3.2d). The flowers have small openings at the apical tip that lead to the narrow stigma pit of the fused styles. Female flowers are visited at night by *Epicephala bipollenella*, which actively pollinates the female flower (Fig. 3.2e), inserts its long ovipositor into the narrow stigma pit (Fig. 3.2f), and lays an egg. Female flowers of *G. obovatum* are columnar like those of *G. acuminatum*, but differ in having distinctly swollen ovaries (Fig. 3.2j). At night the flowers are actively pollinated by *Epicephala obovatella* (Fig. 3.2k), which inserts its abdomen between the style and calyx, and lays an egg into the locules directly through the ovary wall (Fig. 3.2l), rather than through the stigma.

3.2.4 *Seed Infestation by Pollinator Moth Larva*

Fertilized ovules begin to develop, and unpollinated female flowers abscise shortly thereafter. The hatched moth larva bores into the ovary and consumes a few developing seeds within a fruit (Fig. 3.1j). In *G. acuminatum*, a larva usually consumes two seeds to complete larval growth, and escapes from the fruit to pupate on the litter. The life cycles of both the plant and its pollinator moth are inseparably linked (Fig. 3.4). Seed destruction is caused mostly by *Epicephala* moths, but nonpollinating seed-parasitic moths of Pyralidae and Tortricidae also infest seeds. On average, one fruit has 6.1 ovules, of which 1.8 are infested by moth larvae, 3.3 are intact, and 0.9 are sterile or aborted. The overall outcomes are similar among the three *Glochidion* species (Table 3.1).

Fig. 3.2 Flowers and pollinators of *Glochidion zeylanicum* (**a–h**) and *G. obovatum* (**i–l**). (**a**) General appearance of *G. zeylanicum*. (**b**) Flowering branch with male and female flowers toward the base (*right*) and apex (*left*), respectively. (**c**) Male flowers. Bar = 5 mm. (**d**) Female flowers. Bar = 5 mm. (**e**) A female of the pollinator *Epicephala bipollenella* actively depositing pollen on the cryptic stigma. (**f**) A pollinator female ovipositing in the style. (**g**) Fruiting tree. (**h**) Dehisced fruits showing seeds with orange sarcotesta. (**i**) General appearance of *G. obovatum*. (**j**) Flowers and fruits. Male and female flowers are borne toward the base (*left*) and apex (*right*), respectively. (**k**) A pollinator (*Epicephala obovatella*) female actively pollinating the female flower. (**l**) A pollinator female ovipositing through the lateral ovary wall

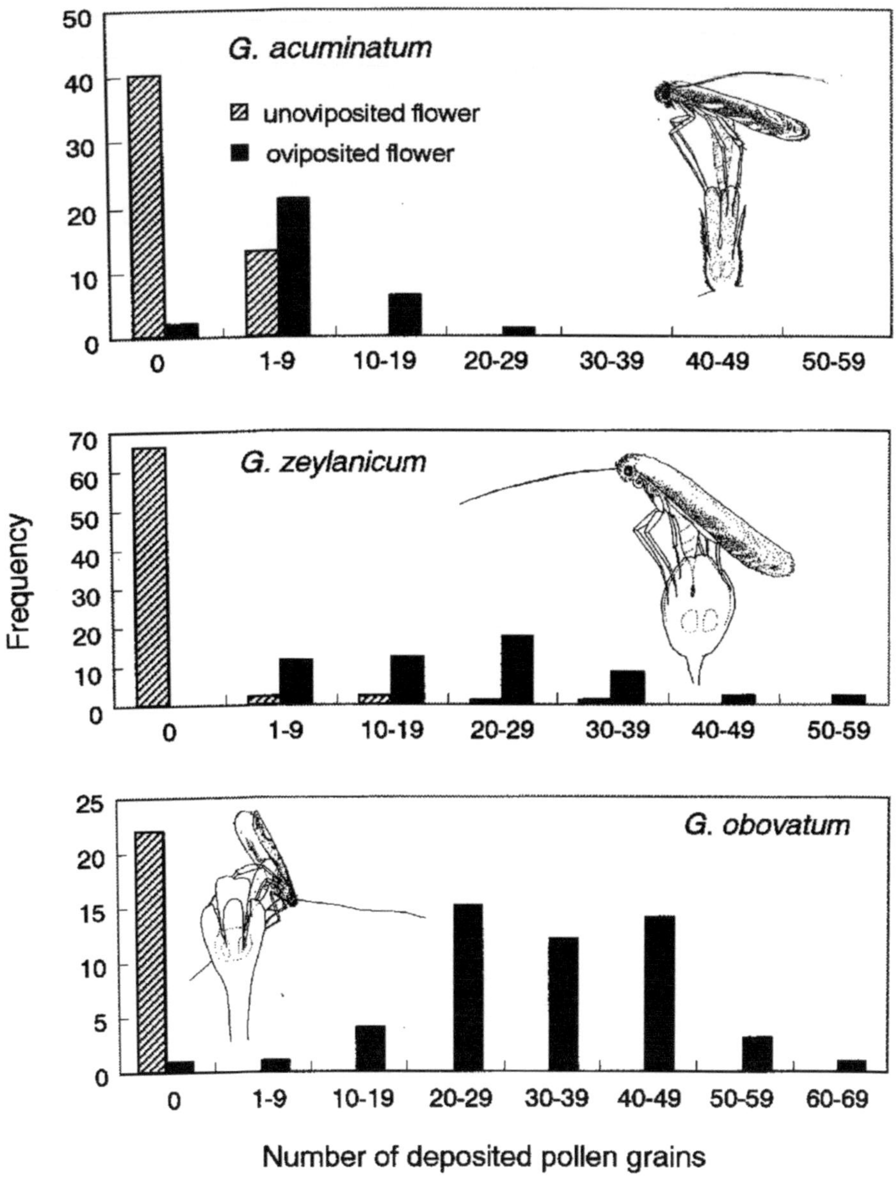

Fig. 3.3 Frequency distributions of the number of pollen grains attached to oviposited (solid) and unoviposited stigmas (shaded) of three *Glochidion* species: *G. acuminatum*, *G. zeylanicum*, and *G. obovatum*. The typical ovipositing postures and oviposited eggs of each *Epicephala* moth species are shown in each inset (Reproduced from Kato et al. (2003). Copyright (2003) National Academy of Sciences)

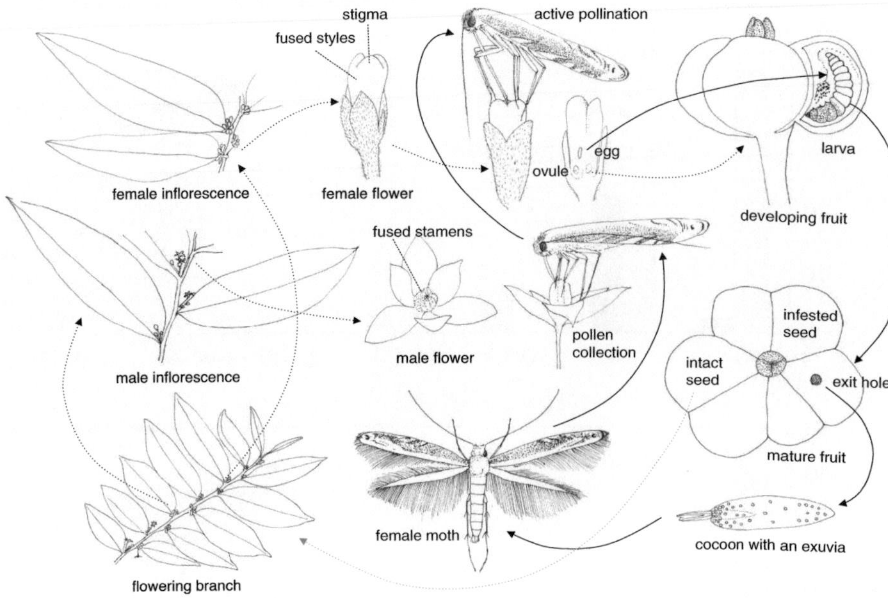

Fig. 3.4 Life cycles of *Glochidion acuminatum* (*broken arrows*) and its pollinator moth, *Epicephala anthophilia* (*solid arrows*)

3.2.5 Host Specificity of Pollinator Moth Species

Surveying other *Glochidion* species in Japan and Taiwan, we found that fruits of all six species harbored an individual (rarely two) of seed-parasitic *Epicephala* species that could be distinguished by its genitalic morphology. Host-specificity of the moths was not surprising because several *Glochidion* species often co-occur at our study sites without apparent hybridization. Host-specificity of the moths was confirmed by investigating nucleotide sequence variation in the mitochondrial cytochrome oxidase subunit I gene (COI) among sampled *Epicephala*, which clearly showed that the sequences of moths collected from different *Glochidion* hosts are highly divergent (3–9% pairwise difference), whereas the divergence between moths using the same *Glochidion* host was minimal (<0.5%).

These observations indicate that at least three *Glochidion* species are pollinated by a species-specific seed-parasitic *Epicephala* species, at the cost of infested seeds. It is notable that the female moth has an exceptionally long ovipositor to insert an egg into a style, and a ciliated proboscis to collect pollen. The moths actively pollinate flowers, similar to fig wasps and yucca moths. The seed-parasitic habit of *Epicephala* is unique in Gracillariidae, most species of which are leaf miners.

3.2.6 Widespread Mutualism in Glochidion *Plants*

Additional *Glochidion* species in New Caledonia, Fiji, Australia, Malaysia, Laos, and Myanmar all had traces of limited seed infestation by the moths. Thus, *Glochidion–Epicephala* mutualism may be widespread among the >300 known *Glochidion* species. This mutualism shares many characteristics with fig–fig wasp and yucca–yucca moth mutualisms, because the reward for the pollinator is the developing ovules or seeds. Furthermore, the sister groups of these pollinator taxa are endophytic herbivores, and the pollinators are females with elongated ovipositors. Outstanding diversification has occurred only in *Ficus* (>750 spp.) and *Glochidion*, both of which are tropical monoecious or gynodioecious woody plants that have highly specialized styles into which small pollinating insects oviposit.

In Malaysia, *Glochidion* is the largest genus (150 spp.) of Phyllanthaceae (Govaerts et al. 2000), and the principal species-diagnostic characteristic is structure of the style (Airy Shaw 1978; Chakrabarty and Gangopadhyay 1995). Because pollinating moths oviposit into styles using diverse and specific methods, the length of their ovipositor and their oviposition behavior are crucial for such specialization. Thus, the specialized structure of the *Glochidion* style and the specialized oviposition behavior of the moths may well serve as barriers against both *Glochidion* hybridization and host-shift by the moths. Plant speciation based on these traits provides a selective basis for speciation and high diversity.

3.3 Mutualism in *Breynia*

3.3.1 Flowers of Breynia vitis-idaea

Breynia vitis-idaea is a monoecious shrub that occurs in forest margins of tropical and subtropical forests in Asia (Fig. 3.5a). The species is distributed from Pakistan to the southern part of Japan, including most parts of tropical Southeast Asia (Govaerts et al. 2000). The flowers lack petals and are dimorphic, with male flowers (Fig. 3.5b) arranged toward the base and female flowers (Fig. 3.5c) at the apex of each branch. Typically, only one or two flowers are borne on axils. Male flowers have fused calyx lobes with inflexed apical ends that make the stamens unlikely to be accessible to opportunistic flower visitors (Fig. 3.5b). Female flowers are campanulate with three short styles fused at the center of the upper surface of the ovary (Fig. 3.5c). Female flowers have three locules, each containing two ovules. Fruits are produced shortly after pollination within 3–4 weeks. In the course of fruit development, pedicels become erect, and the fruit coat eventually turns red to dark purple (Fig. 3.5f). Flowering and fruiting occur throughout the year but typically peak in spring (March to May) and early fall (August to October) in southern Japan.

Fig. 3.5 Flowers and pollinators of *Breynia vitis-idaea* (**a–g**) and *B. fruticosa* (**h–j**). (**a**) General appearance. (**b**) Male flower. (**c**) Female flower. (**d**) A female of the pollinator *Epicephala vitisidaea* actively depositing pollen on the stigma. (**e**) A pollinator female laying an egg in the space between ovary and tepals. (**f**) Mature fruits. One of the fruits has an exit hole excavated by *Epicephala* larva (*arrow*). (**g**) A braconid wasp ovipositing on mature fruit. (**h**) Male flowers. (**i**) Female flowers. (**j**) Capsular fruits containing seeds with fleshy sarcotesta

3.3.2 Active Pollination

Nectar is produced at night on female flowers of *B. vitis-idaea*. Flowers of *B. vitis-idaea* are sometimes visited by ants, but the main pollinator is the moth *Epicephala vitisidaea*. Ants visit female flowers of *B. vitis-idaea* during the day and at night to forage nectar. However, these ants are not observed on male flowers, and they do not carry pollen. At night, female *Epicephala vitisidaea* moths visit female *B. vitis-idaea* flowers, depositing pollen grains with their proboscises (Fig. 3.5d) and subsequently laying an egg within the interspace between calyx lobes and ovary (Fig. 3.5e). The pollination–oviposition behavior is sometimes repeated twice on the same flower. As in *Glochidion*, all the moths that visit female flowers carry numerous pollen grains on their proboscises.

3.3.3 Obligate Pollination Mutualism

Pollinated flowers almost always have moth eggs, whereas unpollinated flowers only rarely have eggs, indicating that *Epicephala vitisidaea* moths are likely exclusive pollinators of *B. vitis-idaea*. Pollen grains are aggregated at the stigmatic part of female flowers as in *Glochidion* (Kato et al. 2003), which is unlikely to occur through passive pollination. Eggs are laid between the ovary and calyx lobes, and individual flowers receive 1–4 eggs with an average of about 1.5 eggs per flower (Kawakita and Kato 2004a).

Of the six seeds contained in a fruit, roughly half are destroyed by moth larvae, and the rest remain intact. However, fruits sometimes remain infested by the moths. These uninfested fruits contain remains of *Epicephala* moth eggs, indicating that egg/larval death of *Epicephala* moths is common in *B. vitis-idaea*. Normally, a single *Epicephala* moth larva does not consume all seeds within a fruit (Fig. 3.6a), but two moth larvae are enough to destroy all seeds of a fruit (Fig. 3.6b). Braconid wasps parasitize early instar *Epicephala* larvae by laying an egg through the fruit wall with their long ovipositors (Fig. 3.5g). When this parasitism occurs, it prevents further seed consumption by *Epicephala* larvae, resulting in greater seed survival (Fig. 3.6c).

Similar mutualism is also observed in another *Breynia* species, *B. fruticosa*, which is distributed from southern China to Indochina (Fig. 3.5h–j). Fruits of this species are dehiscent, and the seeds possess sarcotesta, contrasting with the fleshy fruit of *B. vitis-idaea*. In both *B. vitis-idaea* and *B. fruticosa*, larvae of the moths consumed the developing seeds, but in total, a fraction of the seed crop was left intact, thus resulting in a net benefit to plant reproduction.

The genus *Breynia* currently comprises 35 species distributed in tropical regions of Asia, Australia, and the Pacific Islands (Govaerts et al. 2000). Plants of this genus are characterized by the fused, obconic or turbinate calyx lobes in male flowers and minute styles that are more or less fused in female flowers (Fig. 3.5b,c,h,j; Chakrabarty and Gangopadhyay 1996). These structures likely prevent effective contact with anthers and stigmas by facultative flower visitors and suggest that the specialized *Epicephala* moth pollination is potentially widespread within the genus. Fruits of *B. disticha* in New Caledonia and *B. cernua* and *B. oblongifolia* in Australia are also infested by *Epicephala* moths (A. Kawakita and M. Kato, personal observations), which further supports the widespread occurrence of obligate pollination mutualism in the genus *Breynia*.

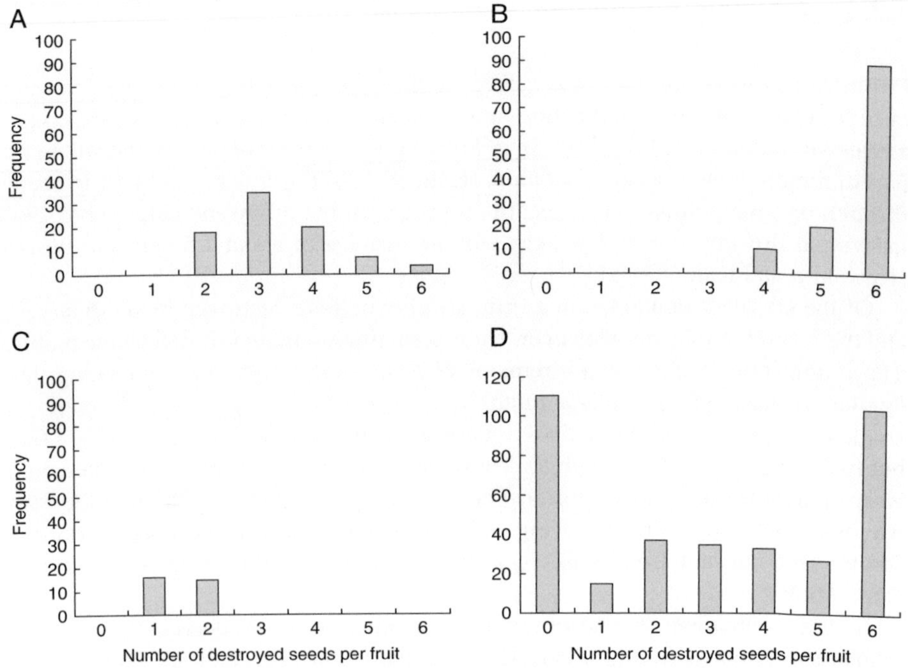

Fig. 3.6 Frequency distributions of the number of destroyed seeds per fruit in *Breynia vitis-idaea*. Fruits of *B. vitis-idaea* have six ovules. Immature larvae of *Epicephala* moths were occasionally parasitized by braconid wasps. (**a**) Fruits with one *Epicephala* larva ($N = 83$). (**b**) Fruits with more than two or more *Epicephala* larvae ($N = 119$). (**c**) Fruits with one or two parasitized moth larvae ($N = 31$). (**d**) All fruits ($N = 365$) (Modified from Kawakita and Kato 2004a)

3.4 Mutualism in New Caledonian *Phyllanthus* (*Gomphidium*)

3.4.1 Flowers of Gomphidium

In 2003, we studied the pollination biology of 25 *Phyllanthus* (all in subgenus *Gomphidium*) species at various localities in New Caledonia, and insect flower visitors were observed for two *Phyllanthus* species: *P. bourgeoisii* and *P. aeneus*. *Phyllanthus bourgeoisii* is a common rheophyte on rocky riverbanks (Fig. 3.7a), and *P. aeneus* is a shrub that is typical of serpentine scrub habitats (Fig. 3.7f). *Phyllanthus aeneus* has open male flowers, whereas those of *P. bourgeoisii* have folded calyx lobes, which make the anthers inaccessible to facultative flower visitors (Fig. 3.8). Female flowers of both species are much reduced and consist of short fused styles that are mostly covered with calyx lobes (Fig. 3.8). The flowers have three locules, each containing two ovules. The two species produce flowers and fruits throughout the year, which is also typical in *Glochidion* and *Breynia*.

Fig. 3.7 Flowers and pollinators of *Phyllanthus bourgeoisii* (**a–e**) and *P. aeneus* (**f–h**). (**a**) Habitat of *P. bourgeoisii* on rocky riverbank. (**b**) Flowering and fruiting branches. (**c**) A pollinator female actively collecting pollen on male flower. (**d**) A pollinator female depositing pollen on the stigma, which is hidden inside the imbricate tepals. (**e**) A pollinator female ovipositing in the style. (**f**) General appearance. (**g**) Flowering branch. A fruit with the exit hole made by *Epicephala* larva is shown in the inlet. Bar = 1 cm. (**h**) Female *Epicephala* moth collected on *P. aeneus* showing its pollen-coated proboscis. Bar =1 mm

3.4.2 Active Pollination

Epicephala moths are the only visitors to the flowers of the two *Phyllanthus* species. In the evening, females of undescribed *Epicephala* species use their proboscises to collect pollen from male *Phyllanthus* flowers (Fig. 3.7c). They then deliberately deposit pollen on the stigma with their proboscises (Fig. 3.7d) and subsequently lay an egg (Fig. 3.7e). Flower-visiting females consistently carry numerous pollen grains on their proboscises (Fig. 3.7h), and their behavior on

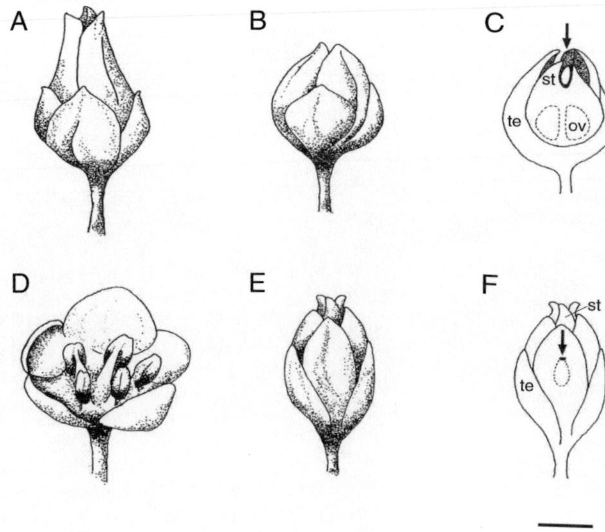

Fig. 3.8 Flowers of *Phyllanthus bourgeoisii* (**a–c**) and *P. aeneus* (**d–f**). (**a**) Male flower. (**b**) Female flower. (**c**) Longitudinal section of a female flower. The *arrow* indicates the location of an *Epicephala* moth egg. (**d**) Male flower. (**e–f**) Female flowers. *Epicephala* eggs are laid within the tissue of the calyx lobes (*arrow*). Abbreviations are: st, style; te, tepal; ov, ovule. Bar = 1 mm (Modified from Kawakita and Kato 2004b)

flowers is very similar to that observed in *Glochidion*- and *Breynia*-pollinating *Epicephala* moths (Kato et al. 2003; Kawakita and Kato 2004a).

In both *Phyllanthus* species, pollen grains are deposited on the inner surface of the fused styles, which unlikely occurs through passive pollination. In *P. bourgeoisii*, moth eggs were laid into the narrow pit of the style apex (Fig. 3.8), whereas in *P. aeneus*, eggs were laid directly into the tissue of the calyx lobes (Fig. 3.8). What is very different from *Glochidion* and *Breynia* is that not all pollinated flowers contain eggs; in both species, only about 50–70% of the pollinated flowers have eggs. *Phyllanthus bourgeoisii* flowers used for oviposition invariably have one egg per flower, whereas *P. aeneus* flowers sometimes contain two and three eggs. Unpollinated flowers do not contain moth eggs.

Overall, infestation by *Epicephala* larvae occurs in less than half of the fruits produced in both species (Fig. 3.9). Each larva consumes all six ovules to complete larval growth and emerges from the fruit (Fig. 3.7g) to pupate on the host leaves or in litter. In *P. bourgeoisii*, *Epicephala* larvae are parasitized by a braconid wasp species. These parasitoids have a significant positive effect on seed set by preventing further seed consumption by the moth larvae (Fig. 3.9).

3.4.3 *Seed Infestation by Pollinator Moth Larva*

Seed destruction by *Epicephala* larvae occurs in most, but not all, species of *Gomphidium* in New Caledonia. When larvae enter the fruits, the seeds within these infested fruits are entirely destroyed, as observed in *P. bourgeoisii* and *P. aeneus*. Curiously, moth eggs are not found in some of the pollinated flowers; for example, moth eggs occurred in proportions ranging from 25% (*P. poumensis*)

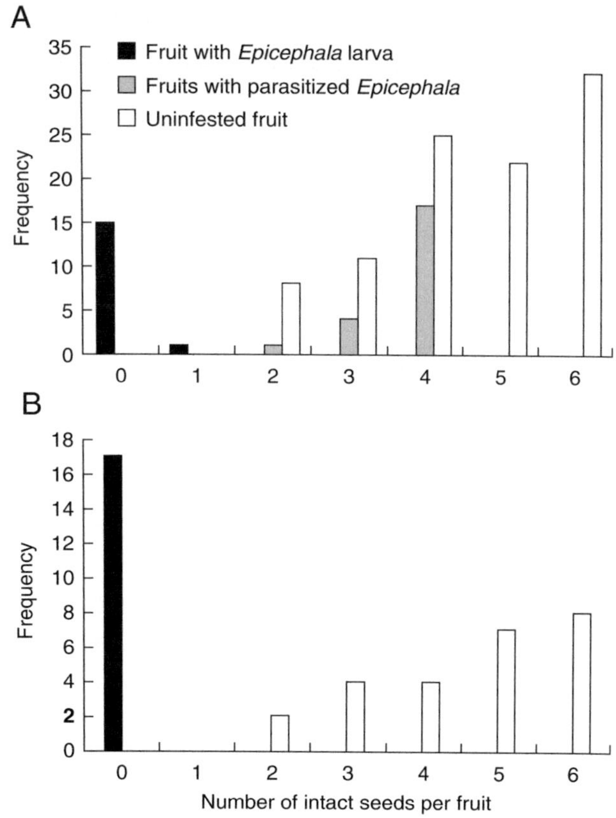

Fig. 3.9 Frequency distribution of the number of intact seeds per fruit. (**a**) *Phyllanthus bourgeoisii*. (**b**) *P. aeneus*. *Epicephala* larvae were parasitized by braconid wasps. Fruits from which moths/wasps had already emerged were assigned to each category based on differences in the exit-hole structure. The number of intact seeds within uninfested fruits ranged from two to six due to the presence of unfertilized/aborted ovules and/or empty, sterile seeds (Modified from Kawakita and Kato 2004b)

to 95% (*P. buxoides*) of the pollinated flowers in one sampling. Eggs are laid on the external surface of the flowers, and thus oviposition by adult moths does not damage the ovary. The *Gomphidium–Epicephala* association is thus probably widespread among other members of this subgenus, but the strength of the interaction may not be as strong as in *Glochidion* and *Breynia*.

Examination of adult moths reared from various *Gomphidium* species (*P. bourgeoisii*, *P. aeneus*, *P. mangenotii*, *P. guillauminii*, *P. chamaecerasus*, *P. koniamboensis*, *P. pilifer*, *P. vulcani*, and *P. pancherianus*) suggested that high species specificity is also the rule in *Gomphidium*. In most cases, individual moths that developed from different hosts are easily distinguishable by wing pattern and relative size. The host specificity of the moths is further supported by nucleotide sequence variations of the COI gene (Fig. 3.10). Sequence differences between individuals collected from different hosts are 3–15%, whereas differences are <1% among individuals parasitizing the same host, despite regional co-occurrence of the host plants (*P. bourgeoisii* and *P. chamaecerasus* at Chutes de Ba, *P. aeneus* and *P. mangenotii* at Cap Bocage, and *P. tiebaghiensis* and *P. guillauminii* at Tiébaghi).

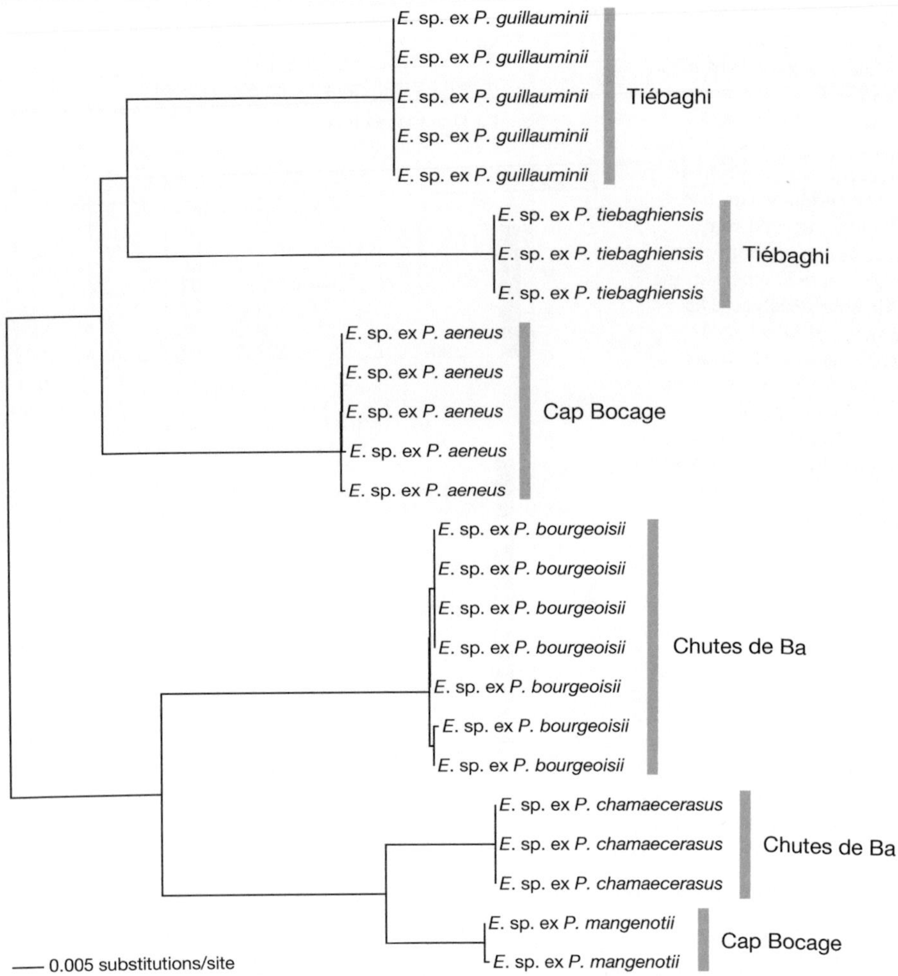

Fig. 3.10 Unrooted neighbor-joining phylogram depicting relative branch lengths within and among *Epicephala* moth individuals collected from different *Phyllanthus* host species. The tree is based on uncorrected pairwise distances within 1317 bp of the mitochondrial cytochrome oxidase subunit 1 gene (COI). All the moths used in the analysis are currently undescribed, and host affiliation of each individual moth is given in parentheses. Locality information is also provided to the right of shaded bars (Modified from Kawakita and Kato 2004b)

3.4.4 Mutualism or Parasitism?

The most critical factor underlying the *Gomphidium–Epicephala* interaction is that a fraction of the fruits is left untouched by the moths. This is most likely brought about by the absence of moth eggs in a fraction of pollinated flowers. One possible explanation for this pattern of egg distribution is that *Epicephala* eggs may be lost from some flowers, possibly by egg predation or strong desiccation. In some yucca–yucca moth interactions, high mortality of eggs and/or early instar larvae is an important process for limiting seed consumption by the moths (Addicott and Bao 1999; Csotonyi and Addicott 2001; Shapiro and Addicott 2003). However, in *P. aeneus*, moths oviposit directly into the tissue of the calyx lobes, thereby scarring the surface of the lobes. Such scars were not observed in flowers without moth eggs, which may allow exclusion of egg mortality as an explanation.

Another possibility is that the moths do not always oviposit in flowers that they pollinate. Such a behavior seems paradoxical, because the moths do not benefit from the pollinating behavior itself. However, this seemingly altruistic pollination behavior can be advantageous to the moth if the presence of uninfested fruits forces the braconid parasitoid to spend excessive time in detecting moth larvae, thus decreasing the probability of successful detection and parasitism. The last possibility is that there are as yet undiscovered copollinators in this system. The flowers of some *Gomphidium* species are reddish, unlike the characteristic greenish flowers of *Glochidion* and *Breynia* (images of flowers are shown in Chap. 4). Also, *Gomphidium* species in the section *Adenoglochidion*, which includes *P. aeneus*, possess female flowers with spread bifid styles, which are also in contrast to the reduced fused styles of *Glochidion* and *Breynia*. More detailed examination of moth pollination and of other potential visitors is clearly needed before these hypotheses can be evaluated robustly.

In some obligate pollination/seed-parasitic mutualisms, plants selectively abscise flowers that contain large numbers of eggs, thereby preventing excessive seed destruction (Pellmyr and Huth 1994; Richter and Weis 1995; Wilson and Addicott 1998; Addicott and Bao 1999; Goto et al. 2010). In light of this, it is paradoxical that *Gomphidium* trees do not abscise flowers containing moth eggs, despite the substantial cost imposed by the larvae. One explanation for the lack of selective abscission in *Gomphidium* is that the potential for such a mechanism is weak because the available resources do not limit seed set and thus need not be allocated to high-quality fruits. However, as hypothesized for some yuccas (Addicott and Bao 1999), *Gomphidium* flowers may not have proximate cues to predict whether their ovules are infested, because oviposition by *Epicephala* moths does not directly damage the ovary. Selective abscission may be more likely involved in the *Glochidion–Epicephala* mutualism, in which the ovipositor of the moth directly cuts through the ovary and/or style tissue, and the reproductive success of the plant strongly depends on the number of eggs laid per flower (Kato et al. 2003; Goto et al. 2010).

Given that *Gomphidium* plants do not possess a mechanism by which to prevent excessive exploitation by *Epicephala* moths, there is also no means by which the pollinators can retaliate against being overexploited by the plant. Once a plant acquires the ability to abscise flowers containing moth eggs selectively, it attains higher relative fitness, which would rapidly lead to pollinator extinction. It is important to note that such a pathway leading to the breakdown of the system is inherently avoided in *Glochidion* and *Breynia*, because the exclusive pollinators of the plants consistently infest the flowers that they pollinate.

Although the proximal process generating seed set in *Gomphidium* plants requires further study, there are major differences in feeding patterns between *Epicephala* moths associated with *Gomphidium* and *Glochidion*/*Breynia* fruits and different mechanisms may be responsible for the evolutionary stability of these specialized interactions.

Chapter 4
Diversity of Phyllanthaceae Plants

Atsushi Kawakita and Makoto Kato

Keywords *Breynia* • *Flueggea* • *Glochidion* • *Margaritaria* • Natural history • Phyllanthaceae • Phyllantheae • *Phyllanthus* • Phylogeny • *Sauropus*

4.1 Taxonomy of Phyllanthaceae

Euphorbiaceae, one of the major components of tropical flora worldwide, has long been considered a heterogeneous group (Webster 1994; Radcliffe-Smith 2001). Recent molecular phylogenetic analyses have clearly shown that Euphorbiaceae is nonmonophyletic, and consequently, Pandaceae, Phyllanthaceae, Picrodendraceae, Putranjivaceae, Peraceae, and Centroplacaceae have been removed from it, although these families all belong to the order Malpighiales together with the updated Euphorbiaceae (Euphorbiaceae s. str.; Angiosperm Phylogeny Group III 2009). Thus, the small unisexual flowers and trilocular capsule that characterize the plants of the former Euphorbiaceae are plesiomorphic or convergent characters. For example, Rafflesiaceae, the family with the world's largest, unisexual but unilocular flowers, is embedded in the former Euphorbiaceae (Davis et al. 2007); the clade sister to Rafflesiaceae is now Euphorbiaceae s. str., and the lineage sister to (Rafflesiaceae + Euphorbiaceae s. str.) is split as Peraceae (Fig. 4.1). Of the seven former Euphorbiaceae families, Euphorbiaceae s. str. and Phyllanthaceae are by far the largest, with the former containing about 300 genera and about 7500 species, and the latter 54 genera and about 2000 species (Table 4.1). Webster (1994) noted an important morphological feature of Phyllanthaceae that clearly distinguishes them from Euphorbiaceae s. str.; the former has two ovules per locule whereas the latter only has one ovule per locule.

A. Kawakita (✉)
Center for Ecological Research, Kyoto University, 2-509-3 Hirano, Otsu, Shiga 520-2113, Japan
e-mail: kawakita@ecology.kyoto-u.ac.jp

M. Kato
Graduate School of Human and Environmental Studies, Kyoto University, Yoshida-Nihonmatsu-cho, Sakyo, Kyoto 606-8501, Japan
e-mail: kato@zoo.zool.kyoto-u.ac.jp

M. Kato, A. Kawakita (eds.), *Obligate Pollination Mutualism*, Ecological Research Monographs, DOI 10.1007/978-4-431-56532-1_4

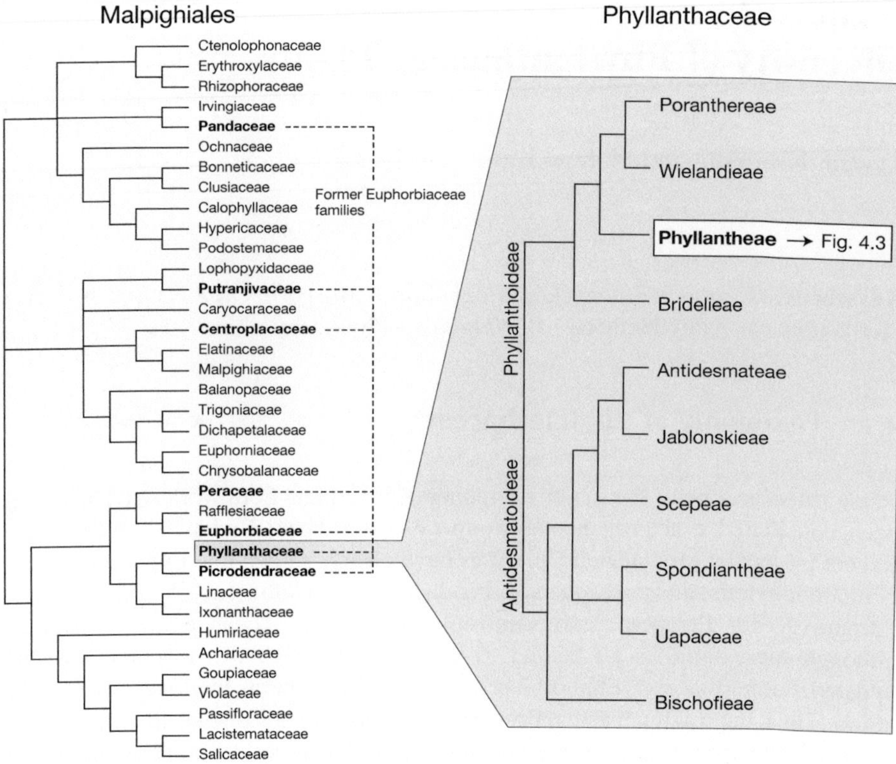

Fig. 4.1 Phylogenies of Malpighiales (*left*) and Phyllanthaceae (*right*). Former Euphorbiaceae families on the Malpighiales phylogeny are indicated in bold. Phyllanthaceae is divided into two subfamilies, Antidesmatoideae and Phyllanthoideae, the latter of which includes the tribe Phyllantheae. Phylogenetic relationships are based on Davis et al. (2005) for Malpighiales and Kathriarachchi et al. (2006) for Phyllanthaceae

Phyllanthaceae is a predominantly tropical family of shrubs and treelets and rarely of herbs and trees. Depending on the author, the family contains 54–60 genera, grouped into 2 monophyletic subfamilies and 10 well-defined, monophyletic tribes (Table 4.1). Of these, association with *Epicephala* moths is limited to Phyllantheae, which has more than half the species in the family (>1200 species) and contains *Glochidion*, *Breynia*, and *Phyllanthus*. Therefore, this chapter focuses entirely on the tribe Phyllantheae and reviews its diversity, phylogeny, and natural history. Recent molecular phylogenetic studies have greatly advanced our understanding of Phyllantheae classification, setting a cornerstone of evolutionary analysis on obligate pollination mutualism.

Table 4.1 Families traditionally classified in Euphorbiaceae and tribal classification of Phyllanthaceae

Family Subfamily Tribe	Genera	Species	Representative genera
Euphorbiaceae	300	7500	*Euphorbia, Croton, Acalypha, Macaranga*
Pandaceae	3	15	*Galearia, Microdesmis, Panda*
Picrodendraceae	24	80	*Picrodendron, Oldfieldia, Austrobuxus*
Putranjivaceae	4	210	*Drypetes*
Peraceae	5	135	*Clutia, Pera*
Centroplacaceae	2	6	*Centroplacus, Bhesa*
Phyllanthaceae	25	2000	
Antidesmatoideae			
Bischofieae	1	1	*Bischofia*
Uapaceae	1	60	*Uapaca*
Spondiantheae	1	1	*Spondianthus*
Scepeae	8	200	*Aporosa, Baccaurea*
Jablonskieae	2	2	*Jablonskia*
Antidesmateae	8	120	*Antidesma*
Phyllanthoideae			
Bridelieae	13	230	*Bridelia, Cleistanthus, Amanoa*
Wielandieae	6	25	*Wielandia*
Poranthereae	8	120	*Actephila, Andrachne, Meineckia*
Phyllantheae	9	1200	*Phyllanthus, Glochidion, Breynia, Sauropus*

4.2 Genera of Phyllantheae

The generic classification of Phyllantheae is still contentious (Hoffmann et al. 2006; Kathriarachchi et al. 2006; Hoffmann 2008; van Welzen et al. 2014), and the number of genera is likely to change as additional progress is made. In this book, we use the following nine genus names to refer to the plants of Phyllantheae, acknowledging the fact that some are unnatural (paraphyletic) groups and should be redefined: *Margaritaria*, *Plagiocladus*, *Lingelsheimia*, *Heterosavia* (formerly a section of *Savia*), *Flueggea* (including former *Richeriella*), *Phyllanthus* (including former *Reverchonia*), *Sauropus*, *Breynia*, and *Glochidion* (Table 4.2). Of these, the last three genera (*Sauropus*, *Breynia*, and *Glochidion*) are clearly embedded within *Phyllanthus* (Kathriarachchi et al. 2005, 2006). *Breynia* is further nested within *Sauropus* (Pruesapan et al. 2008, 2012).

Hoffmann et al. (2006) proposed inclusion of *Sauropus*, *Breynia*, and *Glochidion* in *Phyllanthus* to avoid paraphyletic construction of the latter. By including the three embedded genera, *Phyllanthus* becomes a giant genus of >1200 species. A number of taxonomic and nomenclatural changes have been made for regional flora (Chakrabarty and Balakrishnan 2009; Wagner and Lorence 2011), but many more new combinations are necessary to accommodate fully the >400 species of *Sauropus*, *Breynia*, and *Glochidion* in *Phyllanthus*. Alternatively,

Table 4.2 The genera of Phyllantheae used in this book

Genus	No. of Species	Distribution
Margaritaria	13	Widely distributed in tropics
Plagiocladus	1	Cameroon, Congo, Gabon
Lingelsheimia	7	Africa (Tanzania, Zaire, Gabon), Madagascar
Heterosavia	4	Caribbean
Flueggea	16[a]	Scattered in tropics, temperate east Asia and Europe
Phyllanthus	>800[b]	Widely distributed in tropics and subtropics
Sauropus	>80	Indo-Australian, Mascarene Islands
Breynia	35	Indo-Australian, the Pacific east to Fiji Islands
Glochidion	>300[c]	Indo-Australian, the Pacific east to Pitcairn Islands

[a]Including former *Richeriella* (Hoffmann et al. 2006)
[b]Including former *Reverchonia* (Webster 2007)
[c]Excluding seven Madagascan species transferred to *Phyllanthus* (Hoffmann and McPherson 2003)

van Welzen et al. (2014) proposed dividing *Phyllanthus* into >20 smaller genera because there is no morphological characteristic that defines the expanded *Phyllanthus*. In fact, the flowers, fruits, and growth forms of *Phyllanthus* are unusually diverse for a single genus, particularly if the genus is expanded to include *Sauropus*, *Breynia*, and *Glochidion*. Because the most recent molecular phylogenetic study of Phyllantheae only included approximately 10% of the constituent species (Kathriarachchi et al. 2006), an improved taxon sampling may group the species of *Phyllanthus* into many monophyletic, morphologically recognizable clades that can be recognized at generic rank. However, in the absence of a firm consensus at present, we adopt the traditional use of *Phyllanthus*, *Sauropus*, *Breynia*, and *Glochidion* in this book (Webster 1994; Radcliffe-Smith 2001; but see Chap. 12 for alternative names of Polynesian *Glochidion* species).

The genus *Sauropus* and the embedded *Breynia* have recently been thoroughly reviewed based on morphology and molecular data (Pruesapan et al. 2008, 2012; van Welzen et al. 2014). Based on the results of well-sampled phylogenetic analysis, van Welzen et al. (2014) expanded *Breynia* to include the Southeast Asian species of *Sauropus*, and reinstated the Australian *Synostemon*, a section of *Sauropus*, to generic rank, making each genus monophyletic and morphologically definable. However, we did not reflect these changes here due to the lack of consensus on whether to expand or divide *Phyllanthus*, as discussed above. On the other hand, the inclusion of *Richeriella* in *Flueggea* (Hoffmann et al. 2006) and of *Reverchonia* in *Phyllanthus* (Webster 2007) is generally not contradicted, so we adopted these changes in this book. The name *Phyllanthus* is derived from the Latin *phyll* (leaf) and *anthus* (flower), and its common name is leafflower. This is because the leaves on the lateral, flower-bearing branches of most *Phyllanthus* are flatly arranged on a horizontal plane, resembling the leaflets of compound leaves, and giving an overall impression that flowers are borne on (compound) leaves. Thus, *Phyllanthus* plants are relatively easy to recognize in the field compared to plants of other Phyllanthaceae genera. Phyllanthoid branching is a distinguishing morphological characteristic shared among most, but not all *Phyllanthus* species including

Fig. 4.2 Phyllanthoid (**a**) and nonphyllantoid (**b**) branching. (**a**) Vertical axis of *Breynia retusa* lacks normal leaves (*arrows*), and thus the branching is phyllanthoid. (**b**) Vertical axis of *Flueggea suffruticosa* bears normal leaves and flowers (*arrows*), retaining the ancestral nonphyllanthoid branching

Sauropus, *Breynia*, and *Glochidion* (Webster 1956; Fig. 4.2). In species with phyllanthoid branching, the leaves on the main (vertical) axes are reduced to scales (called cataphylls), and normal leaves are only developed on lateral, flower-bearing axes. This is another reason that the lateral axes of *Phyllanthus* look like compound leaves, particularly in species in which the branchlets are deciduous. Phyllanthoid branching was once gained at an early stage of *Phyllanthus* diversification, probably as an adaptation to herbaceous or shrubby habits in open forest habitats, and was independently lost in several derived lineages, some of which are arboreal. Figure 4.3 summarizes the most recent understanding of Phyllantheae phylogeny based on molecular data (Kathriarachchi et al. 2006; Kawakita and Kato 2009), incorporating the most recent subgeneric arrangements within *Phyllanthus* (Kathriarachchi et al. 2006; Ralimanana and Hoffmann 2011, 2014; Ralimanana et al. 2013). The genus *Phyllanthus* is subdivided into about 15 subgenera and numerous sections, some of which are still poorly defined and require revision. Note that there are many more species of *Phyllanthus*, the phylogenetic positions of which have not been determined; only half of the entire *Phyllanthus* diversity is represented in the figure. An important and rather unexpected finding of molecular phylogenetic analyses was the correspondence between clades and biogeographic boundaries. For example, in *Phyllanthus*, the Neotropical species are grouped into three well-supported clades (although additional clades may appear when additional species are sampled), therefore there were probably only three dispersals from the Old World to the New World and no dispersal in the reverse direction. Therefore, traditional sections or subgenera that spanned both sides of the Pacific are all artificial. The *Sauropus–Breynia* clade is divided into two subclades (corresponding to *Breynia* and *Synostemon* of van Welzen et al. 2014), each occurring exclusively in Australia and Southeast Asia (with the exception of *Sauropus macranthus* that occurs in both regions). In addition, there is a large

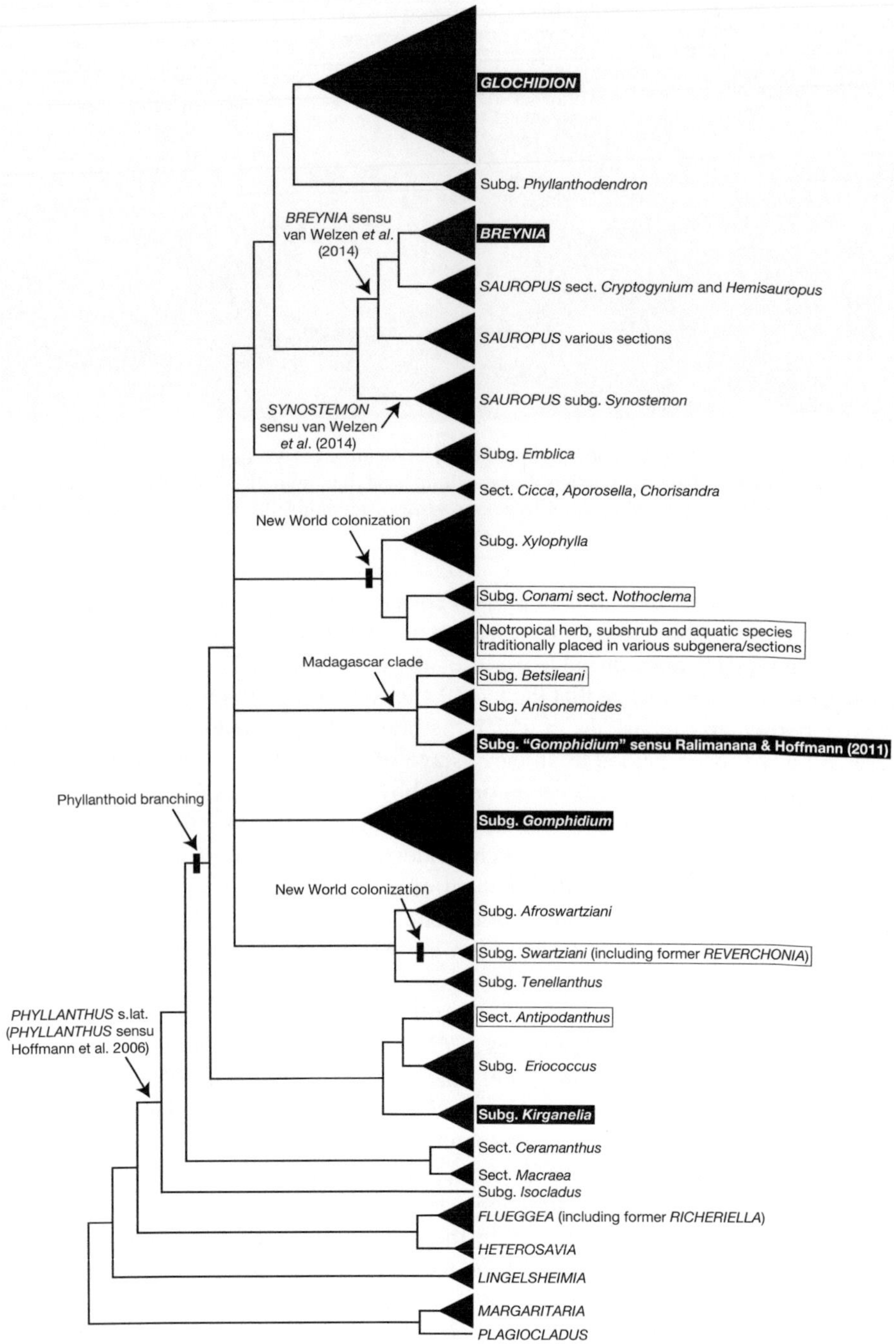

Fig. 4.3 Phylogeny of the tribe Phyllantheae. The tree is based on the most recent phylogenetic analyses of the tribe by Kathriarachchi et al. (2006) and Kawakita and Kato (2009). Species richness of each terminal clade, compiled from various sources, is provided as the area of the clade

monophyletic group of *Phyllanthus* species consisting entirely of species endemic to Madagascar. Thus, long-distance dispersals were probably much less frequent than inferred from traditional taxonomy.

4.3 Global Distribution and Diversity Patterns

The overall distribution of the five basal genera (*Margaritaria*, *Plagiocladus*, *Lingelsheimia*, *Heterosavia*, and *Flueggea*) is highly relictual. For example, *Flueggea* is widespread in the Paleotropics but also has fragmentary distributions in the Iberian Peninsula, Turkey, Caribbean, Ecuador, and northeastern Brazil (Webster 1984; Gemici 1993), which are most likely relicts (there is a single apparent dispersal to Hawaii; Hayden 1987). Similarly, *Plagiocladus*, *Lingelsheimia*, and *Heterosavia* have highly restricted distributions in West Africa, Africa and Madagascar, and the Caribbean, respectively (Table 4.2; Govaerts et al. 2000; Radcliffe-Smith 2001; Lebrun and Stork 2006; Hoffmann 2008), suggestive of range contractions from once broader distributions. On the other hand, distributions of the more derived genera (*Phyllanthus*, *Sauropus*, *Breynia*, and *Glochidion*) are likely the result of more recent dispersals. For example, *Glochidion* includes more than 300 species and ranges throughout the tropical parts of Asia, Australia, and the Pacific, with species colonizing as far east as the Pitcairn Islands (Govaerts et al. 2000). The Southeast Asian tropics have the highest diversity of Phyllantheae plants at both species and genus levels (Fig. 4.4, Table 4.2), although there are also notable diversity centers in New Caledonia (>110 spp.), Madagascar (ca. 60 spp.), Cuba (ca. 50 spp.), and Venezuela (ca. 60 spp.). *Phyllantheae* plants are less common in dense rainforests and thus have low diversity in the Amazonian basin. Much of the current Southeast Asian diversity is attributable to recent diversification of a few derived genera (*Sauropus*, *Breynia*, and *Glochidion*) rather than richness of major *Phyllantheae* lineages. The relative abundance of basal lineages (*Margaritaria*, *Plagiocladus*, *Lingelsheimia*, and *Flueggea*) in Africa, Madagascar, and South America (Table 4.2) may suggest a Gondwanan origin for the tribe as a whole.

4.4 Growth Form and Habitat

The tribe Phyllantheae has remarkable diversity of vegetative form and habitat. The plants range from very small shrubs to canopy-layer trees and a number of *Phyllanthus* lineages have become herbaceous (annual or perennial), which is

Fig. 4.3 (continued) triangle. The five lineages containing *Epicephala* moth-pollinated plants are indicated by *black* shadows on group names. *Phyllanthus* clades that have secondarily lost phyllanthoid branching are indicated by brackets around group names

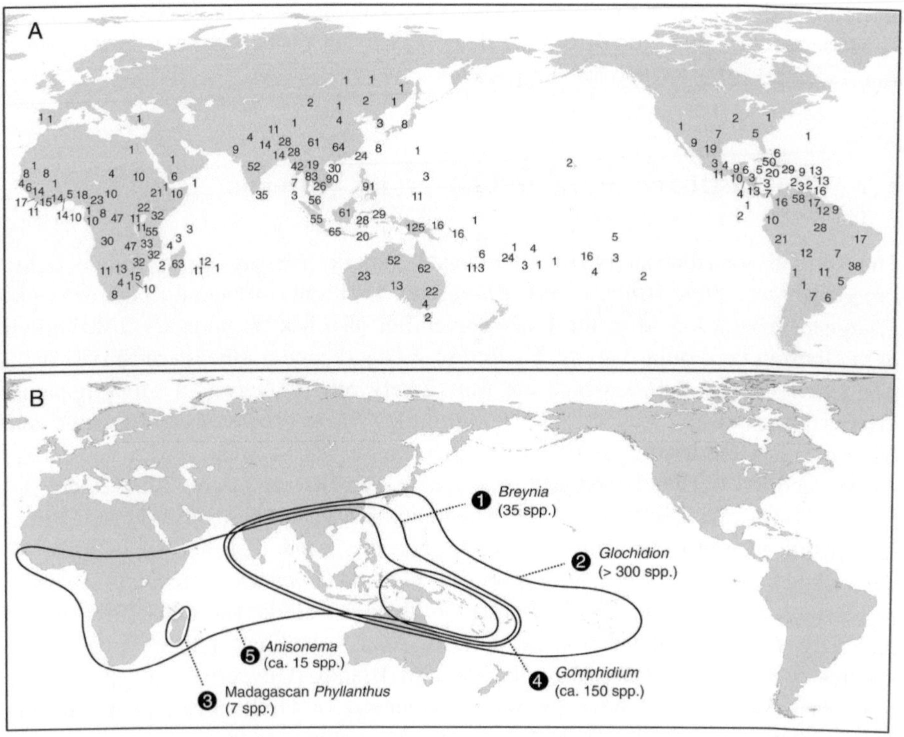

Fig. 4.4 Global distribution of Phyllantheae plants. (**a**) Number of Phyllantheae plants occurring at each geographic region, as defined by Hollis and Brummitt (1992). Data were taken from Govaerts et al. (2000). Note that the geographic regions differ in size, which affects species abundance. (**b**) Approximate distributions of the five Phyllantheae lineages with *Epicephala*-pollinated species

rare among the predominantly woody Phyllanthaceae. Herbaceous species include weeds that have become naturalized in many tropical regions of the world (e.g., *P. amarus*, *P. urinaria*, and *P. tenellus*). One Neotropical species, *P. fluitans*, is a free-floating aquatic (Macbride 1951). There are also a number of rheophytes, including *P. bourgeoisii* in New Caledonia (Schmid 1991), *P. rheophyticus* in Hainan (Li et al. 2008), and *Sauropus heteroblastus* in Indochina (van Welzen 2003). *P. microcarpus* is a scandent shrub that ascends by clinging to the surrounding vegetation (Luo et al. 2011a). *P. mirabilis* is the only succulent species (Chantaranothai 2005).

Phyllantheae plants are most frequently found in open habitats such as forest margins, roadsides, savannas, swamp edges, riverbanks, or coastal scrublands. However, some *Phyllanthus* and *Sauropus* species occur in forest understory, whereas *Margaritaria* and *Glochidion* constitute the canopy of rainforests. Large numbers of species have clear associations with limestone or serpentine vegetation (Webster 1956, 1957, 1958; Schmid 1991; Radcliffe-Smith 1996; van Welzen 2003; Thulin 2003; Gilbert and Thulin 2008; Hoffmann 2008), indicating the propensity of these plants to survive adverse soil conditions. Major radiations of *Phyllanthus* subgenera *Gomphidium* and *Xylophylla* took place in New Caledonia and Cuba, respectively (Webster 1956, 1957, 1958; Schmid 1991), where the land is abundant in serpentine substrates and is barely forested. Some *Phyllanthus* and *Sauropus* species are specialized to limestone cliffs and calcareous rock crevices (Webster 1970; Thulin 2005; Gilbert and Thulin 2008). *P. warnockii* (formerly *Reverchonia arenaria*; Webster 2007) occurs on sand dunes in arid regions of the southwestern United States (Webster and Miller 1963).

Unlike supra-annual flowering of canopy trees in Asian tropical rain forests, the shrubby/herbaceous growth form and open habitat of *Phyllanthus* plants resulted in periodical or continuous flowering, which must have been a prerequisite for the association with pollinator moth partners, because the moths must reproduce periodically and synchronously with their host plants.

4.5 Floral Biology

The flowers of Phyllantheae plants are small (usually <1 cm in diameter) and unisexual. The basal lineages (*Margaritaria*, *Plagiocladus*, *Lingelsheimia*, *and Flueggea*) primarily consist of dioecious plants (although monoecious conditions occur in *Lingelsheimia* and *Flueggea*; Webster 1984; Radcliffe-Smith 2001), whereas the derived genera (*Phyllanthus*, *Sauropus*, *Breynia*, and *Glochidion*) are almost entirely monoecious. Flowers are usually borne singly or in clusters on leaf axils but rarely on the elongate inflorescence axis (e.g., *Flueggea gracilis*, *P. acidus*). Cauliflory occurs in several *Phyllanthus*, *Sauropus*, and *Glochidion* species (Takeuchi 1999, 2003; van Welzen 2003; Thin 2007). Flowers lack petals (except in *Heterosavia*; Hoffmann 2008), as is common to many groups within the family (Webster 1994; Radcliffe-Smith 2001; Hoffmann et al. 2006). Tepals are predominantly greenish regardless of the pollination system, although some non-*Epicephala*-pollinated plants have red-purple or white tepals. Thus, Phyllantheae flowers exhibit relatively poor floral display, and although flowers of *Epicephala*-pollinated plants are hardly conspicuous, inconspicuousness itself is not necessarily their exclusive feature. Nectar production has been confirmed in many taxa (Kawakita and Kato 2004a, 2009), although many, but not all *Epicephala*-pollinated plants are nonnectariferous (Kato et al. 2003). Flowering phenology has not been well studied, but both flowering and fruiting appear to be continuous throughout the year in most species. Clear seasonality occurs in some temperate taxa (e.g., *P. flexuosus*). The most distinctive floral features associated with pollination mode are the structures of the pistils and stamens (Kawakita and Kato 2009; Kawakita 2010). The styles are usually free and bifid at the apices in non-*Epicephala*-pollinated plants, but are reduced to entire tips and are medially fused in most *Epicephala*-pollinated species (Kawakita and Kato 2009; Kawakita 2010). Similarly, filaments and anthers are usually free in non-*Epicephala*-pollinated plants, but are variously fused in species with *Epicephala* pollination. These morphological characteristics are likely associated with the pollination behavior by *Epicephala* females (Chap. 3). In the majority of species, the ovary consists of three locules, each of which has two ovules (Webster 1994; Radcliffe-Smith 2001). Thus, a single fruit contains six seeds. Exceptions include some *Glochidion* and *Phyllanthus* species that have 6, or rarely, 12–15 locules and hence twice the number of seeds. Fruits are predominantly capsules and seeds are dispersed by explosive dehiscence. Fleshy fruits (berries and drupes) occur in a number of groups, including *Flueggea*, *Breynia*, and *Phyllanthus* sections *Anisonema*, *Cicca*, and *Emblica*, which are adapted for bird dispersal. Fruits of *Margaritaria*, *Glochidion*, and some *Sauropus* and *Breynia* are tardily dehiscent, and the seeds are coated with fleshy sarcotesta, which is also consumed by birds (seeds of one New Guinean *Glochidion* are consumed by cassowaries; Takeuchi 2003).

4.6 Natural History of Each Group

To date, *Epicephala* pollination has been uncovered in five separate lineages within Phyllantheae (Kawakita and Kato 2009). In this section, we describe the natural history of each lineage of Phyllantheae along the phylogeny in Fig. 4.3, with a special focus on pollination biology and association with *Epicephala* moths. Overall, there is substantial variation among lineages in the pollination system and mode of association with *Epicephala*, which are both critically important in understanding the origin and evolutionary dynamics of obligate pollination mutualism in Phyllanthaceae.

4.6.1 Margaritaria, Plagiocladus, Lingelsheimia, *and* Heterosavia

One of the most surprising results of recent molecular phylogenetic studies of Phyllanthaceae is the inclusion of *Lingelsheimia* and *Heterosavia* in Phyllantheae. *Lingelsheimia* is a small genus of six species distributed in equatorial Africa (Gabon, Zaire, and Tanzania) and Madagascar, and was traditionally placed in Putranjivaceae (near *Drypetes*) due to its unusually large number of stamens (15–35; Webster 1994), although this placement has been questioned. *Heterosavia*, a genus of five species endemic to the Caribbean, was formerly a section of the genus *Savia*, which was shown to be polyphyletic in recent molecular phylogenetic studies (Kathriarachchi et al. 2005) and now contains only two species in the tribe Bridelieae, with the remaining species transferred to Wielandieae (Hoffmann et al. 2006). *Heterosavia* is unique among Phyllantheae with petals (Hoffmann et al. 2006). Another Phyllantheae lineage that was recently recognized at generic rank is *Plagiocladus*. This genus consists of a single central African species *Plagiocladus*

Fig. 4.5 Genus *Margaritaria*. (**a–c**) *M. indica*. (**d**, **e**) *M. discoidea*. (**a**) Canopy of *M. indica* in subtropical forest of Okinawa Island, Japan. (**b**) Male flowers. (**c**) Female flowers. (**d**) *M. discoidea* in a grassland habitat in Guinea. (**e**) Fruiting branch. Photos (**d**, **e**) courtesy of Ryutaro Goto

diandrus, which was formerly a *Phyllanthus*, but molecular phylogenetic studies clearly placed it as a sister to *Margaritaria* (Kathriarachchi et al. 2006). *Plagiocladus*, *Lingelsheimia*, and *Heterosavia* have never been observed in the field, thus it is still unknown how they are pollinated or whether they are associated with *Epicephala* moths.

Margaritaria (Fig. 4.5) consists of 14 species that are widely distributed in the tropics, with concentrations of species in the Caribbean (4 spp.) and Madagascar (4 spp.). The plants are usually dioecious, with flowers possessing distinct nectaries. Although studies on pollination of the genus have been limited due to their low abundance and canopy-reaching habit, observation of *Margaritaria indica* in Japan indicated that the greenish flowers are frequently visited by nectar-seeking flies (Kawakita and Kato 2009). Information on the fruit-feeding insect fauna is also limited, but gracillariid larvae were not found from fruit samples of *Margaritaria discoidea* collected in Guinea (Kawakita and Kato 2009). Further studies are clearly needed to search more broadly for gracillariid seed feeders in *Margaritaria*. The seeds of some species of *Margaritaria* have a spectacularly metallic blue-green, fleshy sarcotesta, which is likely involved in seed dispersal by birds. The cellular structure of *Margaritaria nobilis* seeds has recently inspired a new fiber material that changes color as it stretches (Kolle et al. 2013).

4.6.2 *Flueggea*

Flueggea (Fig. 4.6) is another lineage of early branching Phyllantheae. The genus consists of 16 species, and as noted earlier (Section 4.4), their distribution is highly relictual. Some of these species are confined to the Iberian Peninsula, Turkey, Indochina, South Africa, the Caribbean, Ecuador, and northeastern Brazil, indicating substantial range contraction of a once more continuous distribution. There is one very widespread species (*Flueggea virosa*) that occurs throughout the tropics of Asia, Australia, and Africa, and another species (*Flueggea suffruticosa*) that occurs widely in the temperate regions of east Asia. *F. virosa* is one of three species in the genus that possess baccate fruits and is dispersed by birds (A. Kawakita, personal observation), which is probably responsible for its wide distribution. *Flueggea neowawraea* is one of the two Phyllantheae species that reached Hawaii (the other being *P. distichus*), although it is unclear how they reached the islands because both plants have dry capsules and the seeds are dispersed by explosive dehiscence. *F. gracilis*, distributed in Southeast Asia, produces flowers in an elongated inflorescence axis, and the species was placed in another genus, *Richeriella*, based on this trait. The genus was subsumed to *Flueggea* after recent molecular phylogenetic studies (Hoffmann et al. 2006). Pollination biology and association with gracillariid moths have been well studied in the Japanese *F. suffruticosa* (Fig. 4.6). This species is dioecious as with most other species of the genus, and both male and female plants produce abundant nectar that is foraged by diurnal bees, flies, beetles, and butterflies (Kawakita and Kato 2009). Similar pollination biology was confirmed for *F. virosa* in Taiwan. Interestingly, *F. suffruticosa* is associated with a gracillariid moth, *Conopomorpha flueggella*, which visits female plants at night and lays eggs in the buds, flowers, or young fruits. The hatched larvae eat the developing seeds in the fruit, and each larva usually consumes all of the six seeds in its natal fruit to complete larval development. None of the moths collected on flowers possessed pollen on the proboscis or displayed pollination behavior on flowers, therefore they are pure seed parasites of *Flueggea*. The moth was initially identified as a species of *Epicephala* (Kawakita and Kato 2009), but subsequent taxonomic study described it under *Conopomorpha*, a genus that includes seed feeders of longan and lychee. However, the species is distantly related to the proper *Conopomorpha* (Kawakita et al. 2010) but is more closely related to *Epicephala*, thus it is best placed in a new genus (also see Chap. 5).

It is interesting to look further for gracillariid seed feeders in other species of *Flueggea*. To date, searches for larvae in *F. virosa* in Taiwan and Laos and *Flueggea jullienii* in Laos have yielded negative results, but examination of herbarium material suggests that the seeds of Ecuadorian *Flueggea elliptica* are

Fig. 4.6 Genus *Flueggea*. (**a–e**) *F. suffruticosa*: (**a**) flowering male individual in a temperate forest in Japan; (**b**) male flowers; (**c**) male flower showing nectariferous gland; (**d**) female flowers; (**e**) fruits. (**f**, **g**) *F. virosa*: (**f**) flowering male individual in tropical Taiwan; (**g**) fruiting branch. (**h**, **i**) *F. jullienii*: (**h**) male and female flowers of *F. jullienii*; (**i**) general habitat on a rocky riverbank in Laos. (**j**) *F. acidoton* in a spiny thicket developed on limestone soil in Jamaica

infested by *Epicephala*-like larvae (see also Chap. 5). Whether the Ecuadorian moth is related to the Asian *Conopomorpha flueggella*, or is alternatively a product of more recent colonization of *Flueggea* by *Epicephala* associated with Neotropical *Phyllanthus*, remains to be determined.

4.6.3 *Subgenus* Isocladus *and Sections* Macraea *and* Ceramanthus

The subgenus *Isocladus* was proposed by Webster (1956) to group all of the *Phyllanthus* species with nonphyllanthoid branching. However, as shown by recent molecular phylogenetic analysis, phyllanthoid-branching taxa are not monophyletic, and consequently, the only species placed in this subgenus is the type species, *P. maderaspatensis*, an herbaceous plant that originally occurred in India but has since been introduced to many parts of the African, Asian, and Australian tropics. Molecular phylogenetic analysis placed *P. maderaspatensis* as sister to the clade including all of the other *Phyllanthus* species (Kathriarachchi et al. 2006). Sections *Macraea* and *Ceramanthus* (Fig. 4.7) are two of the four sections that were separated from former *Isocladus*. Of these, *Macraea* is a group of weedy herbs or

Fig. 4.7 *Phyllanthus* sections *Macraea* and *Ceramanthus*. (**a–c**) *P. virgatus*: (**a**) general habitat on disturbed land in Laos; (**b**) male flowers visited by the ant *Solenopsis* sp.; (**c**) mature fruits. (**d, e**) *P. chrysanthus*: (**d**) general habitat on forest edge of serpentine forest in New Caledonia; (**e**) male flowers. (**f**) *P. myrtifolius* used as hedge plant in Malaysia. (**g**) General habitat of *P. cochinchinensis*

subshrubs distributed in the Old World tropics, but some have colonized the Pacific islands, including *P. chrysanthus*, which is endemic to New Caledonia. *P. myrtifolius*, originally recorded from Sri Lanka, is planted as an ornamental bush in many tropical areas. The most abundant visitors to male and female flowers of *P. ussuriensis* in Japan and *P. virgatus* in Laos are ants that seek nectar in flowers and are likely the most effective pollinators (Kawakita and Kato 2009). Notably, the fruits of *P. ussuriensis* in Japan and *P. chrysanthus* in New Caledonia are infested by larvae of *Epicephala* moths. Although the behavior of adult moths has not been observed in the field, eggs are only found in developing fruits and not in pollination-stage flowers, suggesting that moths only visit young fruits to lay eggs and do not pollinate the plants. A single larva destroys all of the six seeds contained in a fruit. *Ceramanthus* is a Paleotropical section and consists of four species. The group contains *P. cochinchinensis* (Fig. 4.7), a forest understory shrub distributed in Indochina, the flowers of which are pollinated by *Deltophora* moths (Gelechiidae) that are reported to consume *Phyllanthus* pollen by chemically breaking down sporopollenin on the surface of their proboscis (Luo et al. 2011b). The larvae feed on the leaves of *P. cochinchinensis*, therefore this is another form of reciprocally dependent mutualism. A similar mutualism was reported between *P. rheophyticus* and a different species of *Deltophora*, but the phylogenetic position of *P. rheophyticus* has yet to be determined. At present, ecological information is lacking for other members of the section *Ceramanthus*.

4.6.4 Subgenus Kirganelia

The subgenus *Kirganelia* (Fig. 4.8) is a Paleotropical subgenus of about 15 species and is one of the five Phyllantheae lineages currently known to have mutualistic associations with *Epicephala*. The subgenus is most easily recognized by the baccate fruits, which is otherwise only known in *Breynia* within *Phyllanthus* s. lato. Pollination by host-specific *Epicephala* has been studied in *P. reticulatus* and *P. microcarpus* (both in section *Anisonema*), which have particularly wide overlapping distributions from India and Sri Lanka to throughout Southeast Asia (Luo et al. 2011a; their presence in Africa remains to be verified). A detailed natural history of the obligate pollination mutualism is given in Chap. 11. The section *Anisonema* additionally contains four species endemic to Madagascar (Ralimanana and Hoffmann 2011). Although their pollination biology has not been studied, they share the fused style morphology with *P. reticulatus* and *P. microcarpus*, and herbarium specimens of the Madagascan species contained fruits infested by *Epicephala* larvae. Thus, obligate pollination mutualism is likely widespread within the section *Anisonema*.

However, there are species in this subgenus that have conspicuously colored flowers and lack association with *Epicephala*. For example, *P. flexuosus*, which is distributed in Japan, has male flowers with red-purple tepals, and female flowers with widespread styles. Flowers emit an unpleasant odor during the daytime and are

Fig. 4.8 *Phyllanthus* subgenus *Kirganelia*. (**a–c**) *P. flexuosus*: (**a**) flowering individual in the understory of *Cryptomeria* plantation in temperate Japan; (**b**) fruiting individual; (**c**) male (*black arrows*) and female (*white arrow*) flowers. (**d**) Male and female flowers of *P. oligospermus*. (**e**, **f**) *P. deplanchei*: (**e**) general habitat in a dry sclerophyllous forest in New Caledonia; (**f**) female flowers. (**g–i**) *P. reticulatus*: (**g**) general habitat in a disturbed roadside land in tropical Taiwan; (**h**) male flower; (**i**) female flower. (**j**) Fruiting individual of *P. microcarpus* in a forest edge in Laos

visited by dipteran and coleopteran insects (Kawakita and Kato 2009). *Epicephala* moths have not been found in any population studied to date, indicating that this species is free of *Epicephala*. *P. oligospermus*, which occurs in southern Japan and Taiwan, also has an overall similar ecology.

4.6.5 *Subgenus* Eriococcus *and Section* Antipodanthus

Eriococcus (Fig. 4.9) is a group of approximately 30 shrub species with marked diversity in the Indochina region. They are easily distinguished from other *Phyllanthus* by usually having tetramerous male flowers with only two stamens and four distinctly lacerate tepals. Tepals are also often red-purple in color as in *P. flexuous*, which may be an indication of a common pollination system. However, to date, observing flower visitors in *Eriococcus* has been difficult. In *P. liukiuensis*

Fig. 4.9 *Phyllanthus* subgenus *Eriococcus*. (**a–c**) *P. liukiuensis*: (**a**) habitat on a limestone outcrop in subtropical Japan; (**b**) male flower; (**c**) female flower. (**d–g**) *P. pulcher*: (**d**) general habitat on a forest floor along a river in lowland Laos; (**e**) male flower; (**f**) female flower; (**g**) fruits. (**h**, **i**) *P. pulcheroides*: (**h**) fruits; (**i**) longitudinal section of fruit with fully developed seeds and large internal airspace. (**j–l**) *P.* sp.: (**j**) habitat on forest floor of secondary forest in Vietnam; (**k**) male flowers; (**l**) female flowers

Fig. 4.10 *Phyllanthus* section *Antipodanthus*. (**a–d**) *P. dallachyanus*: (**a**) general habitat in understory of fire-prone forest in Queensland, Australia; (**b**) male flower; (**c**) female flower; (**d**) hoverfly nectaring on male flower

occurring in Japan, a few pollen-bearing gall midges were collected on flowers (Kawakita and Kato 2009), so dipterans may pollinate these plants. Another group of species in this subgenus has flowers with long pedicels that are remarkably erect in female flowers. In such species, the tepals of male flowers are imbricate, so they may have a pollination system that is entirely different from species with lacerate and colored tepals. Because *Epicephala* has not been recorded from any *Eriococcus* species, the plants in this subgenus likely have an unknown specialized pollination system. In addition, some *Eriococcus* species have unique, distinctly swollen fruits with a hollow space between the seeds and the endocarp. This is probably an adaptation to escape attack by seed-feeding braconid wasps because their seeds are frequently infested by braconid larvae, and because adult wasps oviposit through the fruit wall using exceptionally long ovipositors. Section *Antipodanthus* (Fig. 4.10) is an entirely Australian group of 13 species (Webster 2002). The plants are subshrubs and occur in fire-prone, open forests. They are notable for the lack of phyllanthoid branching. Both male and female flowers have clearly visible nectaries, and hoverflies were regularly observed visiting flowers to lick nectar during daytime.

4.6.6 *Subgenera* Afroswartziani, Swartziani *(Including Former* Reverchonia*), and* Tenellanthus

The subgenera *Afroswartziani*, *Swartziani*, and *Tenellanthus* (Fig. 4.11) comprise a clade that traditionally contains many herbs or small shrubs placed in various

Fig. 4.11 *Phyllanthus* subgenera *Tenellanthus*, *Swartziani*, and *Afroswartziani*. (**a**) *P. tenellus* (*Tenellanthus*) on roadside in New Caledonia. (**b**) *P. nummulariifolius* (*Tenellanthus*) in rainforest understory in Madagascar. (**c**) *P. amarus* (*Swartziani*) on sandy riverbank in its introduced range in Laos. (**d–g**) *P. warnockii* (*Swartziani*): (**d**) habitat on sandy desert in New Mexico, USA; (**e**) general habitat on sandy dune; (**f**) flowers; (**g**) fruits. (**h–j**) *P. lokohensis* (*Afroswartziani*): (**h**) flowering branch; (**i**) male flowers; (**j**) section of female flower showing airspace between tepals and ovary. (**k**) *P. deblis* on roadside in its introduced range in Japan

sections and subgenera. Ralimanana and Hoffmann (2011) and Ralimanana et al. (2013) reorganized these species into the above three subgenera to reflect phylogenetic relationships. *Afroswartziani* contains shrubs, trees, and rarely herbs that mainly occur in Madagascar and Africa, but many Asian species not sampled in previous molecular phylogenetic studies may also belong to this subgenus. A Madagascan species, *P. madagascariensis*, has male and female flowers with five obovate whitish green tepals and a nectariferous disk, and are visited by nectar-seeking insects such as long-proboscid crane flies. Some Madagascan species (e.g., *P. lokohensis*) have female flowers, the tepals of which form a globe surrounding the pistil that may function to keep the ovary away from *Epicephala* oviposition because their fruits are infested by *Epicephala* larvae (probably nonpollinating species). *P. debilis*, a global weed, also belongs to this subgenus. The subgenus *Swartziani* was revised by Ralimanana et al. (2013) to only include three species native to the New World. These are *P. amarus*, *P. abnormis*, and *P. warnockii* (formerly *Reverchonia arenaria*). *P. amarus* is now a globally distributed weed, and in its introduced populations in Laos their flowers are pollinated by ants, and their fruits are parasitized by nonpollinating species of *Epicephala*. *P. abnormis*, native to the southeastern United States, is also parasitized by an *Epicephala* moth, as demonstrated by the presence of pupae in herbarium specimens. *P. warnockii* is an unusual *Phyllanthus* that grows on sand dunes of the southwestern United States. The flowers are dark red and are visited by flies and small bees in the daytime. *Tenellanthus* is a well-defined subgenus of four species occurring in Madagascar, the Comoro Islands, and Africa, with one species, *P. tenellus*, becoming a pantropical weed. To date association with *Epicephala* has not been found in this subgenus.

4.6.7 *Subgenus* Gomphidium

The subgenus *Gomphidium* is distributed in tropical Australia, New Guinea, and the Pacific Islands as far east as Tonga, with the greatest concentration of species in New Caledonia. The subgenus consists of about 150 species of shrubs or trees, of which 106 species occur in New Caledonia, all of which are endemic except *P. bourgeoisii* (Schmid 1991). New Caledonia (Fig. 4.12) is located in the Pacific Ocean about 1200 km from Australia and consists of the main island (Grande Terre) and a number of surrounding small islands, with a total area of 18,580 km^2. New Caledonia was part of the Gondwana supercontinent and was separated from Australia as early as 66 Ma, although a significant proportion of the land mass was once submerged under the sea around approximately 35 Ma and many of the plants are thought to be descendants of later colonizations (Grandcolas et al. 2008). New Caledonia is renowned for its unique flora; the islands have five endemic angiosperm families (Amborellaceae, Oncothecaceae, Paracryphiaceae, Phellinaceae, and Strasburgeriaceae) and more than 100 endemic genera with the proportion of endemic species exceeding 90%. The earliest branching angiosperm Amborellaceae is only known from New Caledonia, although its fossils are known from other parts of the world. The uniqueness of the flora is exacerbated by the prevalence of serpentine rocks, which cover roughly 40% of the total land area. Serpentine rocks are ultrabasic, and plants occurring on serpentine substrates require particular adaptation to tolerate adverse, magnesium-rich soil conditions, resulting in high endemicity. Many of the notoriously invasive plants, such as *Leucaena leucocephala*, *Aleurites fordii*, and *Psidium littorale*, do not occur on serpentine soil, although they are abundant on nonserpentine regions of New Caledonia. The New Caledonian species of *Phyllanthus* are remarkably diverse in their habitats and vegetative forms, representing one of the most spectacular adaptive radiations in the plant kingdom (Fig. 4.13). It is the largest plant genus

Fig. 4.12 Diverse habitats of New Caledonian *Phyllanthus*. (**a**) *Araucaria*-dominated scrubland on serpentine rock in Poro harboring *P. pancherianus* and *P. bupleuroides*. (**b**) Maquis vegetation near Prony in the southern part of Grande Terre. Note that the soil has the color of iron oxide and lacks introduced plants. (**c**) Rainforest along the valley of Troulala harboring *P. serpentinus*, *P. buxoides*, and *P. pilifer*. (**d**) Rainforest on the foot of Mt. Koghi harboring *P. koghiensis* and *P. yaouhensis*. (**e**) Maquis vegetation along the northwestern coast near Koumac harboring *P. koumacensis*, *P. poumensis*, and *P. peltatus*. (**f**) Humid rainforest of Massif du Panié on the northeastern coast harboring *P. baladensis*, *P. valeriae*, *P. gneissicus*, *P. bourgeoisii*, and *P. loranthoides*

in New Caledonia with 114 recorded species, of which 96% are endemic. The diversity is exceptional for an island that is smaller in size than the state of New Jersey (22,610 km^2) or the island of Sicily (25,710 km^2). Their habitats range from rainforest understory, dry sclerophyllous forest, calcareous forest, riverbank, and maquis (scrubland developed on serpentine soil) with the majority of species being very narrow endemics. *P. bourgeoisii*, *P. loranthoides*, and *P. dracunculoides* are rheophytes with narrow lanceolate leaves. *P. peltatus* is unique with peltate leaves.

Fig. 4.13 New Caledonian *Phyllanthus* (subgenus *Gomphidium*). (**a**) *P. mouensis* in montane forest of Mont Mou. (**b**) *P. peltatus* with peltate leaves in the understory of serpentine scrubland in Koumac. (**c**) *P. poumensis* on serpentine rock in Paagoumène. (**d**) *P. serpentinus* on the understory of serpentine scrub in Troulala. (**e**) *P. valeriae* on the understory of rainforest of Mont Panié. (**f**) *P. mangenotii* along stream in Cap Bocage. (**g**) *P. aeneus* on the understory of serpentine forest in Thio. (**h**) *P. luciliae* in coastal forest of Cap Bocage. (**i**) *P. bourgeoisii* on flood-prone habitat in Cascade de Bâ. (**j**) *P. dracunculoides* on flood-prone habitat in Tiwaka. (**k**) *P.* cf. *montrouzieri* on dry serpentinous scrubland in Kaala-Gomen. (**l**) *P. koghiensis* tree in rainforest habitat of Mont Koghi

Fig. 4.14 Flowers of New Caledonian *Phyllanthus* (subgenus *Gomphidium*). (**a–h**) Section *Adenoglochidion* (male and female flowers on upper and lower column, respectively): (**a**) *P. peltatus*; (**b**, **f**) *P. serpentinus*; (**c**) *P. umbraecola*; (**d**) *P. pronyensis*; (**e**) *P. aeneus*; (**g**) *P. baladensis* (photo courtesy of Eri Yamasaki) and (**h**) *P. dorotheae*. (**i–p**) Section *Gomphidium* (male and female flowers on upper and lower column, respectively): (**i**, **m**) *P. poumensis*; (**j**, **n**) *P. kouaouaensis*, (**k**, **o**) *P. pilifer*; (**l**, **p**) *P. castus*. An *Epicephala* egg is laid externally on the female flower in (**n**)

The subgenus *Gomphidium* contains two large sections, *Gomphidium* and *Adenoglochidion*, and both sections occur in New Caledonia. Male flowers have imbricate folded tepals with narrow apical openings and connate stamens in *Gomphidium*, or spread tepals and free stamens in *Adenoglochidion* (Fig. 4.14). Female flowers have nonbifid styles that are either fused in the middle or spread, although species with bifid styles are known in other parts of the Pacific (Webster 1986). *Epicephala* pollination has been confirmed in two New Caledonian species, *P. bourgeoisii* and *P. aeneus*, belonging to *Gomphidium* and *Adenoglochidion*, respectively (Chap. 3).

4.6.8 Madagascan Clade

There are seven Madagascan endemic *Phyllanthus* species that have until recently been considered members of *Glochidion* (Hoffmann and McPherson 2003; Fig. 4.15). This classification was based on their much reduced and entire styles, a characteristic most likely reflecting the shared *Epicephala* pollination syndrome. Indeed, these species are otherwise widely different from the true *Glochidion* in terms of floral and seed characters as well as growth form (Hoffmann and McPherson 2003). These plants are confined to the cloud forests in the northern part of Madagascar and are pollinated by *Epicephala* moths that feed on their seeds (Kawakita and Kato 2009).

Fig. 4.15 Madagascan *Phyllanthus*. (**a**) *P. marojejiensis*. (**b**) *P. humbertii*. (**c**) *P. sambiranensis*. (**d**) Cloud forest at approximately 2500 m a.s.l. of Mt. Marojeji, Madagascar where *P. marojejiensis*, *P. humbertii*, and *P. sambiranensis* occur in the understory. (**e–g**) *P. marojejiensis*: (**e**) male flower; (**f**) female flower; (**g**) fruit. (**h**, **i**) *P. humbertii*: (**h**) male flower; (**i**) female flower

Ralimanana and Hoffmann (2011) placed the above seven species and an additional three species in the subgenus *Gomphidium*, but molecular phylogenetic analysis suggests that they are distinct from *Gomphidium* known from New Caledonia and Australasia. Thus, the imbricate tepals of male flowers and reduced styles used as the basis for their placement in *Gomphidium* is again a morphological convergence associated with *Epicephala* pollination. These Madagascan species are rather strongly united to subgenera *Betsileani* and *Anisonemoides*, both of which are also confined to Madagascar and contain 3 and 15 species, respectively. Collectively, these three lineages comprise a medium-sized clade that is entirely confined to Madagascar. Pollination biology is unknown for any of the species in *Betsileani* and *Anisonemoides*, but this information is critical to understanding the origin of *Epicephala* pollination in Madagascar. *Epicephala* pollination was studied in *P. marojejiensis* and *P. humbertii* at Mt. Marojeji in northeastern Madagascar. In both species, male flowers have imbricate tepals with either connate or free stamens, and female flowers have very short and entire styles that lean inward (Fig. 4.15). Although flower visitation by *Epicephala* has not been directly observed, pollination by *Epicephala* is suggested by the following observations: (i) fruits of the two species were both infested by species-specific *Epicephala* larvae; (ii) an *Epicephala* female bearing pollen on the proboscis was collected on *P. marojejiensis*; and (iii) there was a near exclusive association between the pollinated status of female flowers and infestation by moth eggs, suggesting that ovipositing moths are the only pollinators (Kawakita and Kato 2009). Eggs are laid internally within the ovary, and a single larva destroys all of the six seeds in each fruit.

4.6.9 The New World Clade

This large clade of entirely New World species consists of the subgenus *Xylophylla*, section *Nothoclema* of subgenus *Conami*, and various Neotropical herb, subshrub, and aquatic species traditionally placed in divergent taxonomic groups. Of these, *Xylophylla* (Fig. 4.16) is a morphologically well-characterized and species-rich group with the center of diversity in the Caribbean Islands. The largest island in the Caribbean is Cuba, which has the greatest concentration of *Xylophylla* species. Its landmass is a mosaic of soil types, including serpentine and limestone; thus, the radiation of *Phyllanthus* in Cuba is reminiscent of *Phyllanthus* diversification in New Caledonia. Many *Xylophylla* species possess flowers with fused tepals and fused styles, and their seeds are infested by *Epicephala* larvae, suggesting that at least some species are pollinated by *Epicephala* moths. However, there are species with well-developed, showy tepals and distinct nectaries (Fig. 4.16), which are probably pollinated by diurnal bees and flies. Two species in section *Cauliflori* have cauline inflorescences. Species of the section *Xylophylla* are characterized by flattened photosynthetic branches (phylloclades) and an associated reduction of leaves (Fig. 4.16). They often occupy exposed habitats on limestone or coastal cliffs

Fig. 4.16 *Phyllanthus* subgenus *Xylophylla*. (**a**) A flowering *P. chamaecrystoides* individual along stream in serpentine forest of Cuba. (**b–d**) *P. williamioides*: (**b**) flowering individual in serpentine scrubland in Cuba; (**c**) male flower; (**d**) female flower. (**e–g**) *P. orbicularis*: (**e**) flowering plant with brightly white flowers in serpentine scrubland in Cuba; (**f**) male flowers; (**g**) female flower. (**h, i**) *P. cauliflorus*: (**h**) cauline inflorescences in limestone forest of Jamaica; (**i**) female flower. (**j–l**) *P. angustifolius*: (**j**) habitat on limestone cliff in Jamaica; (**k**) individual bearing male and female flowers on phylloclades; (**l**) fruits. (**m**) *P. arbuscula* bearing male and female flowers on phylloclades. (**n–p**) *P. nutans*: (**n**) branch bearing erect male flowers; (**o**) branch with female flowers and fruits; (**p**) fruits (one on the lower right has exit hole excavated by *Epicephala* larva)

but are surprisingly robust to hurricane winds. Thus, the loss of leaves and substitution of photosynthesis by branches may be an adaptation to hurricane turbulence. *P. arbuscula*, a species with phylloclades, has bright red flowers, indicative of another unusual pollination system. *Xylophylla* is perhaps the most diverse lineage of *Phyllanthus* in terms of floral and vegetative adaptation, and further studies of the ecology of these plants are required. Section *Nothoclema* is another well-defined group of 11 tree species occurring widely from Mexico to Peru (Webster 2003). They can be easily distinguished from other *Phyllanthus* by their nonphyllanthoid branching, which is otherwise unknown in tree *Phyllanthus*. *Epicephala* moths are frequently found in the herbarium specimens of *Nothoclema* species, but the details of the association are still unknown. The clade sister to *Nothoclema* contains many herbs and subshrubs native to the New World with little known natural history. *P. fluitans* is a floating aquatic native to the Amazon but has also spread commercially as a water plant worldwide. *P. niruri* is used as a medicine to cure bladder stones. The members of the New World clade provide ample opportunity to study *Phyllanthus–Epicephala* association in the New World. The relationship likely varies from obligate pollination mutualism to pure parasitism, the latter involving nonpollinating moth species.

4.6.10 Sections **Cicca, Aporosella,** *and* **Chorisandra**

Sections *Cicca*, *Aporosella*, and *Chorisandra* each contain only one or two species and collectively form a monophyletic group of four species. Production of cauline inflorescences unites the four species morphologically. *P. acidus* (Fig. 4.17) is native to South America but is now widely cultivated in tropical parts of the world, particularly in tropical Asia. It is the most common fruit tree to be planted at home by the Lao people. The plant produces edible yellow drupes, which are eaten fresh or candied in sugar or pickled in salt. *P. acidus* produces abundant fruits under cultivation apparently in the absence of *Epicephala*, although the pollinator of the species is yet unknown. Ecological information is unavailable for species of *Aporosella* and *Chorisandra*.

4.6.11 Subgenus **Emblica**

The subgenus *Emblica* (Fig. 4.18) was traditionally confined to a few Asian shrub species, but Ralimanana et al. (2013) expanded the subgenus to include the herbaceous section *Urinaria*, which has clear phylogenetic affinity to the species traditionally included in *Embli*ca. One such herbaceous species is *P. lepidocarpus*, the flowers of which produce abundant nectar and are effectively pollinated by nectar-seeking ants. The plants are self-compatible and commonly attain 100% fruit even when non-ant visitors are experimentally excluded from flowers. The fruits of

Fig. 4.17 *Phyllanthus* section *Cicca*. (**a–c**) *P. acidus*: (**a**) fruiting tree in a village of Laos; (**b**) cauline inflorescences; (**c**) inflorescence with male flowers and young fruits

P. lepidocarpus are infested by *Epicephala parasitica*, which lacks the pollination behavior and oviposits in young fruits (Kawakita and Kato 2016). The resulting larvae destroy all of the seeds contained in each fruit. *P. emblica*, known as Indian gooseberry, is widely distributed in Southeast Asia and is common along the edges of seasonally dry tropical forests. The plant produces numerous male flowers among which female flowers are embedded, and is visited by honey bees in its native range in China. The plants produce near- spherical drupes about 2 cm in diameter, which is an exceptionally large nondehiscent fruit otherwise unknown in *Phyllanthus*. As with *P. acidus*, the edible fruits are very sour and popularly eaten in many parts of tropical Asia, but the natural consumer of the fruit is unknown, and thus no information is available on the mechanism of seed dispersal. Other members of the subgenus *Emblica* are still poorly known. Because nectaries are well developed in flowers of *Emblica*, they are most likely pollinated by unspecialized, nectar-seeking insects as in *P. lepidocarpus* and *P. emblica*.

4.6.12 Sauropus

The genus *Sauropus* as traditionally circumscribed is clearly a paraphyletic group with respect to the embedded *Breynia* (Pruesapan et al. 2008, 2012). Within the

Fig. 4.18 *Phyllanthus* subgenus *Emblica*. (**a–c**) *P. lepidocarpus*: (**a**) weedy habitat on roadside pebbles in Japan; (**b**) flowering and fruiting branch; (**c**) male flower; (**d**) female flower; (**e**) the ant *Tetramorium* sp. nectaring on male flower. (**f–h**) *P. emblica*: (**f**) flowering branch in Guangzhou, China; (**g**) male flowers; (**h**) fruits sold at market in Vientiane, Laos. (**i–l**): *P.* sp.: (**i**) habitat on limestone forest in northern Laos; (**j**) flowering branch; (**k**) male flowers; (**l**) male flowers with nectar in flower

Sauropus–Breynia clade, the Australian species of *Sauropus* forms a monophyletic group and is the earliest branching lineage. Earlier taxonomists placed Australian *Sauropus* in a separate genus, *Synostemon*, and van Welzen et al. (2014) reinstated *Synostemon* to the generic rank to refer to this Australian clade. The remaining Southeast Asian *Sauropus* and *Breynia* were grouped by van Welzen et al. (2014) in the expanded *Breynia*. The expanded *Breynia* is divided into the subgenera *Sauropus* and *Breynia*, the latter of which is further divided into sections *Cryptogynum* and *Breynia*. However, because *Sauropus* and *Breynia* are deeply embedded in *Phyllanthus*, whether to adopt the generic changes proposed by van

Welzen et al. (2014) is tightly linked to the issue of whether to subsume *Sauropus*, *Breynia*, and *Glochidion* in an expanded *Phyllanthus*. Due to a lack of consensus at present, we followed the traditional circumscription of *Sauropus* and *Breynia*, while acknowledging that the former is a paraphyletic group. *Sauropus* (Fig. 4.19) produces flowers that are among the most unusual of all Phyllantheae.

Fig. 4.19 Genus *Sauropus*. (**a–b**) *S. androgynus*: (**a**) general habitat at forest edge on limestone in Laos; (**b**) fruits. (**c**) *S.* cf. *quadrangularis* in the understory of seasonal dipterocarp-dominated tropical forest in Laos. (**d**) *S. macranthus* in the understory of evergreen broad-leaved forest in mountainous region of Laos. (**e–l**) Flowers of *Sauropus* species (male and female flowers on upper and lower column, respectively): (**e, l**) *S.* cf. *quadrangularis*; (**f, j**) *S. brevipes*; (**g, k**) *S. androgynus*; (**h, j**) *S. macranthus*. Photographs of flowers are on the same scale

In most Southeast Asian species, the tepals of both male and female flowers are spread flat on a horizontal plane, and as a result, the flowers are more or less disclike. In male flowers, the tepals are further inflexed toward the androccium, and together with the short, horizontally spread androecium, form three small cavities on the surface of the disc-shaped flower. As the anthers are held toward the cavities, pollen can only be picked up from inside these small cavities. The bifid styles of female flowers are also spread horizontally, and the stigmatic surface faces outward. Thus, as in male flowers, female flowers can only pick up pollen from flower visitors that approach or hold on to the flower in a particular manner. Several *Sauropus* species produce cauline inflorescences (e.g., *S. racemosus*) or are even geoflorous (*S. discocalyx*). Species that produce flowers in forest understory habitats tend to possess red-purple-colored tepals. Despite these interesting floral morphological characteristics, the pollination system of *Sauropus* is still unknown. None of the *Sauropus* species studied to date host *Epicephala* larvae in the fruit, therefore the moths are not involved in the pollination of *Sauropus*. Their pollination system holds the key to understanding the evolution of *Epicephala* moth pollination in *Breynia*. The ecology of Australian *Sauropus* is much less well known. Most species are rare and have very limited distributions, and new species continue to be found (Telford and Naaykens 2015). One Southeast Asian *Sauropus* species (*Sauropus androgynus*) is cultivated as a leaf vegetable in many parts of tropical Asia.

4.6.13 Breynia

Breynia (Fig. 4.20) contains 35 species distributed throughout the Australasian tropics extending far east to Fiji (Govaerts et al. 2000). Several continental species are very widely distributed, which is likely due to their bird-dispersed fruits. For example, the distribution of *Breynia vitis-idaea* extends from Pakistan in the west to Japan and the Philippines in the east (Chakrabarty and Gangopadhyay 1996). The flowers of most *Breynia* species resemble those of *B. vitis-idaea*, so *Epicephala* moth pollination is likely widespread in the genus. However, at least one species, *Breynia retusa* (Fig. 4.20), does not have an association with *Epicephala* (Kawakita and Kato 2009). The flowers of *B. retusa* are very different from those of *B. vitis-idaea*, suggesting that the plant has an alternative pollination mechanism. Fruit characteristics are also variable in the genus. Most species, including *B. vitis-idaea*, produce berries, whereas others produce dry capsules containing seeds with a fleshy sarcotesta (Fig. 4.20).

B. disticha, a native of New Caledonia, has a horticultural variety that produces strongly variegated leaves, which is known as "Snow Bush" and is popularly cultivated as an ornamental plant in tropical parts of the world.

Fig. 4.20 Genus *Breynia*. (**a–e**) *B. retusa*: (**a**) general habitat on forest edge in Laos; (**b**) flowering branch; (**c**) male flowers; (**d**) female flowers; (**e**) fruiting individual. (**f**) Fruits and seeds of *B. fruticosa*. Note that the fruit is a capsule, and the seeds have an orange sarcotesta

4.6.14 Phyllanthodendron

Subgenus *Phyllanthodendron* (Fig. 4.21) contains approximately 10 species occurring in Southeast Asia. Species of this subgenus have characteristic bell-shaped flowers that are unlike those of any other *Phyllanthus*. In addition, their growth form and leaf traits more resemble those of *Glochidion* than those of most *Phyllanthus*. Thus, the subgenus is often treated at generic rank (Li et al. 2008), and molecular phylogenetic analysis provided support for its unusual placement; *Phyllanthodendron* is sister to *Glochidion* and distantly related to all other *Phyllanthus* (Kathriarachchi et al. 2006).

Fig. 4.21 *Phyllanthus* subgenus *Phyllanthodendron*. (**a**, **b**) *P. roseus*: (**a**) flowering individual on forest edge of seasonal tropical forest in Laos; (**b**) flowers and fruit. (**c–e**) *P.* cf. *caudatifolius*: (**c**) general habitat on rocky riverbank in Laos; (**d**) branch with male flowers; (**e**) close-up of male flower showing the fused anthers

Pollination occurs during the daytime by numerous flies that visit flowers for nectar in *P. roseus*. None of the species studied to date has an association with *Epicephala*, although the seeds of *P. roseus* are consumed by braconid wasp larvae. The braconid wasp has a very long ovipositor that probably coevolved with the swollen fruit of *P. roseus*; its fruit has a large airspace between the fruit wall and ovule that likely distances the ovule from ovipositing wasps. The association between the *Phyllanthus* fruit and the braconid seed parasite reminds us of the evolutionary arms races between balloonlike fruit of a balloon vine (*Cardiospermum*, Sapindaceae) and its seed-sucking long-proboscid soapberry bugs (Carroll and Boyd 1992) and between the thick fruit pericarp of Japanese camellia and its seed-parasitic weevils with long rostrums and long ovipositors (Toju 2009). *P. mirabilis*, native to Thailand, is the only *Phyllanthus* to be caudiciform, or with a succulent trunk. The species is cultivated as an ornamental plant.

4.6.15 Glochidion

Glochidion (Fig. 4.22) contains some of the tallest plants of all Phyllantheae. In the lowland rainforest of Malaysia, *G. arborescens* commonly exceeds 20 m, and others may reach the canopy layer. However, the majority of *Glochidion* species occur along forest edges or roadsides and are typically small-sized trees or treelets. Few are forest understory shrubs. *G. obovatum*, endemic to Japan, occurs in habitat that is prone to deer browsing, and consequently, the stem apices of young plants are modified to thorns. In areas with large deer populations, *G. obovatum* literally dominates the vegetation due to its effective defense against deer browsing provided by the thorns. All *Glochidion* species studied to date are pollinated by *Epicephala* and share the same basic pollination ecology as described in Chap. 3. As the specialized, fused style morphology is shared among all the members of the genus, it is unlikely that other pollination systems exist in the genus. There is considerable variation in the shape of female flowers among different *Glochidion* species (Fig. 4.23). Although the reason for this variation is unknown, morphological differences may partly facilitate host recognition by ovipositing *Epicephala* females. Even greater variation is found in the fruit size and the number of locules per fruit, which should have major consequences on the proportion of seeds consumed by larvae. Some *Glochidion* species, such as *G. littorale*, have fruits exceeding 2 cm in diameter, each containing >20 seeds. There may be selection on *Glochidion* plants to produce larger fruits and more seeds per fruit to escape seed

Fig. 4.22 Genus *Glochidion*. (**a**) *G. sumatranum* on forest edge of lowland rainforest in Malaysia. (**b**) Canopy-reaching tree of *G. philippicum* in Taiwan. (**c**) Forest floor shrub of *G.* sp. in lowland rainforest in Malaysia. (**d**) Thorny young plant of *G. obovatum* in temperate Japan

Fig. 4.23 Flowers and fruits of *Glochidion*. (**a–h**) female flowers: (**a**) *G. lanceolatum* (Japan); (**b**) *G. philippicum* (Taiwan); (**c**) *G. sumatranum* (Sarawak, Malaysia); (**d**) *G. ferdinandii* (Sydney, Australia); (**e**) *G.* sp. (Laos); (**f**) *G.* sp. (Laos); (**g**) *G. sphaerogynum* (Vietnam); (**h**) *G.* sp. (Laos). (**i–m**) fruits: (**i**) *G. obscurum* (Sarawak, Malaysia); (**j**) *G. arborescens* (Sarawak, Malaysia); (**k**) *G. sericeum* (Sarawak, Malaysia); (**l**) *G. philippicum* (Taiwan); (**m**) *G. littorale* (Sarawak, Malaysia). Photographs (**i–m**) are in approximately the same scale. (**n**) *G. rubrum* fruit (Japan) showing dehisced capsule and seeds with orange sarcotesta

predation by *Epicephala* larvae. However, if *Glochidion* fruits evolve to become larger, the moths may adapt by laying more eggs per flower to consume a greater proportion of the seeds available. Therefore, it is interesting to examine whether such coevolutionary escalation is the cause of the large variation in fruit size among *Glochidion* species.

Chapter 5
Diversity of Pollinator Moths

Atsushi Kawakita and Makoto Kato

Keywords Active Pollination • *Epicephala* • Gracillariidae • *Flueggea* • *Glochidion* • Ornixolinae • *Phyllanthus* • Phylogeny • Proboscis • Seed parasite

5.1 Diversity and Classification of Gracillariidae

Gracillariidae, the family to which *Epicephala* belongs, is a large group of miniature moths with roughly 100 recognized genera and 2000 recognized species (De Prins and De Prins 2016). They have a global distribution and are found almost everywhere there are plants, except for extremely harsh environments (e.g., the arctic). Gracillariidae is one of several lepidopteran families that consist almost entirely of leaf-mining species, although the leaf-mining habit itself is known to occur in about 30 moth families (Powell et al. 1999). In most gracillariid species, early-instar larvae have remarkably flat head capsules without chewing mandibles, and feed exclusively on cell sap within the nongreen, epidermal layer of the leaf (sap feeder; Fig. 5.1). Later-instar larvae then feed on the palisade layer and finally the spongy layer of the leaf with functional mandibles, and excrete granular frass (tissue feeder; Fig. 5.1). The larvae of the genus *Phyllocnistis* are exceptional in that they spend all their instars as sap feeders in the leaf epidermal layer. Gracillariid moths are thus unique among insects in that they undergo hypermetamorphosis, a process by which some larval instars become functionally and morphologically distinct from other instars.

Sap feeding is an apomorphy among Gracillariidae that clearly distinguishes them behaviorally from other leaf-mining Lepidoptera. The extent to which their evolutionary success has been facilitated by sap feeding is a fascinating question

A. Kawakita (✉)
Center for Ecological Research, Kyoto University, 2-509-3 Hirano, Otsu, Shiga 520-2113, Japan
e-mail: kawakita@ecology.kyoto-u.ac.jp

M. Kato
Graduate School of Human and Environmental Studies, Kyoto University, Yoshida-Nihonmatsu-cho, Sakyo, Kyoto 606-8501, Japan
e-mail: kato@zoo.zool.kyoto-u.ac.jp

M. Kato, A. Kawakita (eds.), *Obligate Pollination Mutualism*, Ecological Research Monographs, DOI 10.1007/978-4-431-56532-1_5

Fig. 5.1 Various feeding habits of gracillariid moths. (**a**) Early-instar *Gibbovalva quadrifasciata* (Acrocercopinae) larvae mining *Cinnamomum doederleinii*. Sap-feeding larvae mine the epidermal layer of the leaf and produce linear mines, and later tissue-feeding larvae consume the leaf tissue within the expanded blotch mine. (**b**) Late-instar *Psydrocercops wisteriae* (Acrocercopinae) larva-mining a leaflet of *Wisteria floribunda*. (**c**) Mine of *Phyllocnistis* sp. (Phyllocnistinae) on *Ilex pedunculosa*. Pyllocnistinae larvae are sap-feeders throughout the larval stage and only use the epidermal layer of the host leaf. (**d**, **e**) Tentiform mine of *Phyllonorycter lyoniae* (Lithocolletinae) viewed from adaxial (**d**) and abaxial (**e**) sides of the leaf. *Phyllonorycter* larvae line the inner walls of mines with silk on the abaxial side, causing the mine eventually to become tentiform as the silk dries and wrinkles. Note that the initial instars are sap-feeding (*arrow*). (**f**) *Diphtheroptila*

that merits further study. Sap-feeding instars are secondarily lost in *Spinivalva* (leaf miner of *Passiflora*; Brito et al. 2013) and in several non-leaf-mining genera, including *Epicephala*.

The seed-feeding habit of *Epicephala* is unique within Gracillariidae. Nevertheless, deviations from the typical leaf-mining habit occur in a number of groups. *Caloptilia* species, which are common herbivores of *Glochidion* plants (Chap. 7), are sap-feeders as early instars, but later-instar larvae construct leaf rolls in which they live for the rest of the larval period. Larvae inside leaf rolls feed externally on the inner portions of folded leaves (Fig. 5.1). Galling has evolved in a number of gracillariid genera independently, for example, in *Caloptilia* (Kumata 1966; Fig. 5.1), *Borboryctis* (Kumata et al. 1988), and *Parectopa* (Wise 1962). Some species mine plant parts other than the leaf, such as stems or branches (*Marmara*, *Dendrorycter*, *Spulerina*) or fruit peel (*Marmara*, *Spulerina*). Others are plant-borers that use the seed (*Epicephala*, *Conopomorpha*), bud (*Stomphastis*, *Conopobathra*, *Chileoptilia*; Vargas and Landry 2005), or stem gall induced by rust fungi (*Polysoma*; Bashford 2002). Presumably, many more feeding types await discovery. Although the ultimate factor facilitating transitions from leaf mining to alternative feeding habits remains unknown, escape from parasitoid attack is probably a major reason for its evolution, as the structures they construct (leaf roll), induce (gall), or utilize (seed, bud, or fungal gall) are often protective against oviposition by parasitoids.

Gracillariidae species generally have narrow diets, and each specializes upon one or a few species in a single plant genus or, less commonly, in multiple related genera. Exceptions include *Acrocercops transecta*, which feeds on Juglandaceae and Ericaceae (Ohshima 2008); *Calybites phasianipennella*, which uses Polygonaceae and Myrsinaceae (Kumata 1982); and several stem- or fruit-feeding species with exceptionally wide host ranges spanning multiple families (*Marmara*, *Conopomorpha*). Hosts of Gracillariidae are found in 93 diverse angiosperm families, although only three gracillariid species are known to feed on monocots: *Acrocercops maranthaceae* (Maranthaceae; Busck 1934), *Marmara smilacisella* (Smilacaceae; Braun 1909), and *Gibbovalva tricuneatella* (Typhaceae; Sugisima et al. 2005). *Parectopa leucocyma*, the only known gymnosperm leaf-miner, feeds on *Agathis australis* in New Zealand (Wise 1962), and several undescribed gracillariid species are found on *Podocarpus* and *Gnetum* in Japan and Southeast Asia, respectively. No gracillariids are known to attack ferns or bryophytes. For reasons presently unknown, Gracillariidae predominantly use woody hosts, a pattern also observed in certain other groups of leaf-miners (e.g., Nepticulidae,

Fig. 5.1 (continued) scriptulata (Ornixolinae) larva mining *Glochidion acuminatum*. Many species of Ornixolinae expand the mine as they feed on the leaf tissue. (**g**, **h**) *Caloptilia ryukyuensis* (Gracillariinae) on *Glochidion zeylanicum*. Most *Caloptilia* species are sap-feeders as early instars (**g**), whereas late-instar larvae roll the leaf apex and feed on the leaf externally within the rolled leaf (**h**). (**i**) Galls of *Caloptilia cecidophora* (Gracillariinae) on *Glochidion obovatum*. This species is exceptional among Gracillariidae for its gall-inducing habit

Table 5.1 Subfamilies of Gracillariidae. Number of genera and approximate number of described species are given for each subfamily.

Subfamily Genus group	Genera	Species	Representative genera
Acrocercopinae	28	380	*Acrocercops, Dialectica, Spulerina, Eucosmophora*
Gracillariinae	25	700	*Caloptilia, Gracillaria, Aristaea, Calybites*
Lithocolletinae	11	550	*Phyllonorycter, Cameraria*
Marmarinae	2	30	*Marmara, Dendrorycter*
Oecophyllembiinae	6	20	*Metriochroa, Eumetriochroa*
Ornixolinae	28	300	*Epicephala, Conopomorpha, Diphtheroptila, Philodoria*
Parornichinae	4	90	*Parornix, Callisto*
Phyllocnistinae	1	100	*Phyllocnistis*

Information is based on Kawahara et al. (2017)

Heliozelidae). The scarcity of monocot hosts may thus reflect the rarity of woody monocots rather than a true lack of preference.

Gracillariidae species have traditionally been classified into four subfamilies: Gracillariinae, Lithocolletinae, Phyllocnistinae, and Oecophyllembiinae (Kumata et al. 1988a,b; Kumata 1998; note that the circumscription of subfamilies varies among authors). Among these families, Gracillariinae contains by far the largest number of species (>1380), and it has been further divided by Kumata et al. (1988a, b) into four genus groups: *Parornix*-, *Acrocercops*-, *Gracillaria*-, and *Parectopa*-groups. Kawahara et al. (2017) recently provided a robust phylogenetic framework for the family based on the sequences of 22 genes, and proposed a new classification consisting of eight subfamilies: Acrocercopinae, Gracillariinae, Lithocolletinae, Marmarinae, Oecophyllembiinae, Ornixolinae, Parornichinae, and Phyllocnistinae (Table 5.1).

Epicephala belongs to the Ornixolinae subfamily (corresponding to the *Parectopa* group of Kumata et al. 1988a,b), which, unlike other subfamilies that have high species diversity in temperate zones, is concentrated in the tropics. Therefore, the numbers of genera and species are likely to increase dramatically with further taxonomic studies. Ornixolinae is also unique among Gracillariidae in that it contains a disproportionately high number of non-leaf-mining species. In fact, the plant borers listed above (*Epicephala*, *Conopomorpha*, *Stomphastis*, *Conopobathra*, *Chileoptilia*, and *Polysoma*) all belong to Ornixolinae. Thus, there may have been an evolutionary precursor within the Ornixolinae lineage that enabled repeated transitions from leaf-mining to boring feeding habits, giving rise to *Epicephala*.

5.2 Phylogeny of *Epicephala*

The genus *Epicephala* was described by Meyrick (1881) based on specimens collected in Sydney, Australia. Meyrick, the founder of modern Microlepidoptera systematics, made the greatest contribution to *Epicephala* taxonomy, eventually describing more than 20 *Epicephala* species from various parts of the Old World. However, several of the species described by Meyrick (e.g., *Stomphastis chalybacma*) were transferred to other genera after detailed examination of genital morphology by later authors. A few more species currently placed in *Epicephala* will likely be subject to such taxonomic rearrangements. Concurrently with Meyrick, Turner greatly advanced the knowledge of *Epicephala* fauna in Australia, naming 11 currently accepted species. He was the first to note that *Epicephala* are seed-feeders; he observed *Epicephala* adults emerging from the capsules of *Glochidion ferdinandii* in Brisbane, Australia (Turner 1913). He named the moth *E. frugicola*, but later synonymized the species with *E. colymbetella*, the type species of the genus. Later, Vári (1961) described six *Epicephala* species from South Africa and provided the first detailed illustrations of the genitalia for *Epicephala* (as well as for many other genera of Gracillariidae). The study of *Epicephala* slowed down during the late twentieth century, during which time only three species were described (from Russia, Nigeria, and the Marquesas Islands).

The taxonomy of *Epicephala* has attracted renewed interest since the discovery of obligate pollination mutualism (Kato et al. 2003). Li and colleagues, working on *Epicephala* found in China, described 20 species associated with *Glochidion*, *Breynia*, and *Phyllanthus*, and provided detailed descriptions of adult behavior and morphology for some of them. Kawakita and Kato (2016) reviewed the Japanese fauna of *Epicephala* and described seven species. As of April 2016, the genus contains 64 species (Table 5.2; Fig. 5.2). However, ecological, molecular, and biogeographical data suggest that the genus contains several hundred species (Kawakita 2010). For example, some clades of *Epicephala* are confined to Madagascar or New Caledonia, where none of the described *Epicephala* species occurs (Kawakita and Kato 2009). These regions are known for hotspots of *Phyllanthus* diversity, and thus potentially have large numbers of undescribed *Epicephala* species. There is also a high level of *Phyllanthus* diversity in the New World, where *Epicephala* has not been previously recorded; furthermore, as detailed in the following section, *Epicephala* is also prevalent in the Neotropics. Accelerating the taxonomy of *Epicephala* at a global scale is therefore critical for facilitating the ecological and evolutionary study of this model group.

Figure 5.3 shows the most recent analysis of the phylogenetic relationships among *Epicephala* (Kawakita and Kato 2016). Seven clades can be recognized, each of which is generally associated with a particular clade of Phyllantheae. An exception is Clade 2, which consists of species associated with herbaceous *Phyllanthus* belonging to various *Phyllanthus* subgenera. *Conopomorpha flueggella*, which is a nonpollinating seed-feeder of *Flueggea suffruticosa*, is

Table 5.2 The 66 *Epicephala* species named as of April 2016, with known host associations

Species	Distribution	Known Host	References	Notes
E. acinacephora	Australia	–	Turner (1947)	
E. acrobaphes	Australia	–	Turner (1900)	
E. acrocarpa	Samoa Islands	–	Meyrick (1927b)	
E. albifrons	India, Moluccas, Sri Lanka, Thailand	*Phyllanthus* sp.	Stainton (1859) and Robinson et al. (1994)	The larvae feed on the seeds
E. albistriatella	Australia	–	Turner (1894)	
E. ancistropis	Java	*Callicarpa candicans*	Meyrick (1935b) and Robinson et al. (2001)	Doubtful placement in *Epicephala*
E. ancylopa	India, China	*Glochidion* cf. *khasicum*	Meyrick (1918b) and Li and Zhang (2016)	The larvae feed on the seeds
E. angustisaccula	China	*Glochidion wrightii*, *G. sphaerogynum*	Li et al. (2015a)	The larvae feed on the seeds
E. anthophilia	Japan	*Glochidion acuminatum*	Kawakita and Kato (2016)	The larvae feed on the seeds; active pollinator
E. assamica	China	*Glochidion assamicum*	Li and Zhang (2016)	The larvae feed on the seeds
E. australis	Australia	–	Turner (1896)	
E. bathrobaphes	Australia	–	Turner (1947)	
E. bipollenella	China, Japan	*Glochidion zeylanicum*	Zhang et al. (2012a) and Kawakita and Kato (2016)	The larvae feed on the seeds; active pollinator
E. bromias	India	–	Meyrick (1910a)	
E. calasiris	India	–	Meyrick (1908a)	
E. camurella	China	*Glochidion wrightii*, *G. sphaerogynum*	Li et al. (2015a)	The larvae feed on the seeds
E. colymbetella	Australia	*Glochidion ferdinandii*	Meyrick (1881) and Turner (1913, 1940)	Type species of the genus; the larvae feed on the seeds; the adult insect emerges inside the capsule where it remains until liberated by the dehiscence.

(continued)

Table 5.2 (continued)

Species	Distribution	Known Host	References	Notes
E. corruptrix	Japan	*Glochidion rubrum*, *G. obovatum*	Kawakita and Kato (2016)	The larvae induces galls on the female flower; active pollinator
E. daltonii	China	*Glochidion daltonii*	Li and Zhang (2016)	The larvae feed on the seeds
E. domina	China	*Glochidion sphaerogynum*	Li et al. (2015a)	The larvae feed on the seeds
E. duoplantaria	China	*Glochidion lutescens*, *G. daltonii*	Li and Zhang (2016)	The larvae feed on the seeds
E. epimicta	Australia	–	Turner (1913)	
E. eriocarpa	China	*Glochidion eriocarpum*	Zhang et al. (2012a)	The larvae feed on the seeds
E. euchalina	Myanmar	–	Meyrick (1922)	
E. eugonia	Australia	–	Turner (1913)	
E. exetastis	Sri Lanka	–	Meyrick (1908a)	
E. flagellata	Sri Lanka	–	Meyrick (1908a)	
E. frenata	Sri Lanka, Java	*Glochidion* sp.	Meyrick (1908a) and Robinson et al. (2001)	In fruits
E. haplodoxa	South Africa	–	Vári (1961)	
E. homostola	South Africa, Namibia	–	Vári (1961)	
E. impolliniferens	China	*Glochidion sphaerogynum*	Li et al. (2015a)	The larvae feed on the seeds
E. jansei	Zimbabwe	–	Vári (1961)	
E. jianfenglingina	China	*Glochidion* cf. *rubrum*	Li and Zhang (2016)	The larvae feed on the seeds
E. laeviclada	China	*Phyllanthus microcarpus*	Li and Yang (2015)	The larvae feed on the seeds
E. lanceolaria	China	*Glochidion lanceolarium*	Zhang et al. (2012a)	The larvae feed on the seeds
E. lanceolatella	Japan	*Glochidion lanceolatum*	Kawakita and Kato (2016)	The larvae feed on the seeds; active pollinator
E. lativalvaris	China	*Breynia fruticosa*, *B. rostrata*	Zhang et al. (2012a, b)	The larvae feed on the seeds; active pollinator
E. lomatographa	Australia	–	Turner (1913)	
E. microcarpa	China	*Phyllanthus microcarpus*	Li and Yang (2015)	The larvae feed on the seeds

(continued)

Table 5.2 (continued)

Species	Distribution	Known Host	References	Notes
E. mirivalvata		*Breynia fruticosa*, *B. rostrata*	Zhang et al. (2012a, b)	The larvae feed on the seeds; active pollinator
E. nephelodes	Australia	–	Turner (1913)	
E. nudilingua	Japan	*Phyllanthus ussuriensis*	Kawakita and Kato (2016)	The larvae feed on the seeds
E. orientale	India, Malaysia	*Bauhinia* spp.	Stainton (1856a) and Fletcher (1921)	Clearly belongs to *Conopobathra*; the larvae feed on the flower buds
E. obovatella	Japan	*Glochidion obovatum*, *G. rubrum*	Kawakita and Kato (2016)	The larvae feed on the seeds; active pollinator
E. parasitica	Japan	*Phyllanthus lepidocarpus*	Kawakita and Kato (2016)	The larvae feed on the seeds; pollination behavior absent
E. pelopepla	South Africa	–	Vári (1961)	
E. periplecta	New Guinea	–	Diakonoff (1955)	
E. perplexa	Japan	*Glochidion lanceolatum*	Kawakita and Kato (2016)	The larvae feed on the seeds; active pollinator
E. pyrrhogastra	South Africa	–	Meyrick (1908b)	
E. relictella	Russia, China, Korea	*Flueggea suffruticosa*	Kuznetzov (1979), Kawahara et al. (2010) and Hu et al. (2011)	The larvae feed on the seeds
E. scythropis	Myanmar	*Phyllanthus* sp.	Meyrick (1930) and Fletcher (1933)	The adults were bred from galls on *Phyllanthus*
E. sphenitis	India	*Breynia rhamnoides*; *B. vitis-idaea*	Meyrick (1931d) and Robinson et al. (2001)	The larvae feed on the seeds
E. spinula	Marquesas Islands	–	Clarke (1986)	
E. spumosa	Australia	–	Turner (1947)	
E. squamella	Vietnam	–	Kuznetzov and Baryshnikova (2001)	Doubtful placement in *Epicephala*
E. stauropa	India	–	Meyrick (1908a)	

(continued)

Table 5.2 (continued)

Species	Distribution	Known Host	References	Notes
E. strepsiploca	India	–	Meyrick (1918b)	
E. subtilis	India	–	Meyrick (1922)	
E. suttoni	Nigeria	–	Bland (1980)	Treated as a species of *Ectropina* in De Prins and De Prins (2005) but clearly an *Epicephala*
E. tephrostola	South Africa	–	Vári (1961)	
E. tertiaria	China	*Phyllanthus microcarpus*	Li and Yang (2015)	The larvae feed on the seeds
E. trigonophora	Australia	–	Turner (1900)	
E. venenata	Taiwan	–	Meyrick (1935b)	
E. vermiformis	Java	*Cajanus cajan*	Meyrick (1936)	Doubtful placement in *Epicephala*
E. vitisidaea	China, Japan	–	Zhang et al. (2012a) and Kawakita and Kato (2016)	The larvae feed on the seeds; active pollinator
E. zalosticha	Australia	–	Turner (1940)	

depicted as a sister to *Epicephala* in Fig. 5.3. However, the phylogenetic position of *C. flueggella* is contentious because the species is sometimes placed as a sister to Clade 7 (Fig. 5.3). *Conopomorpha flueggella* clearly lacks pollination behavior, and lays eggs in buds or young fruits as well as flowers (Fig. 5.4). It also lacks an ovipositor, placing its eggs on the surface of the ovule (Fig. 5.4). Interestingly, ovipositing females sometimes take nectar with their proboscises prior to oviposition (Fig. 5.4) in a manner similar to *Epicephala*'s use of its proboscis to pollinate. The nectaring behavior of *C. flueggella* may have been the evolutionary precursor for active pollination; however, this question requires a robust understanding of the phylogenetic placement of *C. flueggella* with respect to *Epicephala*. At present, it is equally probable that the behavior of *C. flueggella* represents a secondary loss of pollination behavior. A hatched *C. flueggella* larva usually consumes all six seeds contained in each fruit of *Flueggea suffruticosa* (Fig. 5.4).

Conopomorpha flueggella was described by Hu et al. (2011), who placed it in the genus *Conopomorpha* based mainly on wing morphology. However, the true *Conopomorpha*, which include species that attack seeds of tropical fruit trees such as lychee and longan (Sapindaceae) and cacao (Malvaceae), are distantly related to *Conopomorpha flueggella* or *Epicephala* within Ornixolinae. Therefore, *C. flueggella* can likely be more accurately placed in a separate genus.

Fig. 5.2 Representative specimens of the nine *Epicephala* species in Japan. Variation in wing pattern among species is low, and is of minor importance in morphological identification. Variation in size largely reflects host seed size. The wing pattern of *E. parasitica* is sexually

The basal-most *Epicephala* are the clade of moths associated with the *Phyllanthus* section *Gomphidium* in New Caledonia (Fig. 5.3). They clearly exhibit active pollination behavior (Chap. 3), but lack the sensilla on the proboscis and ovipositor that characterize the derived members of *Epicephala*. As such, they retain the morphology of the earliest *Epicephala* to acquire active pollination behavior. They lay eggs on the surface of the pistils (Fig. 5.5), and a single hatched larva consumes all the seeds contained in each fruit. However, moth mortality is very high, probably owing to desiccation during the egg stage. This high mortality is necessary for a fraction of the seeds to remain intact despite destructive seed-feeding by the larvae (Chap. 3). Because there is a large morphological disparity between moths of this clade and those of the derived *Epicephala*, especially with respect to the female genitalia, the former should probably be placed in a separate genus. However, determining whether it is congeneric with *C. flueggella* requires a closer examination of morphology and a better resolution of the basal phylogenetic relationships.

The remaining moths can safely be placed in *Epicephala*, and share the female ovipositor as a morphological synapomorphy (Clades 1–6; Figs. 5.3, 5.5). Ovipositors are not known in any other Gracillariidae genus; thus, aside from pollination behavior, they represent the trait that best characterizes *Epicephala*. The development of the ovipositor is most likely an adaptation for laying eggs internally in floral tissue, thereby avoiding egg desiccation (Fig. 5.5).

Interestingly, there is considerable variation in egg placement behavior among *Epicephala* species. Most *Epicephala* species associated with *Glochidion* lay eggs either through the apical pit of the stylar column into the stylar tissue or laterally through the ovary wall on the surface of the ovule (Fig. 5.6). This difference in oviposition mode is reflected in morphology; the ovipositors of laterally ovipositing species are distinctly more angular than those of apically ovipositing species (Kawakita and Kato 2016; Fig. 5.6).

Certain species associated with *Phyllanthus* and *Glochidion* lay eggs in the pedicels of female flowers (Chap. 10). In such species, the hatched larvae initially bore through the pedicel to enter the ovary. *Epicephala vitisidaea* and *E. mirivalvata* lay eggs in the narrow space between the sepals and ovary of their host *Breynia* plants, having thus reverted to external oviposition (Kawakita and Kato 2004; Zhang et al. 2012a), although both species have retained functional

Fig. 5.2 (continued) dimorphic, so specimens of both sexes are shown for this species. (**a**) *E. anthophilia* (Amami Island, Kagoshima, ♀, host: *Glochidion acuminatum*). (**b**) *E. bipollenella* (Henoko, Okinawa, ♀, host: *G. zeylanicum*). (**c**) *E. lanceolatella* (Cape Hedo, Okinawa, ♀, host: *G. lanceolatum*). (**d**) *E. perplexa* (Cape Hedo, Okinawa, ♀, host: *G. lanceolatum*). (**e**) *E. obovatella* (Tomogashima, Wakayama, ♂, host: *G. obovatum* and *G. rubrum*). (**f**) *E. corruptrix* (Takae, Okinawa, ♀, host: *G. obovatum* and *G. rubrum*). (**g**) *E. vitisidaea* (Yona, Okinawa, ♀, host: *Breynia vitis-idaea*). (**h**) *E. parasitica* (Yonaguni Island, Okinawa, ♀, host: *Phyllanthus lepidocarpus*). (**i**) *E. parasitica* (Hateruma Island, Okinawa, ♂, host: *P. lepidocarpus*). (**j**) *E. nudilingua* (Watarase-yusuichi, Tochigi, ♀, host: *P. ussuriensis*). Scale bar: 5 mm

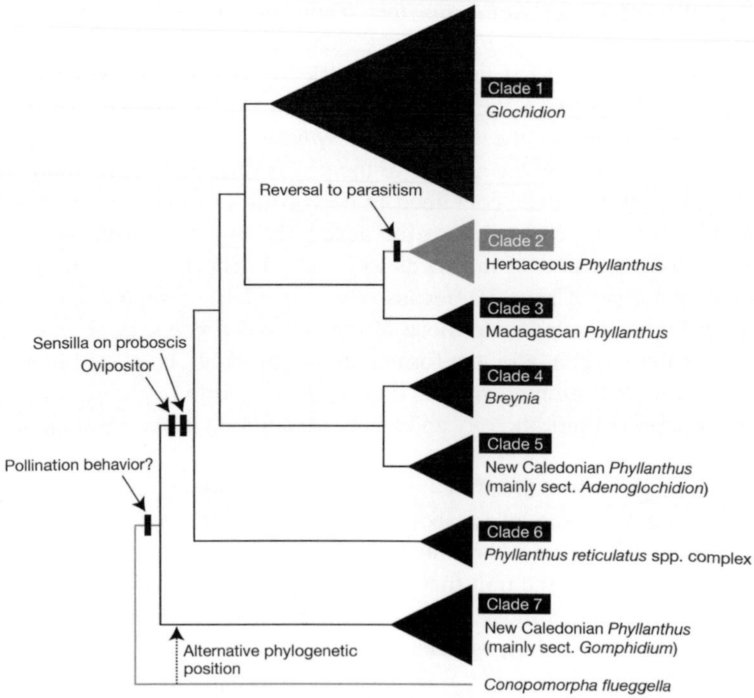

Fig. 5.3 Phylogeny of *Epicephala*. The tree is based on a molecular phylogenetic analysis by Kawakita and Kato (2009). Host plant associations are provided below clade numbers. The phylogenetic position of *Conopomorpha flueggella* is inconsistent among analyses; an alternative placement as a sister to Clade 7 is also indicated. Clade triangle size is proportional to the estimated number of species in each clade. Major evolutionary events are listed on branches. Lineages indicated in *grey* do not possess the pollination behavior

ovipositors. Because Phyllantheae plants are known to abort selectively those flowers with heavy egg loads and abortion is likely based on the extent of mechanical damage to flowers (Chap. 9), external oviposition in these species may have evolved to circumvent the abortion response in their host plants.

However, the adaptive significance of other oviposition modes is less clear. In cases where two or more *Epicephala* species co-occur on the same Phyllantheae host, the different species exhibit different oviposition behaviors (Kawakita et al. 2015; Kawakita and Kato 2016). This may indicate that different oviposition strategies are necessary for stable coexistence on shared hosts.

Another distinguishing characteristic of the core *Epicephala* clade is the numerous sensilla on the proboscises of females (Fig. 5.7). Sensilla are absent in males, and a sensilla-bearing proboscis is not known in any other genera of Gracillariidae. Thus, we can infer that the function of the sensilla is linked to active pollination. Because most pollen is held on the basal area of the proboscis of pollen-carrying females where the density of sensilla is highest, females are able to carry more pollen than would be possible in the absence of such a structure (Fig. 5.7). Support

Fig. 5.4 A nonpollinating, seed-parasitic moth, *Conopomorpha flueggella*. (**a**) Female moth depositing an egg underneath the horizontally spread styles. (**b**) Female moth taking nectar from nectary at the base of the ovary. Note that the moth proboscis lacks sensilla and does not bear pollen (*arrow*). A droplet of nectar collected at the base of the ovary and protruded through the tepals can be seen (*arrow*). (**c**) Moth egg laid externally on the surface of the ovary. (**d**) Mature fruit with exit hole excavated by moth larva. Seeds are usually entirely destroyed in such fruits

for the interpretation that the sensilla assist in pollination comes from the absence of sensilla in species that have secondarily lost the pollination behavior. For example, of the six major lineages of the core *Epicephala* clade, the herbaceous *Phyllanthus* clade consists of species that lay eggs in young fruits of herbaceous *Phyllanthus* and thus do not pollinate flowers. Accordingly, the sensilla on their proboscises are lost entirely or are rudimentary (Fig. 5.7). The *Phyllanthus reticulatus* clade represents another *Epicephala* lineage that lost the sensilla. Of the six species associated with plants of the *Phyllanthus reticulatus* species complex in Taiwan, three induce gall formation on female flowers, although two of them retain pollination behavior. Sensilla are completely lost in the three gall-inducing species, probably because selection to transfer pollen is relaxed or absent owing to their ability to induce galls (Chap. 11).

Fig. 5.5 Difference in the placement of eggs by *Epicephala* moths with and without ovipositors. Roughly half of the species associated with New Caledonian *Phyllanthus* (Clade 7 in Fig. 5.3) do not possess ovipositors and oviposit externally (**a, b**) whereas species of derived clades have well-developed ovipositors and lay eggs internally in floral tissue (**c, d**). (**a**) Female flower of *Phyllanthus kouaouaensis* with an *Epicephala* moth egg laid externally on style surface (*arrow*). (**b**) Female flowers of *P.* cf. *koniamboensis* with *Epicephala* eggs. (**c**) Female *E. eriocarpa* with extended ovipositor (*arrow*). (**d**) Vertical section of *Glochidion zeylanicum* female flower containing an *Epicephala* moth egg (*arrow*). *Ova*, ovary; *ovu*, ovule; *st*, style; *te*, tepal

As mentioned in the previous section, gall-inducers have arisen repeatedly in the course of gracillariid evolution, but the evolution of galling may be particularly common in *Epicephala*. In the Ryukyu Archipelago of southern Japan, *E. corruptrix* is associated with two *Glochidion* species (*G. obovatum* and *G. rubrum*). On both hosts, *E. corruptrix* exhibits pollination behavior, but the resulting fruits do not develop normally. Instead, the locule in which the larva develops grows irregularly and eventually becomes a gall (Fig. 5.8). Populations of *G. obovatum* and *G. rubrum* associated with *E. corruptrix* suffer very low seed production. This contrasts sharply with populations elsewhere in their ranges, where the plants are pollinated by non-gall-inducing *E. obovatella*, and produce large numbers of normal fruits and seeds. *Epicephala corruptrix* is distantly related to the gall-inducers of the *Phyllanthus reticulatus* species complex, so the galling habit has independently evolved at least twice in *Epicephala*. Similar gall-like development is found in *P. humbertii* in Madagascar and *P. cuscutiflorus* in Australia. It is therefore interesting to consider how many other lineages of

Fig. 5.6 Oviposition behavior (**a–f**) and ovipositor morphology (**g–l**) of six *Epicephala* species associated with *Glochidion* in Japan. (**a**) *E. anthophilia* ovipositing through stylar pit of *G. acuminatum* flower. (**b**) *E. bipollenella* ovipositing through stylar pit of *G. zeylanicum* flower. (**c**) *E. lanceolatella* ovipositing through stylar pit of *G. lanceolatum* flower. (**d**) *E. perplexa* ovipositing through lateral ovary wall of *G. lanceolatum* flower. The ovipositor of this species penetrates both the tepal and the ovary wall. (**e**) *E. obovatella* ovipositing through lateral ovary wall of *G. obovatum* flower. The ovipositor of this species is inserted directly into the ovary and does not penetrate the tepal. (**f**) *E. corruptrix* ovipositing through ovary wall of *G. rubrum* flower. Similar to that of *E. obovatella*, the ovipositor of this species also does not penetrate the tepal. (**g**) *E. anthophilia*. (**h**) *E. bipollenella*. (**i**) *E. lanceolatella*. (**j**) *E. perplexa*. (**k**) *E. obovatella*. (**l**) *E. corruptrix*. Note that the ovipositors of species that oviposit through lateral ovary walls are distinctly angular

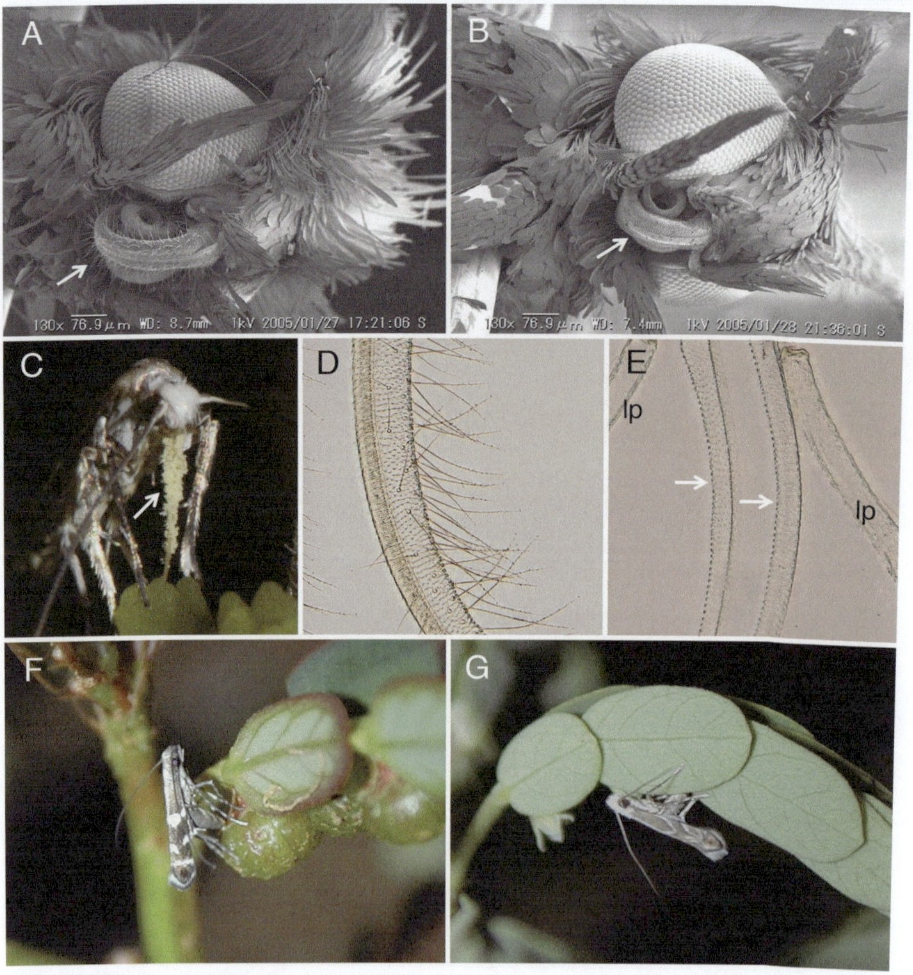

Fig. 5.7 Proboscis sensilla. (**a, b**) Scanning electron micrographs of female (**a**) and male (**b**) proboscises of actively pollinating *Epicephala* species (*Epicephala* sp. associated with *Phyllanthus reticulatus*). Note that numerous sensilla are present on the female proboscis, whereas they are absent on the male proboscis. (**c**) *E. anthophilia* female actively depositing pollen on *G. acuminatum* stigma in Amami Island, Japan. Sensilla on moth proboscises allow more pollen to be held on the proboscis. (**d**) Proboscis of female *E. bipollenella* bearing sensilla. (**e**) Proboscis of female *E. nudilingua*, a nonpollinating species, lacking sensilla. (**f**) *E. parasitica*, a nonpollinating species, ovipositing in a young fruit of *P. lepidocarpus* in Ishigaki Island, Japan. (**g**) Another nonpollinating species (*E.* sp.) ovipositing in a very young fruit of *P. amarus* through leaves folded at night in Thakhek, Laos. Scale bar: 0.1 mm

Epicephala evolved the gall-inducing ability. The adaptive significance of galling is still unclear, but is probably linked to escape from parasitoid attack (Chap. 11).

Fig. 5.8 Fruits and galls produced by *Epicephala* species on *Glochidion obovatum*. (**a**) Fruit produced after pollination by *E. obovatella*. (**b**) Gall induced on female flower by *E. corruptrix*. (**c**) Cross-section of the gall induced by *E. corruptrix*. *Arrow* indicates the galled locule with feeding traces of *Epicephala* larva. Note that the irregularly developed ovules of the galled locule have merged and become indistinguishable from septa. Scale bar: 2 mm

5.3 Global Diversity of *Epicephala*

The genus *Epicephala* has thus far been described only in the Old World tropics, but they are also prevalent in the Neotropics. Observation of *Phyllanthus* in Cuba and Jamaica indicate that several species in the subgenus *Xylophylla* are associated with seed-feeding *Epicephala*. Some of the these *Phyllanthus* species (e.g., *P. chamaecrystoides*, *P. myrtilloides*) bear female flowers with fused styles characteristic of *Epicephala*-pollinated plants in the Old World, whereas others possess bifid and spread styles indicative of pollination by nonspecialized insects (e.g., *P. angustifolius*, *P. nutans*). Whether obligate pollination mutualism occurs in *Xylophylla* is still unclear, but the outcome of association between *Xylophylla* and *Epicephala* appears to be highly variable among species. There are roughly 60 species of *Xylophylla*, with the highest concentration of species being in the Caribbean islands, although the subgenus occurs as far south as Andean Peru.

The widespread occurrence of *Epicephala* in the New World is further demonstrated by the presence of *Epicephala* larvae and pupae on herbarium specimens (Fig. 5.9). Because plant specimens bearing fruits may sometimes contain *Epicephala* larvae at the time of collection, the larvae are occasionally found on *Phyllanthus* specimens, especially those in which the capsules dehisced while being dried. The larvae of most *Epicephala* species have a characteristic red color with narrow white bands, and cannot be mistaken for those of other Lepidoptera. Mature larvae inside fruits may also spin cocoons on the edges of leaves before they are completely dried. While spinning cocoons, the larvae of *Epicephala* excrete bubblelike balls from the anus, grab them with the mandibles, and attach them to the surface of the cocoon through a hole punched from inside the cocoon (Fig. 5.9). Although the adaptive role of ball production is unknown, such a habit is known only in *Epicephala* and several related genera of the Ornixolinae. Therefore, the

Fig. 5.9 Larva and cocoon of *Epicephala* moth. (**a**) Larva of *Epicephala* sp. associated with *P. reticulatus* in Taiwan. The number and position of white bands varies among species. (**b**) Cocoon of *E. bipollenella* whose surface is decorated with bubbles. The moth has emerged from the cocoon, and the exuvia can be seen. (**c**) Herbarium specimen of *Phyllanthus mocinianus* at the herbarium of the University of California, Davis. The herbarium sheet is attached with an envelope (*arrow*) containing detached plant parts (mostly dehisced capsules and seeds), among which *Epicephala* larvae are sometimes found. (**d**) Dried *Epicephala* larva found inside the envelope. (**e**) An *Epicephala* cocoon found on a detached leaf in the envelope

presence of such cocoons on herbarium specimens provides reliable evidence of an association with *Epicephala*.

Epicephala larvae and cocoons are often found on specimens of *Phyllanthus* section *Nothoclema* (subgenus *Conami*; Fig. 5.9; Table 5.3). The group contains 10 species distributed from Mexico to Argentina, and is often a prominent

Table 5.3 List of *Flueggea* and *Phyllanthus* species associated with *Epicephala*

Genus	Subgenus	Section	Species	Geographic Location	Evidence	*Epicephala* Species
Flueggea			*F. suffruticosa*	Temperate east Asia	f, l	*E. relictella*, *Cono-pomorpha flueggella*
			F. elliptica	**Ecuador**	**h**	
Phyllanthus	*Kirganelia*	*Anisonema*	*P. reticulatus*	China, Taiwan, Laos	f	
			P. microcarpus	China, Taiwan, Laos, Malaysia	f, l	*E. microcarpa*, *E. laeviclada*, *E. tertieria*
			P. ciccoides	**Australia**	**f**	
			P. duplanchei	**New Caledonia**	**f**	
			P. casticum	**Madagascar**	**h**	
			P. pervilleanus	**Madagascar**	**h**	
	Swatziani		*P. amarus*	Laos	f	
			P. abnormis	**USA**	**h**	
	Afroswartziani		***P. lokohensis***	**Madagascar**	**f**	
			P. madagascariensis	**Madagascar**	**f**	
	Gomphidium	*Gomphidium*	*P. bourgeoisii*	New Caledonia	f	
			P. buxoides	New Caledonia	f	
			P. cuscutiflorus	**Australia**	**f**	
			P. koumacensis	New Caledonia	f	
			P. mangenotii	New Caledonia	f	
			P. pancherianus	New Caledonia	f	
			P. poumensis	New Caledonia	f	
			P. tenuipedicellatus	New Caledonia	f	
			P. tiebaghiensis	New Caledonia	f	
			P. caudatus	New Caledonia	f	
			P. chamaecerasus	New Caledonia	f	
			P. koghiensis	New Caledonia	f	

(continued)

Table 5.3 (continued)

Genus	Subgenus	Section	Species	Geographic Location	Evidence	*Epicephala* Species
			P. koniamboensis	New Caledonia	f	
			P. pilifer	New Caledonia	f	
		Andenoglochidion	*P. aeneus*	New Caledonia	f	
			P. gneissicus	New Caledonia	f	
			P. guillauminii	New Caledonia	f	
			P. vulcani	New Caledonia	f	
	Gomphidium sensu Ralimanana and Hoffmann (2011)		*P. marojejiensis*	Madagascar	f	
			P. humbertii	Madagascar	f	
	Conami	***Nothoclema***	***P. graveolens***	**Peru**	**h**	
			P. mcvaughii	**Mexico**	**h**	
			P. mocinianus	**Mexico**	**h**	
	Xylophylla	***Elutanthos***	***P. botryanthus***	**Venezuela**	**h**	
			P. huallagensis	**Peru**	**h**	
			P. nutans	**Jamaica**	**f**	
		Orbicularia	***P. chamaecristoides***	**Cuba**	**f**	
			P. myrtilloides	**Cuba**	**f**	
		Xylophylla	***P. angustifolius***	**Jamaica**	**f**	
	Emblica		*P. lepidocarpus*	Japan	f	*E. parasitica*
			***P.* sp.**	**Laos**	**f**	
	Unplaced	*Macraea*	*P. ussuriensis*	Japan	f	*E. nudilinguae*
			P. chrysanthus	**New Caledonia**	**f**	
	Unplaced	***Microglochidion***	***P. chimantae***	**Venezuela**	**h**	
			P. duidae	**Venezuela**	**h**	
			P. obfalcatus	**Venezuela**	**h**	
	Unplaced	***Paraphyllanthus***	***P. liebmannianus***	**USA**	**h**	

Data in bold indicate those newly reported here

component of the local flora. The female flowers of the *Nothoclema* species have spread styles and do not appear to be specialized to *Epicephala*; thus, it would be interesting to clarify the pollination system of *Nothoclema* plants to understand whether obligate pollination mutualism occurs in the New World. One surprising finding was an association between *Epicephala* and *Flueggea elliptica* (Table 5.3), a plant that is only known from a small coastal area in southernmost Ecuador. The Old World *Flueggea suffruticosa* is host to *Conopomorpha flueggella* and *Epicephala relictella*, but other *Flueggea* species, such as the widespread and abundant *F. virosa*, have never been found hosting *Epicephala*. The *Flueggea–Epicephala* association in the New World is probably phylogenetically independent from that found in Asia, but determining where the Ecuadorian species belong within the *Epicephala* phylogeny is critical for the global understanding of *Epicephala* diversity and evolution.

Finally, one exciting possibility is the occurrence of obligate pollination mutualism on the tepuis of the Guiana Highlands. The tepuis are table-top mountains of granite arenite sandstone rising abruptly from the Amazonian rainforest that are host to a spectacular array of endemic plants and animals. The majority of the tepuis occur in Venezuela and Western Guyana, but some are also found in Colombia, Suriname, French Guiana, and in northernmost Brazil.

The *Phyllanthus* species found in the Guiana Highlands are grouped into a well-defined section, *Microglochidion* (Webster and Carpenter 2002, 2008). Many species possess elongated, nonbifid, and fused styles, which strikingly resemble those of *Glochidion* or New Caledonian *Phyllanthus*. Although neither larvae nor cocoons of *Epicephala* were found on herbarium specimens, the seeds of several *Microglochidion* species have holes that are typical of those made by *Epicephala* larvae, suggesting that the association with *Epicephala* is widespread among the ~10 species of Guiana Highland *Phyllanthus* (Table 5.3). It is exciting to think of the possibility that obligate pollination mutualism, which probably originated in the Old World, has reached some of the most exotic biota on earth and produced an impressive number of endemic species.

Chapter 6
Origin of Active Pollination and Mutualism

Atsushi Kawakita and Makoto Kato

Keywords *Breynia* • *Epicephala* • *Flueggea* • Gracillariidae • *Glochidion* • Phyllantheae • *Phyllanthus* • Phylogeny

6.1 Evolution of Pollination by Seed Parasites

The mutualisms between plants and their seed-parasitic pollinators, such as the fig–fig wasp, yucca–yucca moth, and leafflower–leafflower moth mutualisms, provide textbook examples of specialized pollination systems (Weiblen 2002; Cook and Rasplus 2003; Pellmyr 2003; Kato et al. 2003). Remarkably, in all three systems, the pollinator insects actively collect and transport pollen between flowers in order to ensure food for their seed-feeding larvae. Reciprocal adaptation by plants to restrict floral access by other visitors resulted in extreme mutual dependence between plants and insects. Consequently, these mutualisms served as principal model systems for the studies of coevolution and mutualism.

Despite a wealth of documented examples of specialized pollination systems in angiosperms, however, pollination by obligate seed parasites is rare. This is because seed parasitism inflicts a heavy cost on plants, whereas abundant copollinators swamp the mutualistic effect of pollination by seed parasites (Thompson and Pellmyr 1992; Thompson and Cunningham 2002). In fact, exclusion of pollinators has not occurred in plants that were more recently found as being pollinated by seed parasites, including *Lithophragma* plants pollinated by *Greya* moths (Thompson and Pellmyr 1992), senita cactus pollinated by senita moths (Fleming and Holland 1998), and *Silene* plants pollinated by *Hadena* and *Perizoma* moths (Kephart et al. 2006). Nevertheless, there are other plant–seed parasite associations that have

A. Kawakita (✉)
Center for Ecological Research, Kyoto University, 2-509-3 Hirano, Otsu, Shiga 520-2113, Japan
e-mail: kawakita@ecology.kyoto-u.ac.jp

M. Kato
Graduate School of Human and Environmental Studies, Kyoto University, Yoshida-Nihonmatsu-cho, Sakyo, Kyoto 606-8501, Japan
e-mail: kato@zoo.zool.kyoto-u.ac.jp

M. Kato, A. Kawakita (eds.), *Obligate Pollination Mutualism*, Ecological Research Monographs, DOI 10.1007/978-4-431-56532-1_6

evolved into reciprocal specialization (globeflower–globeflower fly and *Rheum nobile*–fungus gnat mutualisms; Jaeger and Després 1998; Song et al. 2014). Understanding of how and under what circumstances obligate pollination mutualisms evolve would thus benefit from exploring the origin of active pollination and mutualism in the Phyllantheae–*Epicephala* association.

In this chapter, we describe the pollination systems and associations with *Epicephala* of 26 Phyllantheae species studied during 2002–2007 in Southeast Asia, New Caledonia, Australia, Madagascar, Guinea, and North America (Table 6.1). Based on this information, we explore the origin of the Phyllantheae–*Epicephala* mutualism using robust molecular phylogenies for 46 species of Phyllantheae and associated *Epicephala* moths. Although the present analysis focuses on only a small proportion of the global diversity of Phyllantheae, the sampled species cover the entire range of taxonomic diversity within the tribe (Hoffmann et al. 2006; Kathriarachchi et al. 2006), allowing an overview of broad coevolutionary history of the Phyllantheae–*Epicephala* association. Overall, the results reveal an unexpectedly complex origin of the Phyllantheae–*Epicephala* pollination mutualism and provide important general insights into how a combination of evolutionary innovation and partner shifts shapes the evolutionary dynamics of mutualism in coevolving species interactions.

6.2 Diversity of Pollination Systems in Phyllantheae

As detailed in Chaps. 3 and 4, there are five Phyllantheae lineages that are obligately pollinated by host-specific *Epicephala* moths (*Glochidion*; *Breynia*; and New Caledonian *Phyllanthus*; *Phyllanthus* section *Anisonema*; and an unclassified group of *Phyllanthus* endemic to Madagascar). The remaining species are pollinated by diurnal insects that visit flowers for nectar and pollen, and do not have associations with pollinating *Epicephala*. However, *Flueggea suffruticosa* is parasitized by *Conopomorpha flueggella*, and three herbaceous *Phyllanthus* species are parasitized by seed-parasitic *Epicephala* species that do not pollinate the flowers (Chap. 5).

The plants that are not pollinated by *Epicephala* employ a variety of insects as pollinators. Species of *Flueggea* have the broadest range of flower visitors, including bees, flies, beetles, and butterflies, which all probably contribute to pollination. Although observation is limited, dipteran insects appear to be important pollinators of many other Phyllantheae, such as *Phyllanthus flexuosus*, *P. oligospermus* (both subgenus *Kirganelia*), *P. liukiuensis* (subgenus *Eriococcus*), *P. roseus* (subgenus *Phyllanthodendron*), *Breynia retusa*, and *Sauropus quadrangularis*. Notably, most herbaceous *Phyllanthus*, which are phylogenetically spread across the entire Phyllantheae phylogeny (Chap. 4), are pollinated by ants that visit flowers for nectar. In a controlled experiment where only ants were allowed to visit flowers of *P. lepidocarpus* grown in cages, plants regularly attained full fruit set (Fig. 6.1), whereas they produced no fruits when insects were fully excluded. Because

Table 6.1 List of species studied

Species sampled[a]	Abbreviation	Study site	*Epicephala* as pollinator	Criteria for pollinator determination[b]	Style spreading
Margaritaria					
M. discoidea	Mdis	Guinea: Bossou	No	E, M	7.07
M. indica	Mind	Japan: Okinawa Is.	No	M	5.46
Flueggea					
F. jullienii	Fjul	Laos: Mahaxai	No	M	4.99
F. suffruticosa	Fsuf	Japan: Hyogo/Hiroshima/Amami Island	No	E, M	4.82
F. virosa	Fvir	Laos: Vieng Xai/Taiwan: Fangliao	No	E, M	4.47
Phyllanthus					
P. (Mc.) ussuriensis	Puss	Japan: Tokyo/Kyoto	No	E, M	7.01
P. (Mc.) virgatus	Pvir	Laos: Vientiane	No	E, M	5.97
P. (Er.) liukiuensis	Pliu	Japan: Okinawa Island	No	E, M	7.87
P. (Er.) pulcheroides	Ppul	Laos: Mahaxai	No	E, M	8.69
P. (Ki.) reticulatus	Pret	Taiwan: Henchun	Yes	E, M	0.45
P. (Ki.) microcarpus	Psp	Laos: Laksao	Yes	E, M	0.5
P. (Ki.) flexuosus	Pfle	Japan: Kyoto/Hyogo/Miyazaki	No	E, M	4.87
P. (Ki.) oligospermus	Poli	Japan: Yonaguni Island	No	E, M	4.96
P. (Tn.) tenellus	Pten	Japan: Okinawa Island	No	E, M	6.41
P. (Sw.) amarus	Pama	Japan: Ishigaki Island/Laos: Thakhaek	No	E, M	4.09
P. (Sw.) warnockii	Pwar	USA: New Mexico	No	E, M	1.87
P. (Af.) debilis	Pdeb	Japan: Ishigaki Island	No	E, M	4.32

(continued)

Table 6.1 (continued)

Species sampled[a]	Abbreviation	Study site	*Epicephala* as pollinator	Criteria for pollinator determination[b]	Style spreading
P. (Go.) aeneus	Paen	New Caledonia: Cap Bocage	Yes	L, M	1.08
P. (Go.) gneissicus	Pgne	New Caledonia: Mt. Panié	Yes	L	n. a.
P. (Go.) guillauminii	Pgui	New Caledonia: Tiébaghi	Yes	L	n. a.
P. (Go.) vulcani	Pvul	New Caledonia: Riviere Bleue	Yes	L, M	0.62
P. (Go.) bourgeoisii	Pbou	New Caledonia: Cap Bocage	Yes	L, M	0.38
P. (Go.) chamaecerasus	Pcha	New Caledonia: Chutes de Ba	Yes	L	n. a.
P. (Go.) caudatus	Pcau	New Caledonia: Riviere Bleue	Yes	L	n. a.
P. (Go.) cf. *koniamboensis*	Pkon	New Caledonia: Tinip	Yes	L	n. a.
P. (Go.) mangenotii	Pman	New Caledonia: Cap Bocage	Yes	L, M	0.49
P. (Ci.) acidus	Paci	Laos: Vientiane (cultivated)	No	L, E, M	2.5
P. (Em.) emblica	Pemb	Laos: Ban Chomesy	No	L, E	n. a.
P. (Em.) lepidocarpus	Plep	Japan: Kyoto/ Miyako Island/ Ishigaki Island	No	E, M	3.12
P. (Pd.) roseus	Pros	Laos: Phialat	No	E, M	1.99
P. marojejiensis	Pmar	Madagascar: Mt. Marojeji	Yes	E, M	0.18
P. humbertii	Phum	Madagascar: Mt. Marojeji	Yes	E, M	0.39
Sauropus					
S. androgynus	Sand	Laos: Thakhaek	No	E, M	2.03
S. brevipes	Sbre	Laos: Vientiane	No	E, M	2.14
S. granulosus	Sgra	Laos: Vientiane	No	E, M	2.04

(continued)

Table 6.1 (continued)

Species sampled[a]	Abbreviation	Study site	*Epicephala* as pollinator	Criteria for pollinator determination[b]	Style spreading
S. quadrangularis	Squa	Laos: Vientiane	No	E, M	2.53
Breynia					
B. disticha	Bdis	New Caledonia: Koumac	Yes	M	0.25
B. fruticosa	Bfru	Laos: Vientiane	Yes	L, M	1.45
B. oblongifolia	Bobl	Australia: Windsor Tableland	Yes	M	0.2
B. retusa	Bret	Laos: Vientiane	No	E, M	3.02
B. vitis-idaea	Bvit	Japan: Amami Island	Yes	L, M	0.43
Glochidion					
G. acuminatum	Gacu	Japan: Amami Island	Yes	L, M	0.86
G. lanceolatum	Glan	Japan: Ishigaki Island	Yes	L, M	0.31
G. obovatum	Gobo	Japan: Wakayama	Yes	L, M	0.93
G. rubrum	Grub	Japan: Ishigaki Island	Yes	L, M	0.87
G. zeylanicum	Gzey	Japan: Okinawa Island	Yes	L, M	0.24

[a]*Phyllanthus* sections and subgenera are abbreviated as follows: *Mc.*, *Macraea*; *Er.*, *Eriococcus*; *Ki.*, *Kirganelia*; *Tn.*, *Tenellanthus*; *Sw.*, *Swartziani*; *Af.*, *Afroswartziani*; *Go.*, *Gomphidium*; *Ci.*, *Cicca*; *Em.*, *Emblica*; *Pd.*, *Phyllanthodendron*. Placement of *Phyllanthus marojejiensis* and *P. humbertii* is uncertain

[b]Each species was judged as either *Epicephala*- or non-*Epicephala*-pollinated based on literature information (L), ecological data (E), and/or style morphology of the female flower (M)

P. lepidocarpus is self-compatible, ants are probably sufficient to pollinate this species fully in wild conditions as well.

Whether a plant species is pollinated by *Epicephala* is most clearly reflected in style morphology. In species pollinated by the moths, styles are reduced and fused to form a narrow apical cavity into which moths insert the proboscis to deposit pollen. By contrast, species diurnally pollinated by various nectar-seeking insects usually have bifid styles that are spread horizontally, which facilitates passive pollen receipt from insect bodies (Fig. 6.2). Overall, species with different

Fig. 6.1 Ant pollination of *Phyllanthus lepidocarpus*. (**a**) An ant, *Paratrechina flavipes*, consuming nectar on male flower of *P. lepidocarpus*. (**b**) *P. lepidocarpus* pollen attached to the head of *Tetramorium* sp. (**c**) Result of selective exclusion experiment in *Phyllanthus lepidocarpus*. Fruit set of the caged treatment is significantly lower than the fruit sets of other three treatments (Kruskal–Wallis test, $\chi^2 = 40.01$, df $= 3$, $P < 0.001$). Numbers inside bars are sample sizes. Error bars are too small to be seen

pollination syndromes have nonoverlapping degrees of style spreading (Fig. 6.3); thus, pollination systems can be reliably assigned to plant species for which sufficient ecological data are not available.

6.3 Phylogeny of Phyllantheae and *Epicephala*

To investigate the origin of the Phyllantheae–*Epicephala* mutualism, pollination systems were mapped onto the molecular phylogeny of 46 Phyllantheae species. The phylogeny is based on the combined chloroplast *matK*, *ndhF*, *atpB*, and nuclear *PHYC* gene dataset for 92 species of Phyllanthaceae including the above 46 Phyllantheae species. Maximum parsimony, likelihood, and Bayesian analyses all produced a highly resolved and well-supported phylogeny for Phyllantheae (Fig. 6.4). Similarly, the phylogeny of 26 *Epicephala* species associated with the above Phyllantheae species were reconstructed based on the combined mitochondrial *COI*, nuclear *ArgK*, *EF-1α*, *Wg*, and the *18S rDNA* gene dataset. This produced a well-resolved phylogeny, although the phylogenetic placement of *Conopomorpha flueggella* with respect to *Epicephala* remained ambiguous (Fig. 6.5).

Fig. 6.2 Floral morphology of Phyllantheae plants with *Epicephala* (**a–h**) and non-*Epicephala* (**i–p**) pollinators. (**a**, **e**) *Glochidion acuminatum*. (**b**, **f**) *Breynia vitis-idaea*. (**c**, **g**) *Phyllanthus reticulatus*. (**d**, **h**) *Phyllanthus marojejiensis*. (**i**, **m**) *Flueggea suffruticosa*. (**j**, **n**) *Phyllanthus flexuosus*. The *arrowheads* indicate female flowers. (**k**, **o**) *Breynia retusa*. (**l**, **p**) *Sauropus quadrangularis*. For each species, male flowers are shown *above* female flowers (Reproduced from Kawakita 2010)

These phylogenies provide important insights into the origin of mutualism and active pollination. First, Phyllantheae species pollinated by *Epicephala* are not monophyletic, indicating that there have been multiple shifts in pollination systems. Reconstruction of ancestral character states for the pollination system along the Phyllantheae phylogeny suggest that there are five independent origins of the obligate pollination mutualism in Phyllantheae, with a single reversal to non-*Epicephala* pollination in *Breynia retusa* (Fig. 6.6). The pollinator *Epicephala* species are also nonmonophyletic, and ancestral character state reconstruction indicated a likely single origin of pollination behavior with a single event of secondary loss (Fig. 6.6). Major clades of *Epicephala* generally have specific

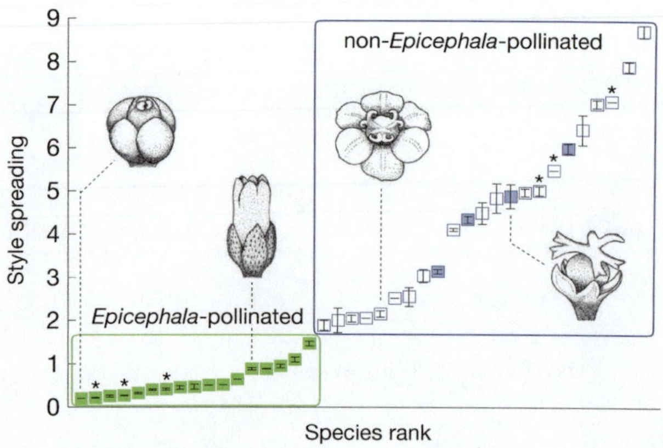

Fig. 6.3 Distribution of style spreading in Phyllantheae, quantified as the ratio of apical to basal style width. Species pollinated by *Epicephala* (*green*) have reduced styles that are medially fused, whereas non-*Epicephala*-pollinated species (*blue*) have horizontally spread, bifid styles. *Filled* and *empty boxes* indicate species with and without associations with *Epicephala*, respectively. Ecological data were not available for species with asterisks, but because species with different pollination syndromes had nonoverlapping distributions of style spreading, their pollination modes could be assigned reliably. Female flowers are drawn for *Phyllanthus marojejiensis*, *Glochidion acuminatum*, *Sauropus brevipes*, and *Flueggea suffruticosa* (from *left* to *right*). Error bars, ±1 SE (Reproduced from Kawakita and Kato 2009)

associations with well-defined taxonomic groups of Phyllantheae, but relationships at higher levels were largely incongruent, indicating that host shifts have occurred repeatedly (Fig. 6.6).

The above analysis of ancestral character state reconstruction indicates that *Epicephala*-pollinated Phyllantheae plants evolved multiple times independently. However, because the taxon sampling was limited to 46 species amid the global diversity of Phyllantheae (>1200 species), results of ancestral state reconstruction might change with the addition of more taxa. Therefore, divergence times for the Phyllantheae and *Epicephala* phylogenies were estimated to test whether the multiple origins hypothesis is in fact the preferred scenario. If the age of the most recent common ancestor of moth-pollinated plants is contemporary to that of *Epicephala*, a single origin of the mutualism followed by multiple losses would still be a viable hypothesis. Alternatively, evolution of pollinating behavior postdating initial host divergence would provide strong support for the multiple origins hypothesis.

A major obstacle when estimating divergence times is the scarcity of fossils, which is also the case for Phyllantheae and *Epicephala*. Nevertheless, there are several fossils of Phyllantheae and plants in other tribes of Phyllanthaceae that can be used to provide minimum age constraints on Phyllanthaceae phylogeny. The fossils used are *Bischofia*-type pollen from Bartonian, Middle Eocene (37.2 mya); *Actephila*-type pollen from Late Eocene (33.9 mya); *Phyllanthus*-type pollen from Early Eocene (48.6 mya) (Gruas-Cavagnetto and Köhler 1992); and *Glochidion* leaf

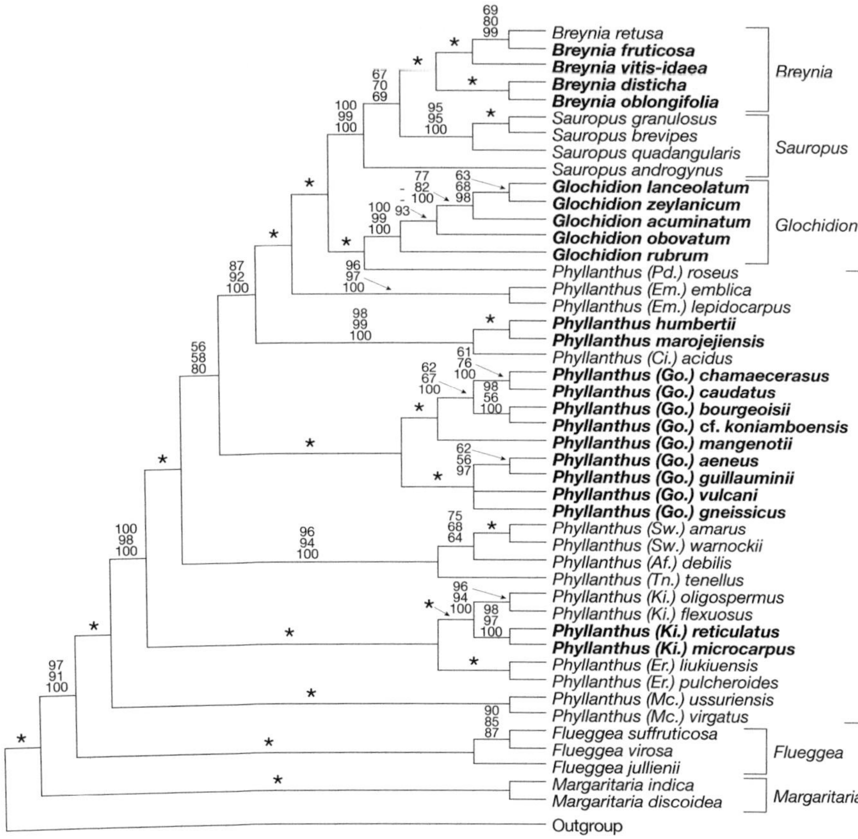

Fig. 6.4 Bayesian majority consensus cladogram of 46 Phyllantheae species based on sequences of combined plastid *matK*, *ndhF*, *atpB*, and nuclear *PHYC* genes. *Numbers* indicate maximum parsimony and likelihood bootstrap values, and Bayesian posterior probability (from *top to bottom*; shown only when >50). *Asterisks* indicate maximal nodal support (100 for all three measures)

impressions from Middle Miocene (11.6 mya; Prasad 1994; Antal and Prasad 1996). The root node (i.e., the node splitting Phyllanthaceae and Picrodendraceae) was assumed to be no older than 108 mya, which is the oldest estimate of the corresponding node in a study of Malpighiales radiation (Davis et al. 2005). Because attribution of some of the Phyllanthaceae fossils may still need refinement (Gruas-Cavagnetto and Köhler 1992), caution may be necessary when taking the precise dates resulting from this analysis. Because gracillariid moths are extremely scarce in the fossil record (Lopez-Vaamonde et al. 2006), *Epicephala* divergence times were obtained assuming a molecular clock of the COI gene. Only the COI clock was used because it is generally conserved across arthropod taxa (Gaunt and Miles 2002), has been widely used for dating in insects (Kandul et al. 2004; Quek et al. 2007; Ueda et al. 2008), and clusters at approximately 1.5% myr^{-1} in several arthropod groups (Farrell 2001; Quek et al. 2004; Sota and Hayashi 2007).

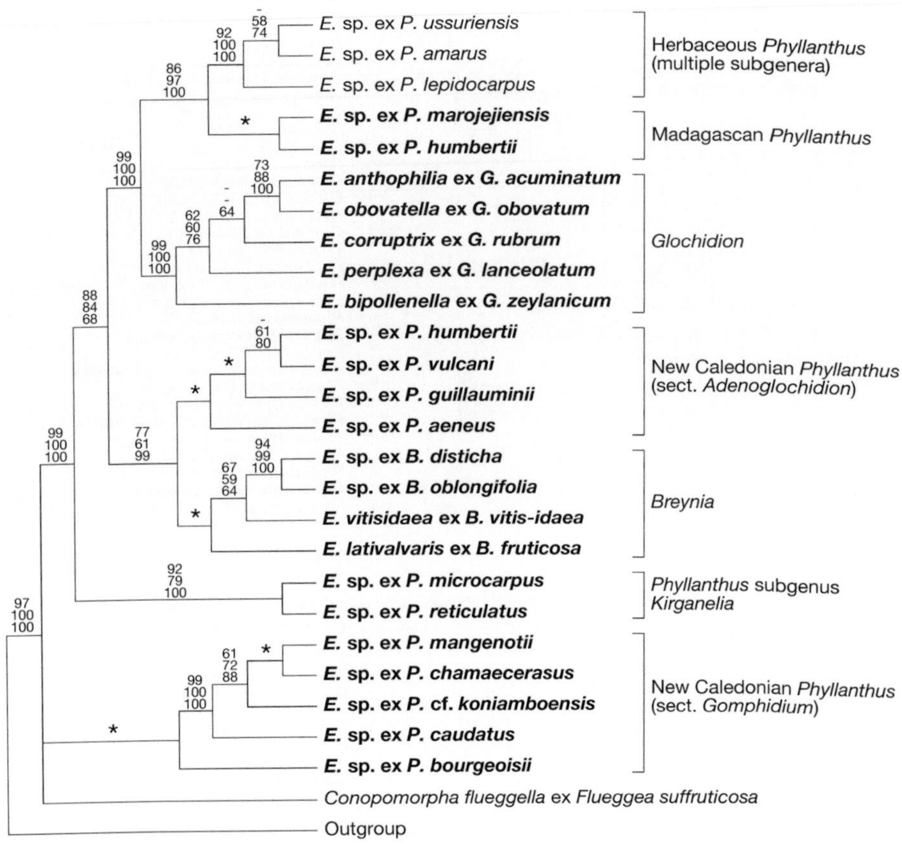

Fig. 6.5 Bayesian majority consensus cladogram of 25 *Epicephala* species and related *Conopomorpha flueggella* based on sequences of combined mitochondrial *COI* and nuclear *ArgK*, *EF-1α*, *Wg*, and *18S rDNA* genes. *Numbers* indicate parsimony and likelihood bootstrap values, and Bayesian posterior probability (from *top to bottom*; shown only when >50). *Asterisks* indicate maximal nodal support (100 for all three measures). Associated host taxonomic groups are given

The analysis of divergence times indicates that the most recent common ancestor of *Epicephala*-pollinated plants occurred 41.0 mya (95% credibility interval, 39.3–48.3 mya; Fig. 6.6). In contrast, estimated ages of the split between *Conopomorpha flueggella* and *Epicephala* clustered within a timeframe between 20 and 30 mya. These estimates for the age of active pollination postdates initial host divergence by roughly 10–20 myr (Fig. 6.6), which is consistent with delayed radiation of *Epicephala* and hence multiple origins of the obligate pollination mutualism in Phyllantheae. Although the estimate of the timing of *Epicephala* divergence depends largely on the accuracy of the COI molecular clock, the assumed 1.5% myr^{-1} is among the slowest of known rates for the arthropod COI gene (1.3–2.3% myr^{-1}; Brower 1994; Quek et al. 2004), and using higher rates would only give younger estimates for the age of the *Epicephala* root node; thus, the method employed is conservative with respect to providing young ages.

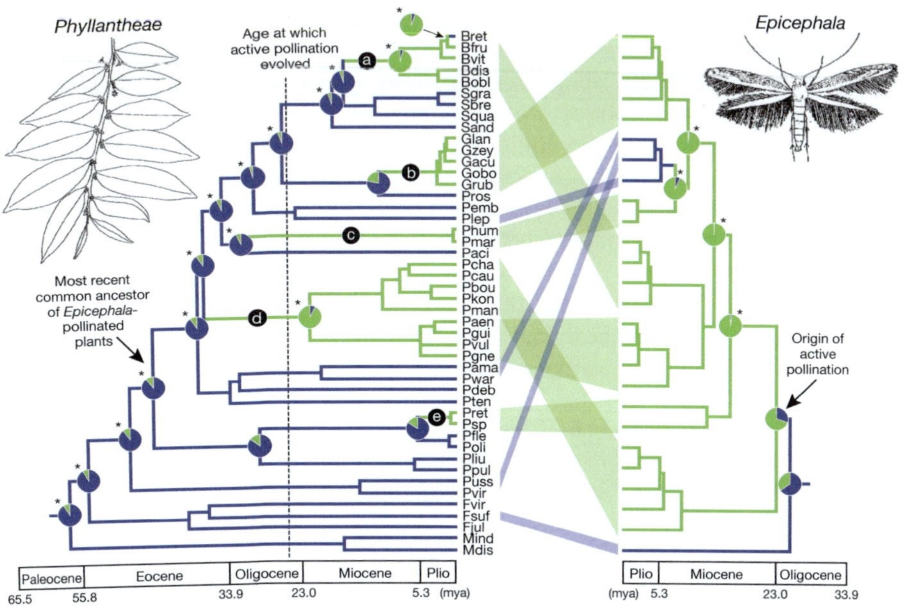

Fig. 6.6 Origin of the Phyllantheae–*Epicephala* obligate pollination mutualism. Chronograms for Phyllantheae plants (left) and associated *Epicephala* moths (right). *Pie charts* indicate the probabilities of *Epicephala* and non-*Epicephala* pollination systems (Phyllantheae) or the presence/absence of active pollination behavior (*Epicephala*) occurring at ancestral nodes. *Asterisks* indicate significant difference in likelihoods. Mutualism is represented in *green*, and associations with major plant and moth clades are indicated

6.4 Origin of Active Pollination and Mutualism

The above phylogenetic analyses and divergence time estimations allowed a general overview of the evolutionary history of the Phyllantheae–*Epicephala* association. Because the taxon sampling was limited to 20% of the global diversity of Phyllantheae at the section level (Kathriarachchi et al. 2006) and less than 5% at the species level, the entire picture of the evolutionary history of *Epicephala* pollination in Phyllantheae is probably much more complex than as depicted here. However, inclusion of other lineages would likely only strengthen the conclusion of repeated independent evolution because these plants generally have bifid, horizontally spread styles that are characteristic of non-*Epicephala*-pollinated plants (Fig. 6.2). Exceptions are the New World *Phyllanthus* subgenus *Xylophylla*, which consists of approximately 60 species having reduced columnar styles (Webster 1958) and section *Microglochidion*, which consists of approximately 10 species occurring on the tepuis of the Guiana Highlands (Chap. 5). Field observation and examination of herbarium specimens indicate that they are also associated with seed-feeding *Epicephala*. It is thus tempting to clarify the pollination systems of these plants and phylogenetic positions of associated *Epicephala*, as they may represent additional origins of *Epicephala* moth pollination in Phyllantheae.

Our finding that the obligate pollination mutualism arose repeatedly in Phyllantheae is in stark contrast with the situations in the fig–fig wasp and yucca–yucca moth mutualisms. Coevolutionary analyses in the fig and yucca systems indicate that these associations arose only once in each partner lineage 40–60 mya (Pellmyr and Leebens-Mack 1999; Rønsted et al. 2005). An exception is *Hesperoyucca whipplei*, which is phylogenetically distant from the rest of the yuccas and independently established the mutualism with a yucca moth (Bogler et al. 1995; Pellmyr et al. 2007; Smith et al. 2008a). In the Phyllantheae–*Epicephala* system, major lineages of Phyllantheae had already emerged when *Epicephala* colonized these plants ~30 mya. Sequential radiation of *Epicephala* on an already diverged host lineage has likely provided opportunities for the moth pollinators to establish new mutualistic associations in distant host lineages. Thus, specialization to moth pollination occurred multiple times independently in Phyllantheae as *Epicephala* spread onto a broad range of the Phyllantheae lineage.

Our results also indicate that colonization of new host lineages by the pollinators sometimes results in a loss of mutualistic traits. A derived clade of *Epicephala* has completely lost the pollinating behavior after colonizing herbaceous species of *Phyllanthus*. These plants regularly attain full seed set through ant pollination (Fig. 6.1); thus, time and energetic costs required during pollination probably outweighed the benefit of assuring seed set in these moth lineages. At the same time, effective pollination by ants probably swamped the mutualistic effect of pollination by moths; thus, selection did not favor these *Phyllanthus* to specialize to moth pollination.

Taken together, the overall evolutionary history of Phyllantheae and *Epicephala* provides two general implications for the coevolutionary dynamics of mutualisms. First, although species associations are phylogenetically conserved in most coevolving interactions (Thompson 2005), rare shifts by a partner possessing the mutualistic trait can give rise to new mutualisms in phylogenetically distant partner lineages. In this sense, the active pollination behavior in *Epicephala* has been of critical importance for the establishment and maintenance of the Phyllantheae–*Epicephala* mutualism and thus represents a key innovation in this association. Second, the outcome of a species interaction can vary greatly depending on the community context in which it occurs (Thompson and Pellmyr 1992; Thompson and Cunningham 2002; Westerbergh 2004); thus, transitions between mutualism and antagonism can occur repeatedly within a single phylogenetic lineage. This parallels findings in other mutualisms where derived parasitic taxa are nested within ancestrally mutualistic clades (Pellmyr et al. 1996b; Machado et al. 2001; Als et al. 2004). Of particular relevance to future studies is our finding that the mutualism arose independently in several Phyllantheae lineages, which provides outstanding opportunities for comparative analyses of character evolution, diversification rates, and factors affecting mutualism establishment and stability.

Part II
Ecology

Chapter 7
Reinforced Specificity of Pollinator Moths

Atsushi Kawakita and Makoto Kato

Keywords *Caloptilia* • *Cuphodes* • *Diphtheroptila* • *Epicephala* • *Glochidion* • Gracillariidae • Host specificity • Phylogeny

7.1 Partner Specificity in Mutualisms

The remarkably high level of partner specificity is a hallmark feature of the leafflower–leafflower moth mutualism. Together with the fig–fig wasp and yucca–yucca moth systems, obligate pollination mutualisms provide some of the best examples of highly species-specific plant–insect associations. However, the evolutionary processes underlying these patterns are poorly understood. The high degree of specificity in pollinating seed parasites is often regarded as the fortuitous result of specialization in their ancestors because these insects are derived from endophytic herbivores that are themselves highly host-specific. This chapter focuses on the comparison of the level of host specificity in *Epicephala* to those of purely parasitic gracillariid relatives as a test of whether mutualism reinforces partner specificity. When interpreted with what is known in the fig and yucca systems, such an analysis serves as a useful approach to determine how partner specificity is shaped in coevolved mutualisms.

Parasitic lifestyles, in general, favor extreme specialization to one or few host species because they require complex adaptations to circumvent host defenses and sustain life on a single host (Ehrlich and Raven 1964; Price 1980; Thompson 1994; Strauss and Zangerl 2002). Although specific mechanisms underlying host specialization may vary among taxa, the broad general understanding is that host–parasite coevolution promotes specialization in parasitic organisms (Thompson 1994,

A. Kawakita (✉)
Center for Ecological Research, Kyoto University, 2-509-3 Hirano, Otsu, Shiga 520-2113, Japan
e-mail: kawakita@ecology.kyoto-u.ac.jp

M. Kato
Graduate School of Human and Environmental Studies, Kyoto University, Yoshida-Nihonmatsu-cho, Sakyo, Kyoto 606-8501, Japan
e-mail: kato@zoo.zool.kyoto-u.ac.jp

M. Kato, A. Kawakita (eds.), *Obligate Pollination Mutualism*, Ecological Research Monographs, DOI 10.1007/978-4-431-56532-1_7

2005). Strict host-specificity of parasites is often linked to high species diversity because specialization to different hosts can result in host-associated speciation (Mitter et al. 1988; Farrell 1998; Schluter 2000; Coyne and Orr 2004).

In contrast, the evolutionary processes that determine the level of specialization in mutualisms are far less understood. Although many mutualisms do not evolve to exhibit high degrees of specificity (e.g., most plant–pollinator and plant–seed disperser interactions), reciprocal partner specialization is often found in intimate mutualisms, such as those between myrmecophytic plants and their resident ants (Davidson and McKey 1993; Heil and McKey 2003; Guimarães et al. 2007), ants/termites and their cultivated fungi (Mueller et al. 1998; Aanen et al. 2002; Currie et al. 2003), or various invertebrates and their endosymbiotic microorganisms (Moran and Telang 1998; Hosokawa et al. 2006). Both ultimate and proximate causes of specialization have been proposed, including selection for elimination of less-cooperative partners (Heil et al. 2005; Poulsen and Boomsma 2005) and chemical or physical mechanisms of partner discrimination (Federle et al. 1997; Brouat et al. 2001; Edwards et al. 2006; Grangier et al. 2009). However, the general understanding of the evolutionary conditions favoring specialization in mutualisms is still very limited (Thompson 1994, 2005), and molecular approaches continue to refine our view of how mutualists are associated with one another on both local and broad geographic scales (Molbo et al. 2003; Mikheyev et al. 2006; Quek et al. 2007; Visser et al. 2009).

A currently estimated 500 species of leafflowers are each pollinated by host-specific *Epicephala* moth species (Chap. 6). Similarly, figs and yuccas have diversified into more than 700 and 40 species, respectively, and a corresponding high diversity of pollinator species has evolved, each of which is obligately mutualistic with one or few fig/yucca hosts (Weiblen 2002; Pellmyr 2003; Herre et al. 2008). This level of specificity is unusual among pollination mutualisms because, although selection may favor plants to depend on specialized visitors for effective conspecific pollination, pollinators are generally expected to maximize the range of plants they visit to optimize resource use (Pellmyr 2002; Gómez and Zamora 2006). The high specificity of pollinating seed parasites is therefore considered to be the result of their inherently parasitic lifestyle (Thompson 1994, 2005) because seed-feeding insects commonly specialize to a narrow range of host plants. Indeed, ecological and phylogenetic studies of the yucca moth family Prodoxidae have found that close relatives of the pollinators are also highly host-specific herbivores (Pellmyr and Thompson 1992; Pellmyr 1999; Pellmyr et al. 2006), suggesting that the high degree of pollinator specificity is driven by the parasitic part of the interaction and cannot be attributed to mutualistic selection (Thompson 1994, 2005).

However, observations suggest that the above view of host specificity in pollinating seed parasites may require revision. Within the yucca moth lineage, two cheater species have independently lost their pollinating behavior and oviposit in young fruits to exploit the seeds that other yucca moth species have pollinated (Pellmyr et al. 1996a, b; Pellmyr 1999). In contrast to their pollinating relatives, each of these cheater species evolved to utilize 4–6 yucca hosts (Pellmyr 1999), suggesting that host specificity in the pollinators may not be determined solely by the herbivorous habit of the moths (Pellmyr 2003). In the fig system, nonpollinating agaonid wasps that are closely related to and co-occur with pollinating fig wasps tend to be less host-specific than are the pollinators (Weiblen and Bush 2002; Marussich and Machado 2007; but see Lopez-Vaamonde et al. 2001; Jousselin et al. 2006, 2008). In addition, fig herbivores in general are dominated by insects that feed on several locally available fig hosts. Given that shared pollinators can result in hybridization among closely related, co-occurring figs (Machado et al. 2005; Wang et al. 2016), selection may favor figs to rely on specialist pollinators to achieve effective conspecific pollination. Thus, these observations indicate that pollinating seed parasites may in fact attain a higher degree of host specificity than that of their parasitic ancestors due to coevolutionary selection arising after the evolution of pollination mutualism.

Glochidion plants and associated *Epicephala* moths provide an ideal system for testing whether host specificity is greater in pollinating seed parasites than in their herbivorous ancestors because *Glochidion* is host to two other genera of Gracillariidae that are purely parasitic, namely *Diphtheroptila* and *Caloptilia* (Fig. 7.1). *Diphtheroptila*, which belongs to the subfamily Ornixolinae together with *Epicephala*, are leaf miners that utilize young *Glochidion* leaves, whereas *Caloptilia* (subfamily Gracillariinae) are leaf miners as early instar larvae and, as they develop into late instars, construct leaf rolls or induce leaf galls, depending on the species (Chap. 5). Phylogenetic analyses of Gracillariidae focused on Ornixolinae and Gracillariinae indicate that neither *Diphtheroptila* nor *Caloptilia* is the direct sister of *Epicephala* (Fig. 7.2); thus, *Epicephalae* are not derived from leaf herbivores with which they share host plants, and *Diphtheroptila* and *Caloptilia* likely each colonized *Glochidion* plants independently. Nevertheless, the shared use of *Glochidion* by the three genera allows for a rigorous test of how different life histories affect patterns of host specificity by controlling for the effect of host-plant species. Furthermore, *Glochidion* plants are attacked by three other genera of seed-parasitic moths in the families Tortricidae, Pyralidae, and Carposinidae (Fig. 7.1), whose host specificity may be determined by a common mechanism with that of *Epicephala* due to their shared larval diet.

The geographic region of focus is southwestern Japan and Taiwan, where there are five common *Glochidion* species (Fig. 7.3; two additional *Glochidion* species occur in Taiwan, but both are relatively uncommon). Of the five species, only *G. obovatum* ranges as far north as mainland Japan; the remaining four species occur in the Ryukyu Archipelago (the southern island chain of Japan) and Taiwan (and elsewhere in subtropical and tropical Asia). As many as four *Glochidion* species coexist on several of these islands. Any pair of *Glochidion* species can be found growing side by side (Fig. 7.4), although each species has more or less distinct microhabitat requirements. For example, *G. zeylanicum* is often found near streams and wetlands, and *G. obovatum* tends to occur along edges of coastal forest. Because most *Glochidion* species flower continuously from spring to autumn, there is a large overlap in the flowering period between sympatric *Glochidion* species.

Fig. 7.1 Gracillariid moths associated with *Glochidion*. (**a**) A mature *Glochidion obovatum* leaf mined by *Diphtheroptila scriptulata* in Amami Island, Japan. (**b**) A young *G. obovatum* leaf mined by *Diphtheroptila* sp. 1 in Amami Island, Japan. Note that the leaf used by this species is distinctly younger than that used by *D. scriptulata*. Also, the larvae of this species cut the margins of the mine to defecate feces, a behavior not seen in *D. scriptulata*. (**c**) Leaf rolls produced by *Caloptilia ryukyuensis* on *G. lanceolatum* in Yonaguni Island, Japan. (**d**) Leaf galls induced by *Caloptilia cecidophora* on *G. acuminatum* in Okinawa Island, Japan. (**e**) *Epicephala obovatella*. (**f**) *Diphtheroptila scriptulata*. (**g**) *Caloptilia ryukyuensis*. (**h**) *Tritopterna* sp. (Tortricidae). (**i**) *Cryptoblabes* sp. (Pyralidae). (**j**) *Peragrarchis syncolleta* (Carposinidae)

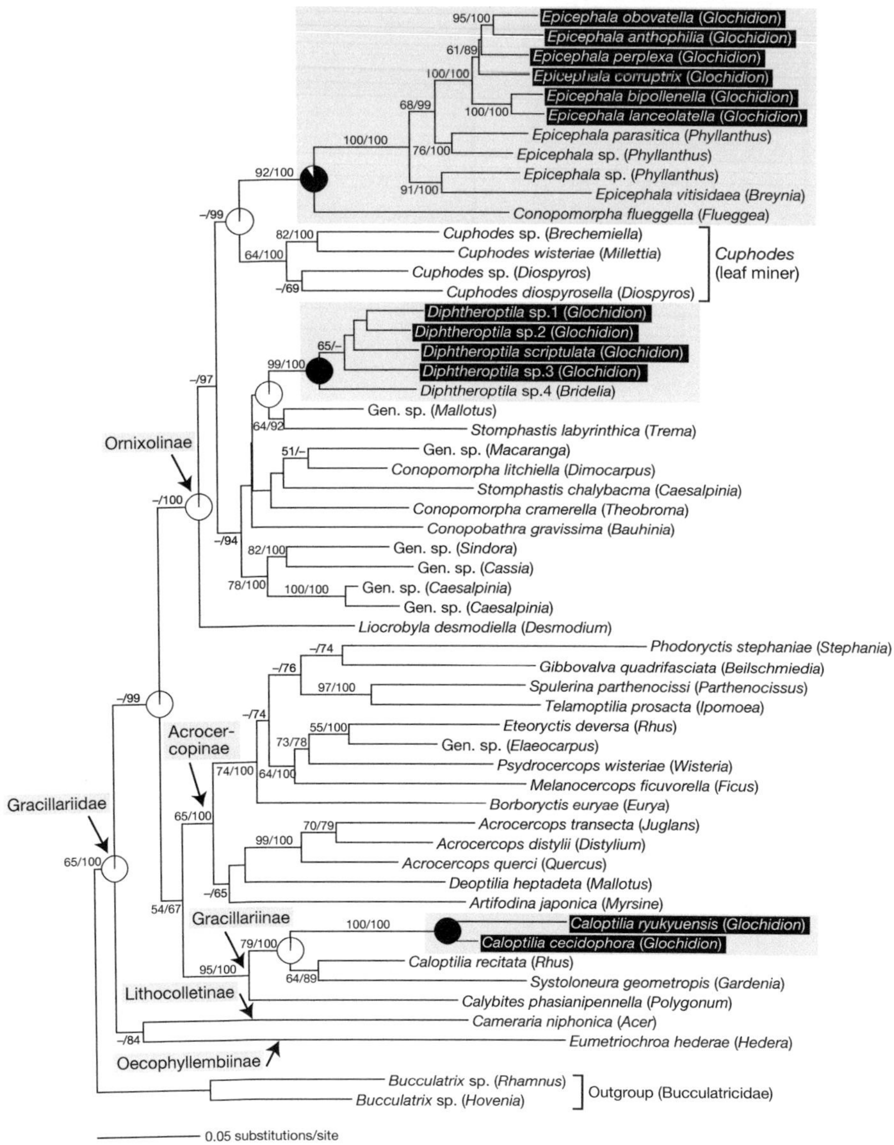

Fig. 7.2 Maximum likelihood phylogeny of Gracillariinae based on 2548 bp of the combined mitochondrial *COI* and nuclear *EF-1α*, *ArgK*, and *18S rRNA* genes. *Numbers* at nodes indicate maximum likelihood bootstrap values followed by Bayesian posterior probabilities. Clades boxed in *grey* are those feeding on Phyllanthaceae plants, and species highlighted individually in *black* are those associated with *Glochidion*. *Pie graphs* show the relative likelihoods of alternative host associations at selected ancestral nodes: *black*, Phyllanthaceae host; *white*, non-Phyllanthaceae host. Taxon names in *parentheses* indicate host plant genera

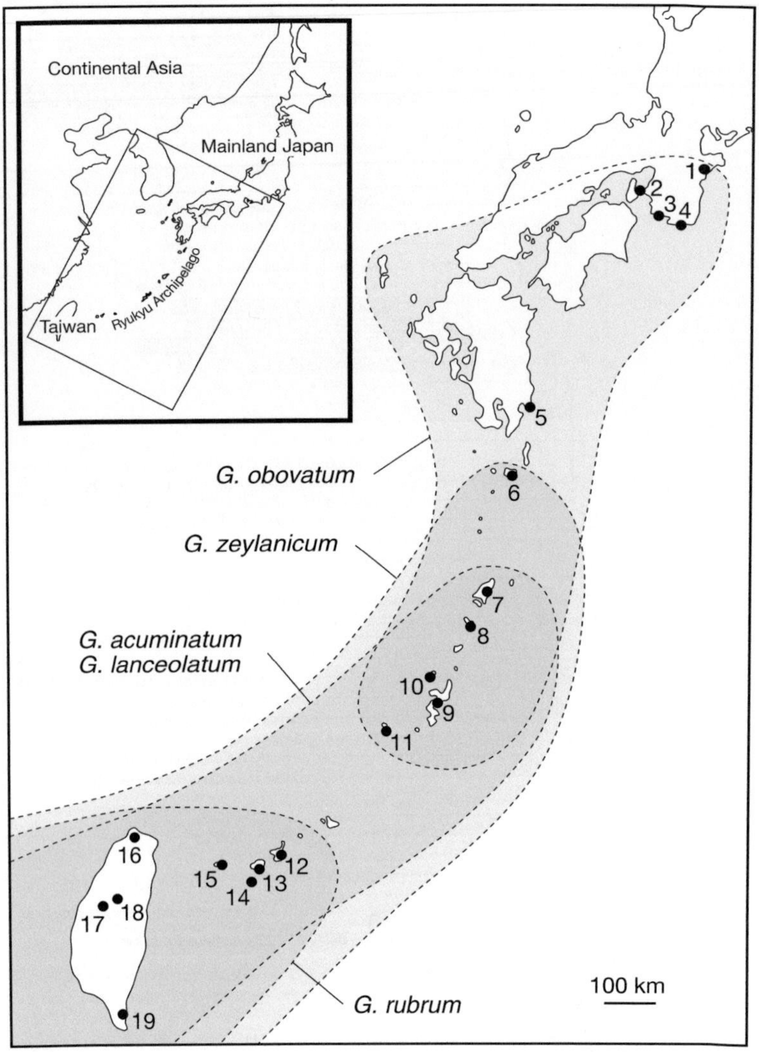

Fig. 7.3 Map of southwestern Japan and Taiwan showing moth sampling localities and distribution ranges of host *Glochidion* species. Locality numbers are: (*1*) Nantou; (*2*) Tomogashima; (*3*) Yura; (*4*) Koza; (*5*) Cape Toi; (*6*) Yaku Island; (*7*) Amami Island; (*8*) Tokuno Island; (*9*) Okinawa Island; (*10*) Iheya Island; (*11*) Kume Island; (*12*) Ishigaki Island; (*13*) Iriomote Island; (*14*) Hateruma Island; (*15*) Yonaguni Island; (*16*) Taipei; (*17*) Sun Moon Lake; (*18*) Wushe; (*19*) Nanren

Fig. 7.4 *Glochidion obovatum* (*left*) and *G. lanceolatum* (*right*) growing side by side in Amami Island, Japan

7.2 Host Specificity of *Epicephala*

Epicephala moths are generally specific to a single host species. However, an often-assumed one-to-one specificity is not perfectly met, as many Phyllantheae plants host more than one *Epicephala* species, or a single *Epicephala* species uses two or more host species either locally or in different parts of their ranges. The *Epicephala* moths associated with the five *Glochidion* species in Japan and Taiwan consist of six species that are clearly distinguishable by morphology and sequences of mitochondrial and nuclear DNA (Fig. 7.5). Of the six species, two species are specific to single host species; *E. anthophilia* is specific to *G. acuminatum*, and *E. bipollenella* uses only *G. zeylanicum*, throughout this region (Fig. 7.6). Two species, *E. lanceolatella* and *E. perplexa*, are associated with *G. lanceolatum* and are regularly found co-occurring on the same trees (Fig. 7.6). Lastly, two species, *E. obovatella* and *E. corruptrix*, share two closely related parapatric hosts (*G. obovatum* and *G. rubrum*; Fig. 7.6). Behaviorally, all six species possess the actively pollinating habit, although the larvae of *E. corruptrix* probably induce gall formation and provide little benefit to the host as compared to the other five species (Chap. 5).

Within this geographic scale, genetic variation within each *Epicephala* species is minimal (Fig. 7.5). However, individuals of *E. obovatella* in the Wushe population, which is located at 1400 m in the central mountain range of Taiwan, exhibit consistent divergence in both mitochondrial and nuclear gene sequences from

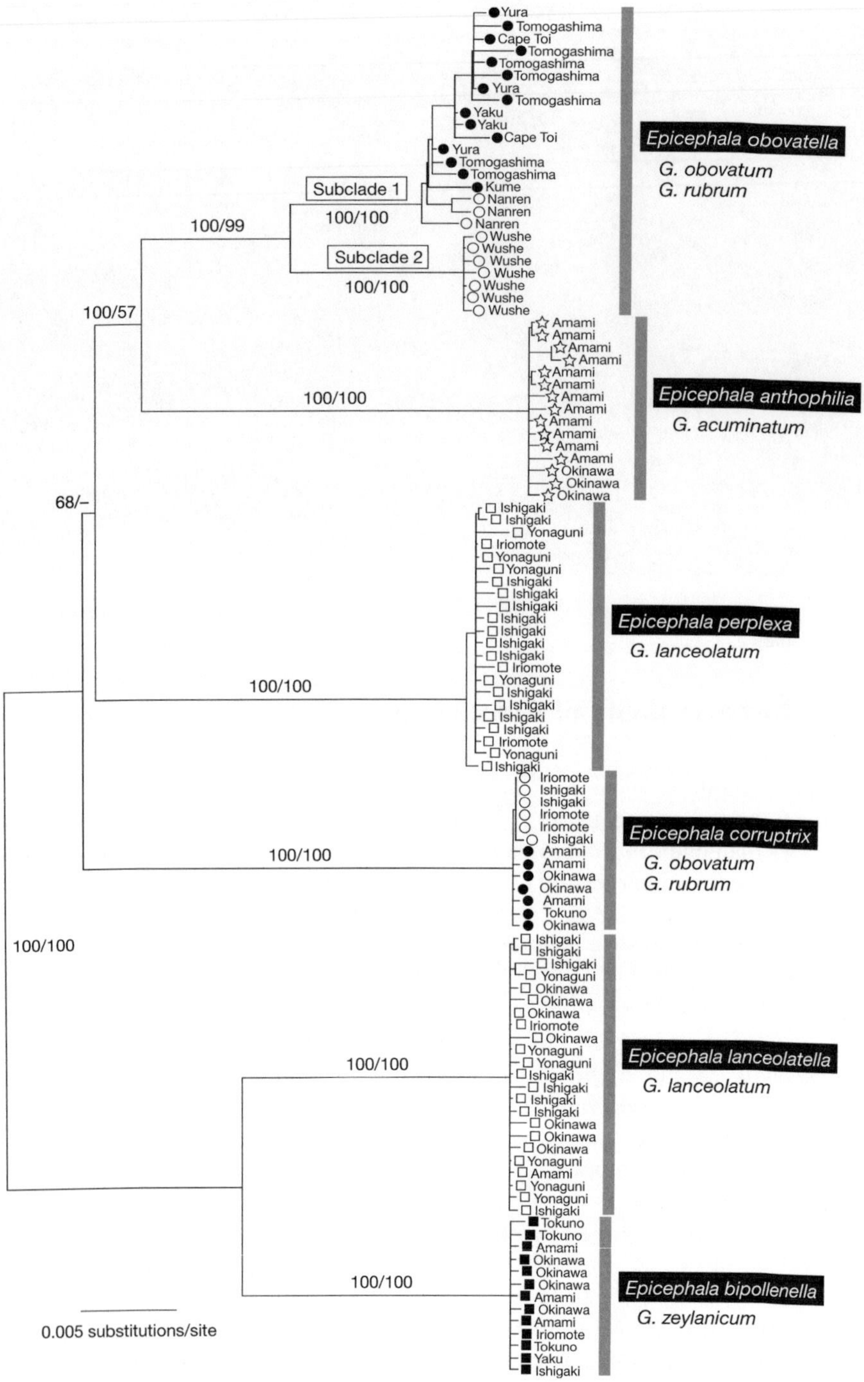

Fig. 7.5 Bayesian 50% majority rule consensus phylogram obtained by simultaneous analysis of *COI*, *ArgK* , and *EF-1α*. Terminal symbols represent host *Glochidion* species followed by locality names: *filled circle*, *G. obovatum*; *open circle*, *G. rubrum*; *filled square*, *G. lanceolatum*; *open*

individuals of other populations located at lower elevations (subclades 1 and 2 in Figs. 7.5 and 7.6), although genital morphology is uniform across all *E. obovatella* populations. Notably, *G. rubrum* trees of the Wushe population tend to have shorter pedicels and larger fruits than plants occurring elsewhere; thus, although a further quantitative study is necessary, this may represent an incipient stage of simultaneous speciation in plant and pollinator.

Phylogenetic analysis also indicates that the two *Epicephala* species co-occurring on *G. lanceolatum* (*E. laceolatella* and *E. perplexa*) are not sister taxa, so a host shift has occurred at least in one of the two species. Whether the coexistence of two pollinator species on a shared host is evolutionarily stable is unknown, because the age at which the two species started to coexist cannot be inferred from available data. Nevertheless, the prevalence of similar situations in figs, yuccas, and leafflowers (Pellmyr 1999; Molbo et al. 2003) may suggest that long-term coexistence on a shared host can occur.

In contrast, the two species associated with *G. obovatum* and *G. rubrum* (*E. obovatella* and *E. corruptrix*) were not found within the same population (Fig. 7.6). Although the number of moths sampled is limited to rule out coexistence, the pattern is in clear contrast with that observed for the two species pollinating *G. lanceolatum*. The disjunct distribution of *E. obovatella* in mainland Japan, Taiwan, and a few small islands (Fig. 7.6), coupled with a relatively high level of intraspecific sequence variation (Fig. 7.5), may indicate ancient widespread distribution of *E. obovatella* and subsequent extinction through competitive exclusion by *E. corruptrix* in most of the Ryukyu Archipelago. Because *E. corruptrix* is probably the less beneficial species (Chap. 5), it is interesting to identify the mechanism that has shaped the current distribution pattern of the two species.

Overall, the pattern of association between *Glochidion* and *Epicephala* is far more complex than a simple one-to-one relationship. However, an important pattern consistently observed throughout this region is that, at any given location, these moths are all specific to only one of multiple co-occurring *Glochidion* species. This likely prevents interspecific pollen transfer and helps maintain reproductive isolation of sympatric *Glochidion* species. For example, on Amami Island of southern Japan, four *Glochidion* species (*G. acuminatum*, *G. obovatum*, *G. zeylanicum*, and *G. lanceolatum*) co-occur and all flower simultaneously during May–June. Three of the four species other than *G. acuminatum* have a prolonged flowering season lasting from spring to autumn, so phenological isolation is virtually absent among these *Glochidion* species. Although studies are needed to quantify the strength of pollinator isolation, local host specificity of *Epicephala* is probably a necessary condition for multiple *Glochidion* species to coexist stably.

Fig. 7.5 (continued) square, *G. zeylanicum*; *open star*, *G. acuminatum*. *Numbers above branches* indicate Bayesian posterior probabilities followed by parsimony bootstrap values (shown when >50%). Species name is given to the *right* of each clade on *black background*; host plant species are given immediately *below*

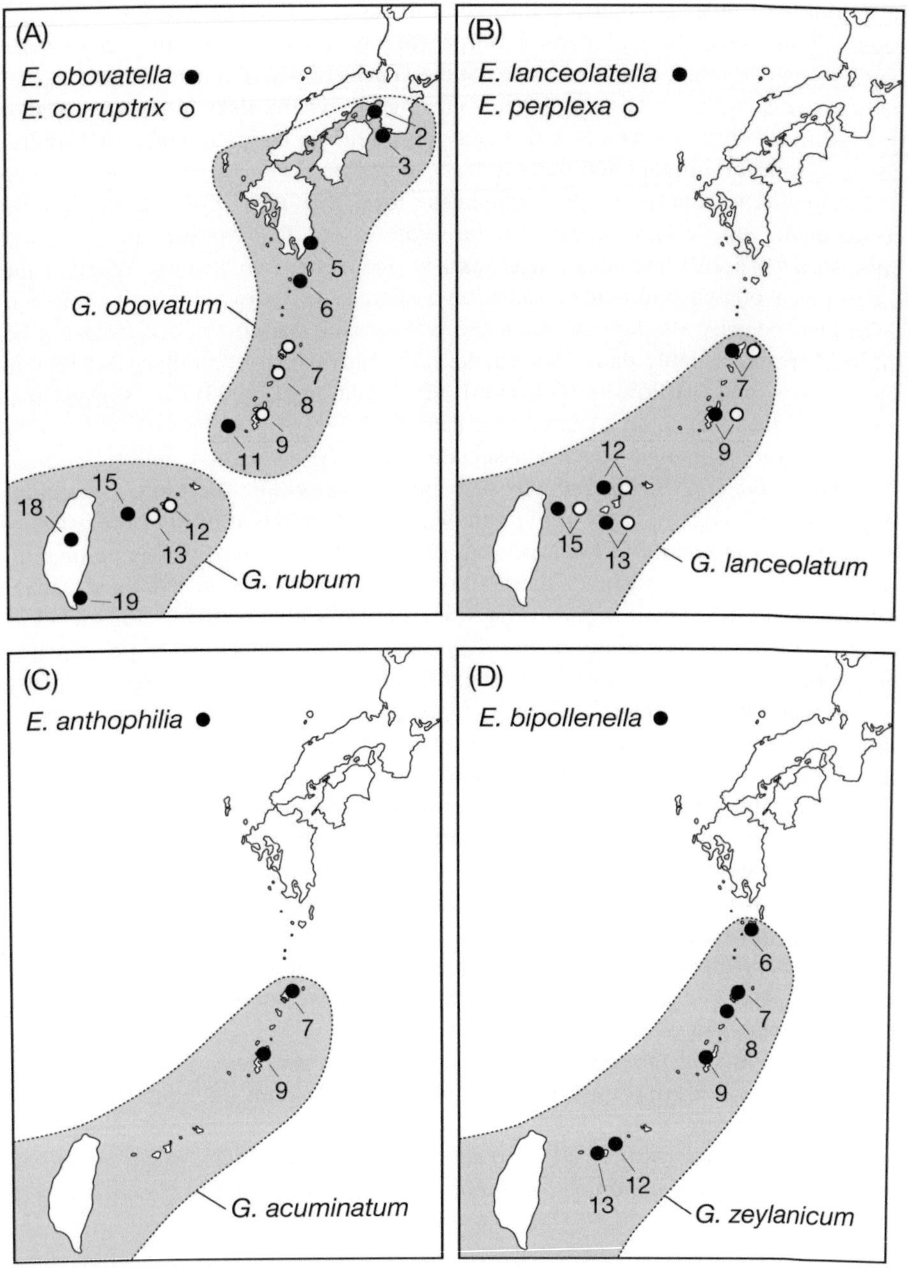

Fig. 7.6 Geographic distribution of the six *Epicephala* species in Japan along with range of their host plants. *Numbers* on the map correspond to those in Fig. 7.3. *Epicephala lanceolatella* and *E. perplexa* co-occur at each location

7.3 Host Specificity of *Diphtheroptila* and *Caloptilia*

Diphtheroptila and *Caloptilia*, sampled within precisely the same geographic range as the above *Epicephala* moths, each contained four distinct species that are widely separated genetically (Fig. 7.7). They are also distinguishable by wing pattern, male genitalia morphology, and larval feeding habit (Fig. 7.2). Notably, each *Diphtheroptila* and *Caloptilia* species commonly utilizes more than one coexisting *Glochidion* species (Fig. 7.7). Although the possibility of hidden divergence associated with *Glochidion* species cannot be ruled out, it is unlikely that all the *Diphtheroptila* and *Caloptilia* species under consideration are at incipient stages of such host-associated divergence. Moreover, the level of host-associated differentiation, if any, is considerably lower than that found in *Epicephala*, in which individuals associated with different *Glochidion* hosts in any population are morphologically distinct and divergent by at least 4% uncorrected pairwise sequence difference in the COI gene (Fig. 7.5). Therefore, available evidence suggests that *Epicephala* are more highly host-specific than are their leaf-feeding relatives that utilize the same sets of *Glochidion* hosts.

7.4 Host Specificity of Other Seed-Feeding Moths

The observed increase in the level of host specialization in *Epicephala*, however, may simply be the result of a shift to seed feeding, rather than coevolutionary selection resulting from being a pollinator. This possibility can be evaluated by comparing the level of host specificity in seed-infesting lepidopterans that share the same larval food with *Epicephala* moths. Non-gracillariid moths that emerge from *Glochidion* fruits are morphologically identified as either *Peragrarchis syncolleta* (Carposinidae) or as undescribed species of *Tritopterna* (Tortricidae) or *Cryptoblabes* (Pyralidae). However, there is only one species in each genus, as judged by the negligible divergence found in mitochondrial and nuclear gene sequences, and each species attacks 4–5 different *Glochidion* hosts. Although it is not straightforward to compare host specificity directly between moths of different families, the level of host specialization found in these groups is at the opposite extreme from the pattern expected if seed feeding is to promote higher host specificity. Therefore, there is no positive evidence that seed feeding favors a higher degree of host specialization; thus, *Epicephala* host specificity is likely determined by factors other than larval diet.

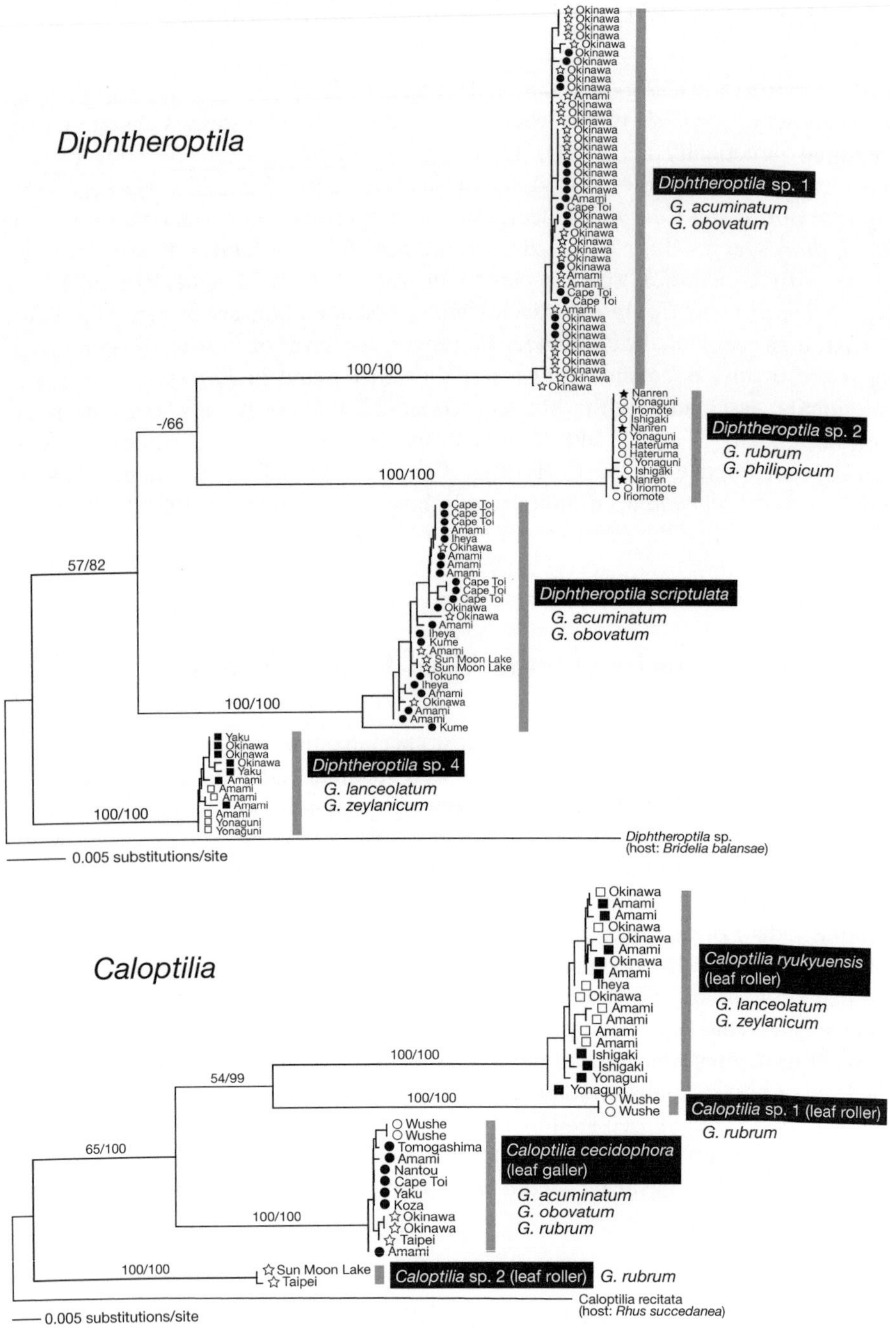

Fig. 7.7 Maximum likelihood phylogeny of *Diphtheroptila* and *Caloptilia* moths based on 1058 bp of the combined mitochondrial *COI* and *EF-1α* genes. Terminal symbols represent host *Glochidion* species followed by locality names: *filled circle*, *G. obovatum*; *open circle*, *G. rubrum*;

7.5 Host Specificity of *Cuphodes*

Another explanation for strict host specialization in *Epicephala* is that such high host specificity is a common feature among all the closest relatives of *Epicephala*, and that pollinator habit evolved against a background of high host specificity. The closest relative of *Epicephala* on the Ornixolinae phylogeny (Fig. 7.3) is *Cuphodes*, which uses species of Fabaceae, Ebenaceae, and Rhamnaceae as hosts. Although support for this relationship is low, adults of *Epicephala* and *Cuphodes* rest with their abdomens raised (Fig. 7.8), a feature not otherwise found in any genera of Ornixolinae, suggesting that *Cuphodes* is likely one of the closest relatives of *Epicephala*.

Cuphodes moths sampled from roughly the same geographic region as the above- sampled *Glochidion* feeders consisted of eight putative species (Fig. 7.8), which can be distinguished by wing pattern, male genitalia morphology, and larval mining pattern. As with *Diphtheroptila* and *Caloptilia*, single *Cuphodes* species regularly use 2–4 closely related plants (Fig. 7.8), suggesting that the closest relatives of *Epicephala* do not show the same degree of host specificity as *Epicephala*. Although the use of different host plant families in *Epicephala* and *Cuphodes* may make direct comparison difficult, available evidence suggests that *Cuphodes* species exhibit much broader host ranges than do the species of *Epicephala*. For example, *C. wisteriae* utilizes *Wisteria* and *Millettia*, which are distantly related genera within Fabaceae, having diverged at least 50 Ma (Lavin et al. 2005), whereas the age of the *Glochidion* crown group is estimated to be only <10 Ma (Chap. 6). Similarly, *Cuphodes* sp. 4 feeds on two genera (*Berchemia* and *Berchemiella*) of the Rhamnaceae family, although the antiquity of their divergence is unknown. Thus, the high degree of host specialization found in *Epicephala* is also likely not an ancestral condition predating the evolution of pollinator habit.

7.6 Reinforced Specificity of *Epicephala*

Taken together, the pattern of host specificity of *Diphtheroptila*, *Caloptilia*, *Cuphodes*, and seed-feeding non-gracillariid moths all indicate that the level of host specialization in *Epicephala* is higher than would be expected if host specificity were determined solely by the herbivorous habit of the moths. A more likely view is that pollinator habit favors higher host specificity than the ancestral parasitic lifestyle. What, then, is the ultimate cause driving the strict host specificity of *Epicephala*? As discussed in Chap. 8, there are clear differences in the chemical

Fig. 7.7 (continued) filled square, *G. lanceolatum*; *open square*, *G. zeylanicum*; *filled star*, *G. philippicum*; *open star*, *G. acuminatum*. *Numbers above branches* indicate maximum likelihood bootstrap values followed by Bayesian posterior probabilities. Species name is given to the *right* of each clade on black background; host plant species are given immediately *below*

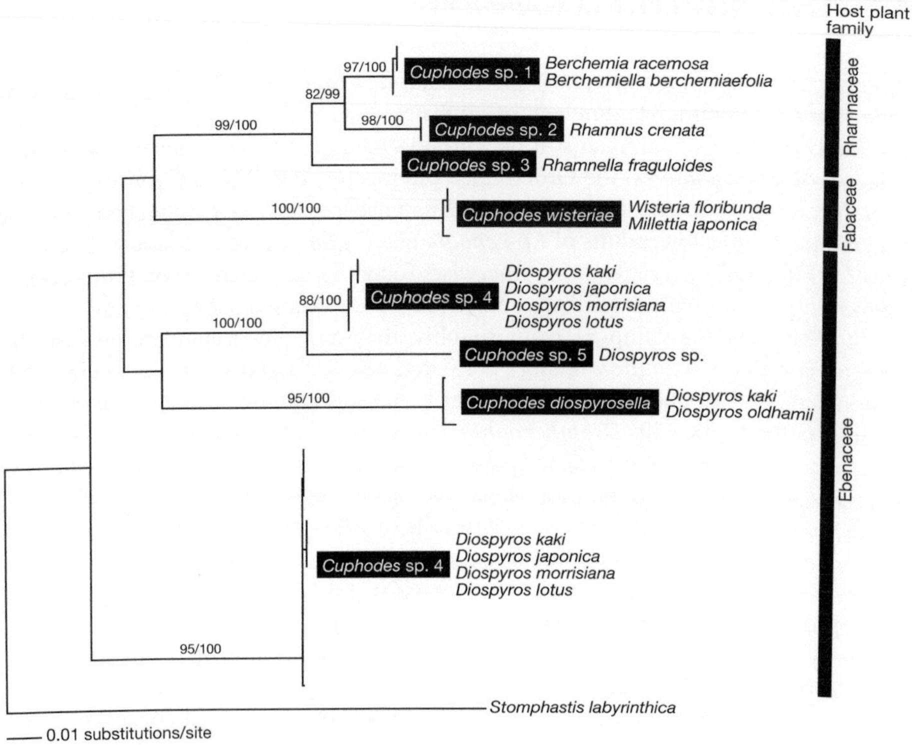

Fig. 7.8 Maximum likelihood phylogeny of *Cuphodes* based on 1601 bp of the combined mitochondrial *COI* and nuclear *EF-1α* and *ArgK* genes. Species name is given to the *right* of each clade on *black background*; host plant species are given immediately *below*. Host plant families are indicated using *bars* on the *right*. *Numbers above branches* indicate maximum likelihood bootstrap values followed by Bayesian posterior probabilities

composition of floral volatiles among coflowering *Glochidion* species. These differences are perceived by host-seeking *Epicephala* females and likely facilitate the attraction of species-specific pollinators. Thus, selection may operate on *Glochidion* plants to produce distinct floral scents and attract specific pollinators and thereby avoid hybridization. Although further experimentation is needed to determine whether interspecific crosses result in fruit production, any decrease in quantity and/or quality of hybrid fruits is likely also to facilitate specialization by *Epicephala* to species-specific floral volatiles. Thus, it is possible that the high plant–pollinator specificity in obligate pollination mutualisms is driven by a plant's interest to avoid less-advantageous hybridization.

Overall, reinforced specificity of *Epicephala* contrasts with what is known in the yucca moth lineage (Pellmyr and Thompson 1992; Thompson 1994, 2005). The closest relative of the pollinating yucca moths, *Prodoxus*, feed on inflorescence stalk, fruit, or, rarely, leaves of yucca plants and have very similar degrees of host specificity with the pollinating yucca moths (Pellmyr et al. 2006). This difference is probably due to contrasting patterns of flowering phenology between yuccas and *Glochidion*. Because both pollinating (*Tegeticula* and *Parategeticula*) and nonpollinating (*Prodoxus*) yucca moths are short lived and their life histories are strongly associated with yucca flowers (Pellmyr 1999, 2003; Pellmyr et al. 2006), the adult moths must emerge during a short period when host flowers are available. However, phenological isolation is strong between coexisting yucca species (Pellmyr 2003); thus, there is little opportunity for both pollinating and nonpollinating yucca moths to select among multiple hosts within a single population. In contrast, *Glochidion* plants produce flowers and leaves continuously from spring to autumn, and different *Glochidion* species commonly flower at the same time. Under such circumstances, both leaf-feeding and flower-infesting moths are provided with multiple available hosts, but the latter are more selective in their host choice probably due to a stricter chemical coadaptation with which they are constrained. The occurrence of multiple coflowering host species is also the case in figs, for which pollinating fig wasps are more host specific than the nonpollinators (Weiblen and Bush 2002; Marussich and Machado 2007; but see Lopez-Vaamonde et al. 2001; Jousselin et al. 2006, 2008).

Although *Epicephala* exhibits higher degrees of host specialization than do their parasitic ancestors, a more direct test of host specificity would be to include nonpollinating gracillariid seed parasites in the analysis. *Conopomorpha flueggella*, a nonpollinating seed feeder of *Flueggea* that is very closely related to *Epicephala* (Chap. 5), may be specialized to *Flueggea suffruticosa*, but a rigorous test is necessary in regions where multiple *Flueggea* species occur. Also, a derived clade of *Epicephala* has secondarily lost the pollinating habit, and presently there are three species that are each specific to a single *Phyllanthus* host (Kawakita and Kato 2009). However, closely related *Phyllanthus* hosts are rarely available within the same population, which precludes a direct comparison of host specificity with pollinating *Epicephala* in this case as well. Within the yucca moth lineage, two derived species have independently lost their pollinating behavior and oviposit in young fruits to exploit the seeds that other yucca moth species have pollinated (Pellmyr et al. 1996b; Pellmyr 1999). These cheater species evolved to utilize 4–6 yucca hosts (Pellmyr 1999, 2003), which is consistent with the view that pollinator habit promotes host specificity in pollinating seed parasites. The cheater yucca moths are likely to have a broader phenological window for successful oviposition (Pellmyr 2003); thus, selection for host specialization may have been relaxed in these derived nonpollinators.

Although further research is required to identify coevolutionary forces driving pollinator specificity, a viable hypothesis is that mutualistic selection reinforces host specificity of pollinating seed parasites in obligate pollination mutualisms. Pollinator specificity is likely to impact strongly patterns of gene flow between coexisting plant species and play an important role in facilitating reproductive isolation between diverged populations (Machado et al. 2005; Smith et al. 2008b, 2009). Thus, identifying the mechanism that shapes partner specificity is the key to understanding the role of coevolution in promoting speciation and diversification in obligate pollination mutualisms.

Chapter 8
Species-Specific Floral Scents as Olfactory Cues in Pollinator Moths

Tomoko Okamoto

Keywords Active pollination behavior • *Epicephala* • *Glochidion* • Phyllantheae • Sexual dimorphism • Species specificity

8.1 Floral Scents as Pollinator Attractants

Within the diverse taxa in Phyllanthaceae, many species are obligately pollinated by host-specific seed-parasitic moths. There must be some mechanism by which a plant species attracts a pollinator partner because several Phyllanthaceae plants can co-occur in tropical regions.

To attract pollinators, flowers generally use visual and olfactory signals (Proctor et al. 1996; Ollerton et al. 2011). Although diurnal butterflies mainly use visual cues to detect flowers, olfactory cues together with visual cues are used in pollination mutualism by oligolectic bees (Dötterl et al. 2005), pompilid wasps (Shuttleworth and Johnson 2009), bumblebees (Byers et al. 2014), euglossine bees (Ackerman 1989), fungus gnats (Okamoto et al. 2015), and thrips (Terry et al. 2007). Another well-known case in which plants strongly depend on olfactory signals to attract pollinators is the deceptive pollination system, in which flowers emit floral scents mimicking pollinators' brood sites (Stensmyr et al. 2002; Oelschlägel et al. 2014) or virgin females of the pollinator species (Peakall and Whitehead 2014). Chemical mimicry of insect brood sites has been observed in 61 plant species in 11 families (Jürgens et al. 2013), the flowers of which emit floral scents mimicking decaying plant material, fungi, carrion, or dung and are pollinated by saprophagous, fungivorous, or coprophagous beetles and flies (Jürgens et al. 2013).

In contrast to diurnal pollination, some plants are pollinated only during the night when visual cues are essentially unavailable. Due to the darkness, nocturnal animal-pollinated flowers depend on chemical signals, and many nocturnal insect pollinators have developed an olfactory sense, as reported in fig wasps (Chen et al. 2009) and moths (Kato and Inoue 1994; Singer 2002; Raguso et al. 2003; Sugiura and Yamazaki 2005; Okamoto et al. 2008). Plants pollinated by nocturnal moths

T. Okamoto (✉)
Faculty of Applied Biological Sciences, Gifu University, 1-1 Yanagido, Gifu 501-1193, Japan
e-mail: tokamoto@gifu-u.ac.jp

M. Kato, A. Kawakita (eds.), *Obligate Pollination Mutualism*, Ecological Research Monographs, DOI 10.1007/978-4-431-56532-1_8

usually have a conspicuous perianth with a white, pale yellow, or light pink color and heavy scents described as "pleasant" or "sweet" (Dobson 2006; Waser 2006). These pale flowers stand out against the background on moonlit and starlit nights, and olfactory signals guide the moth pollinators to the flowers regardless of how dark it is. In *Datura* (Solanaceae) flowers, the pollinator moth, *Manduca sexta*, has been observed feeding on flowers only when both visual and olfactory cues are available, even at night (Raguso and Willis 2005).

As discussed in other chapters, the Phyllanthaceae plants adopting obligate pollination mutualism are pollinated by gracillariid moths only during the night. This suggests that the mutualism is underpinned by chemical communication between the flowers and pollinator moths. This chapter focuses on the floral scent and the olfactory sense of moths, both of which contribute to the high host-specificity of the mutualism.

8.2 Species Specificity Mediated by Floral Scents

The *Epicephala* moth-pollinated plant genus *Glochidion* includes more than 300 species distributed from tropical Asia to Australia and Polynesia (Govaerts et al. 2000). All members of *Glochidion* are thought to be pollinated exclusively by species-specific *Epicephala* moths. Five species commonly occur at forest edges in southern Japan, *G. acuminatum*, *G. lanceolatum*, *G. obovatum*, *G. rubrum*, and *G. zeylanicum*, and a few *Glochidion* species can grow sympatrically. In such populations, flowering periods often overlap and the sympatric *Glochidion* species do not grow in spatial isolation. The inconspicuous flowers and their nocturnal flowering suggest that the flowers attract moth pollinators via scent. Therefore, the high host-specificity of the pollinator moths suggests that (1) the floral scent composition differs among *Glochidion* species, and (2) pollinator *Epicephala* moths can distinguish between the floral scents of their own host and nonhost species.

The first hypothesis was tested by chemical analyses of the floral scents of five *Glochidion* species using gas chromatography-mass spectrometry (GC-MS). The floral scents of *Glochidion* flowers were blends of 42 volatiles, mainly monoterpenes and sesquiterpenes, and were dominated by linalool or β-ocimene. To detect intra- and interspecific variation in the volatile profiles obtained from GC-MS, the profile data were transformed using chord-normalized expected species shared distances (CNESS) and nonmetric multidimensional scaling (NMDS). The CNESS–NMDS analysis of the volatile profile data of the five *Glochidion* species revealed that the floral scents could be clearly distinguished among plant species based on the relative compositions of volatile chemicals, especially the composition of minor compounds (Fig. 8.1a). Although the composition of floral volatiles was also variable among conspecific flowers on different trees, marked interspecific differences were observed in the volatile composition. In CNESS analyses, sensitivity to the abundance of dominant compounds can be adjusted by altering the analysis parameters, that is, by emphasizing the importance of major or minor compounds (Trueblood et al. 1994). Under the parameter set sensitive to both dominant and minor compounds, NMDS scatterplots revealed that the floral samples were divided into three main groups (Fig. 8.1b). *Glochidion lanceolatum*, *G. obovatum*, and *G. rubrum* were not separated from each other in the scatterplot, probably because they shared a dominance of (*R*)-(−)-linalool. In comparison, under the parameter set most sensitive to minor compounds, the five *Glochidion* species were largely distinguished by volatile composition, although there were slight overlaps among the *G. lanceolatum*, *G. obovatum*, and *G. rubrum* samples, suggesting that the presence of minor compounds or their amounts relative to other compounds is also important in attracting specific *Epicephala* moths. Among the three *Glochidion* species, *G. obovatum* and *G. rubrum* are closely related and have allopatric distributions: the former appears on mainland Japan and Okinawa Island and the latter on the Yaeyama Islands. These two *Glochidion* species do not have different scent profiles because they never co-occur (Fig. 8.2). The CNESS/NMDS analysis did not use an equal number of samples from flowers of each plant species, but a similar analysis based on samples from the same numbers of flowers clearly differentiated the floral scent profiles among the five *Glochidion* species.

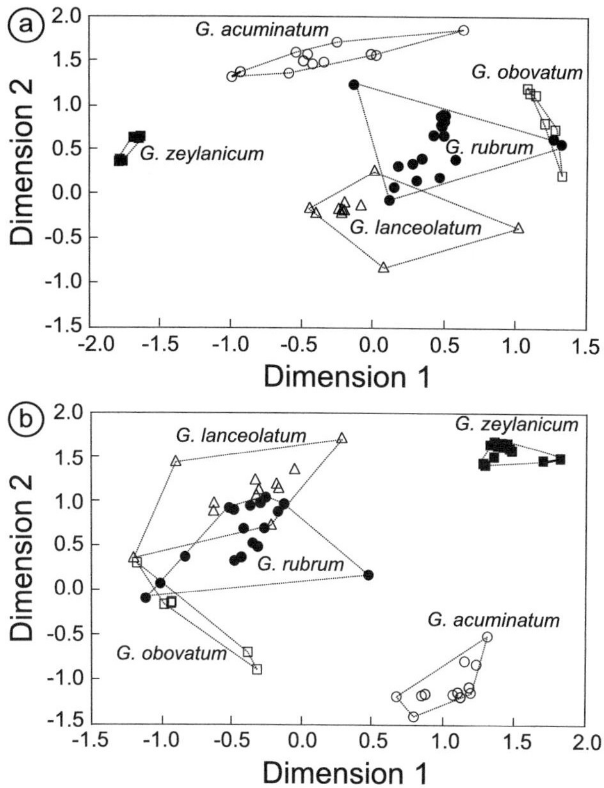

Fig. 8.1 Scatterplots resulting from the NMDS analysis based on CNESS distance index of floral scents in the five Japanese *Glochidion* species. CNESS dissimilarity was calculated by either (**a**) assuming a greater weight on minor compounds and (**b**) weighing all compounds equally

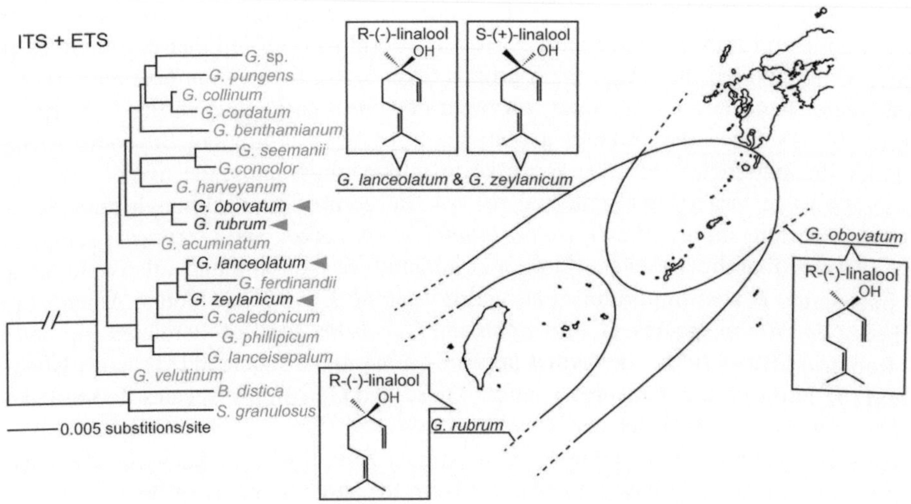

Fig. 8.2 Relationship between dominant floral scent compounds and distribution pattern of four *Glochidion* species. Phylogenetic tree based on ITS and EST shows two pairs of closely-related species: *G. obovatum*/*G. rubrum* and *G. lanceolatum*/*G. zeylanicum*. The former species pair occurs in allopatry and share R-(-)-linalool as the dominant compound. The letter species pair occurs in sympatry, and each has different dominant compound

To test the hypothesis that a pollinator *Epicephala* moth can distinguish between the floral scents of its own host and nonhost species, a bioassay was used to investigate the response of pollinator moths to floral scents. Pollinator moths reared from the seeds of *G. lanceolatum* and *G. rubrum* were used in a preference test using a Y-shaped olfactometer. In this bioassay, an unmated male/female moth introduced into a Y-shaped tube was allowed to select its way in response to airborne stimuli of floral volatiles of the two *Glochidion* species. The preference test revealed that *Epicephala* moths could detect their host plants by floral scents alone (Okamoto et al. 2007). Interestingly, the fact that both males and females responded to the floral scent suggests that the moths may use the floral scents as cues to visit flowers not only for pollination but also for mating.

8.3 Not a Private Channel, But a Blend

A "private channel" involves exclusive, system-specific chemicals that mediate species-specific interactions between plants and insects and between fungi and spore dispersers (Steinebrunner et al. 2008; Chen et al. 2009; Franke et al. 2009; Soler et al. 2010). Well-known examples of private channels are found in sexually deceptive orchids: the floral scents of *Ophrys* and *Chiloglottis* species mimic the sex pheromones produced by female pollinator insects (Ayasse et al. 2003; Schiestl and Peakall 2005). Species-specific olfactory signals also mediate the association between *Ficus* plants and fig wasps, and the chemical that attracts single pollinator species has recently been identified in *Ficus semicordata* (Chen et al. 2009). The receptive male and female syconia of *F. semicordata* emit scents consisting primarily of 4-methylanisole, which is attractive to its sole pollinator wasp, *Ceratosolen gravelyi*. Although 4-methylanisole is not rare as a floral scent compound, and has been detected in flowers of other angiosperms, it has not been found in flowers of other *Ficus* species (Proffit et al. 2008, 2009).

To detect the presence or absence of private channels in the obligate pollination mutualism in Phyllanthaceae, chemical analyses of floral scents and electrophysiological analyses of the pollinator's olfactory receptors in the *Breynia–Epicephala* obligate pollination mutualism were conducted. In contrast to *Glochidion*, *Breynia* is characterized by shrubby habitat, saccate male flowers, discal or clavate female flowers with connate sepals and bifid stigmas, red/black fleshy fruit, and nocturnal leaf-folding behavior. In *Breynia vitis-idaea* from Ishigaki Island, a mixture of two floral volatiles, 2-phenylethyl alcohol and 2-phenylacetonitril, was as attractive as real flowers to the females of specific *Epicephala* moth species, although each compound alone was less attractive (Svensson et al. 2010). The two volatiles are common floral compounds frequently found in various angiosperms (Knudsen et al. 2006), including male flowers of *Glochidion lanceolatum* (Okamoto et al. 2013).

Gas chromatography with electroantennographic detection (GC-EAD), coupled GC-MS, and olfactometer bioassays were used to determine which compounds elicit responses from the antennal olfactory receptors of the pollinator *Epicephala* moths. The results revealed that five compounds, including the two above-mentioned volatiles, elicited responses from the antennal olfactory receptors of both males and females of the pollinator *Epicephala* moths (Fig. 8.3). These results

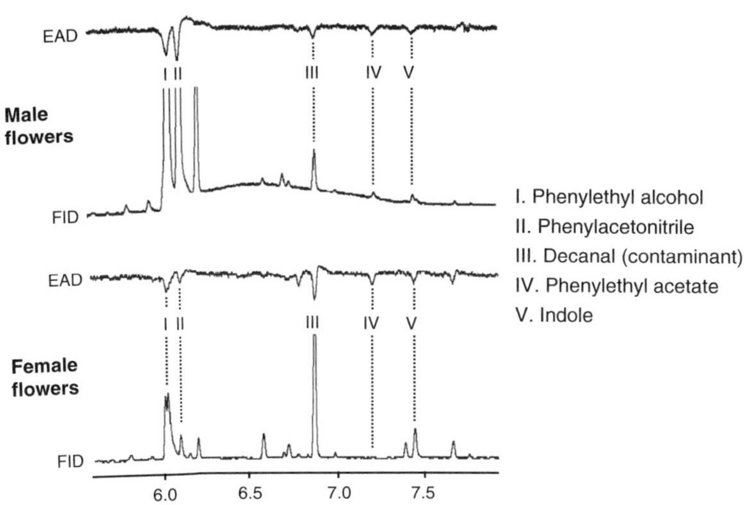

Fig. 8.3 Simultaneous responses of the flame ionization detector (*FID*) and electroantennographic detection (*EAD*) using the antennae of female *Epicephala* moths to male and female floral scents of *Breynia vitis-idaea*

indicate that *B. vitis-idaea* uses a blend of conventional floral scent compounds as an attractant for its exclusive *Epicephala* pollinator, and that system-specific chemistry is not a necessity for efficient host location by the moth partners.

8.4 Evolution of Sexual Dimorphism of Floral Scent

Like some members of *Ficus*, but unlike *Yucca*, most members of Phyllanthaceae are monoecious; that is, male and female flowers grow on a single plant. In *Epicephala* moth-pollinated species, male and female flowers are visited by the same female moth, although with different behavior on each: pollen collection from male flowers and active pollination and oviposition on female flowers. This begs the question, do the floral scents differ between male and female flowers?

Most angiosperms are hermaphrodites bearing bisexual flowers, although about 30% of angiosperm species are monoecious or dioecious, producing unisexual flowers (Lloyd and Webb 1977). Among these monoecious and dioecious plants, wind- and water-pollinated unisexual flowers exhibit extensive sexual dimorphism in floral traits, which are specialized for pollen release in males and pollen reception in females. However, because animal-pollinated plants must attract the same pollinator species to both male and female flowers to secure conspecific pollen transfer, the flowers of both sexes tend to resemble each other in morphological and olfactory traits. Consequently, many animal-pollinated unisexual flowers exhibit low sexual divergence in floral signals, even if they have different floral rewards (Willson and Ågren 1989; Fenster et al. 2004). However, this is not the case in the obligately moth-pollinated plants in Phyllanthaceae because the pollinator moths execute different missions on flowers of different sexes.

Female *Epicephala* moths deliver pollen grains from male to female flowers using their proboscis, and then lay eggs in ovaries because the larvae consume part of the developing seed until pupation. Therefore, active pollination by *Epicephala* moths is a specialized trait to secure food for larvae. Because active pollination behavior involves the collection of pollen from male flowers and then depositing it on female flowers, *Epicephala* moths must be able to distinguish male and female flowers and to visit male flowers before visiting female flowers. A Y-tube preference test indicated that mated female *Epicephala* moths without prior experience of pollen collection preferred the scent of male flowers over that of female flowers (Okamoto et al. 2013). Virgin female moths first mate before visiting flowers; the mating enhances their motivation to visit male flowers to collect pollen. Having collected pollen, the moths are attracted by female flowers, which they pollinate and oviposit.

In *Epicephala* moth-pollinated plants of the genera *Glochidion*, *Breynia*, and *Phyllanthus* (section *Anisonema*), the floral scents of the male and female flowers of each plant species are composed of similar compounds, although the scent profiles clearly differ between male and female flowers. In contrast, in the Phyllanthaceae

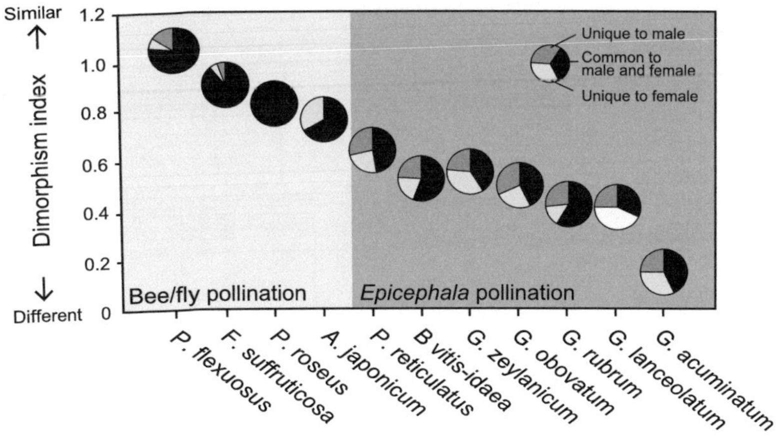

Fig. 8.4 Pattern of sexual dimorphism in floral scent of *Epicephala* moth-pollinated and bee/fly-pollinated species. Dimorphism index approaches 1 as the male and female floral scents become more similar. Pie graphs indicate the proportion of compounds unique to either male or female floral scents. Black color indicate compounds shared between male and female flowers, dark gray indicates compounds unique to male flowers, and light gray indicates compounds unique to female flowers

generally pollinated by diurnal bees and flies, for example, *Antidesma*, *Flueggea*, and *Phyllanthus* (subgenera *Kirganelia* and *Phyllanthodendron*), the floral scent profiles are very similar between male and female flowers (Fig. 8.4). Sexual dimorphism of the floral scent profiles was investigated using a dimorphism index (*D*) based on the Bray–Curtis dissimilarity index (Bray and Curtis 1957), which is a positive value that approaches 0 as the floral scents become more sexually dimorphic, and approaches 1 as male and female floral scents become more similar. The *D* values of 11 Phyllanthaceae species (Fig. 8.4) suggest that the obligately moth-pollinated plants have low *D* indices, indicating sexual dimorphism in floral scents. Marked sexual dimorphism was observed in floral scents between male and female flowers in multiple Phyllanthaceae lineages that have independently evolved via *Epicephala* pollination, strongly indicating that the sexual dimorphism of floral scent is associated with *Epicephala* pollination (Fig. 8.5).

Interestingly, the difference between male and female floral scents involves major qualitative differences in volatile blends. For example, the dominant compounds are derived from different biosynthetic pathways: in *Glochidion zeylanicum* and *G. lanceolatum* the dominant compounds of the female floral scent are terpenoids synthesized by the 2-*C*-methyl-D-erythritol 4-phosphate/1-deoxy-D-xylulose 5-phosphate pathway (MEP/DOXP pathway), whereas those of the male floral scent are simple aromatic compounds synthesized by the shikimic acid pathway (Fig. 8.6). In another case, male and female flowers emit different enantiomers of the same compound as dominants from each sex: in *G. obovatum* and *G. rubrum*,

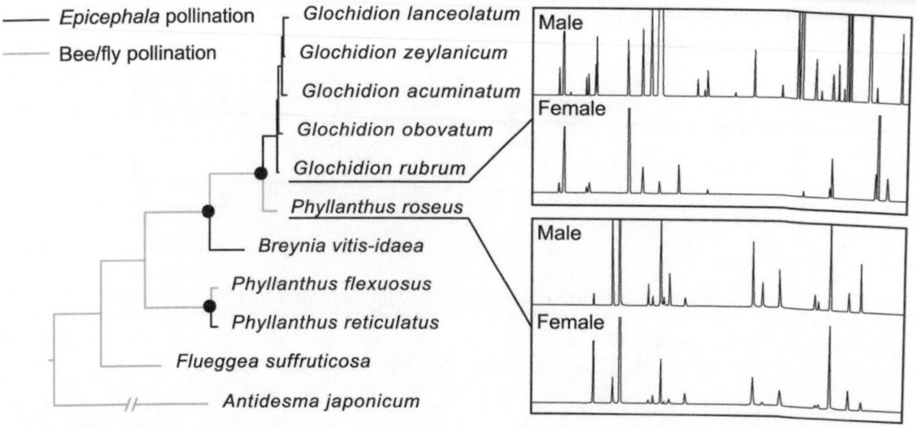

Fig. 8.5 Evolution of floral scent sexual dimorphism in Phyllantheae. Phylogenetic relationships of the 11 Phyllanthaceae species are based on the maximum-likelihood tree of Kawakita and Kato (2009). Black circles on the tree indicate evolutionary transitions from bee/fly pollination to *Epicephala* moth-pollination

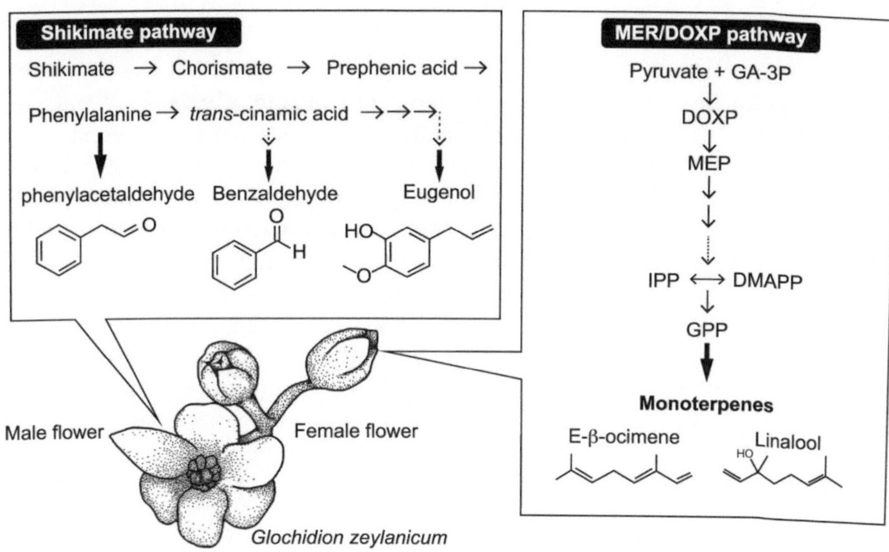

Fig. 8.6 Compounds dominating in the male and female floral scent of *G. zeylanicum* and their synthetic pathways. Simple aromatic compounds are synthesized via the shikimate pathway in the chloroplast. Monoterpenes are synthesized by the *MEP/DOXP* pathway in the chloroplast

(R)-(−)-linalool in male flowers and (S)-(+)-linalool in female flowers. Overall, the average proportion of compounds unique to one sex is 36.5% in *Epicephala*-moth pollinated species, whereas it is only 8.9% in bee/fly-pollinated species (Fig. 8.4).

The sexual dimorphism of floral scents is uncommon in animal-pollinated plants, especially in plants whose female flowers attract pollinator animals without offering rewards. Soler et al. (2012) demonstrated evidence for intersexual mimics of floral scents in dioecious *Ficus* species. Almost one half of *Ficus* species are gynodioecious, producing "functionally male" (hermaphrodite) trees and female trees (Cook and Rasplus 2003). Pollinating fig wasps use only the "female" flowers of male syconia for their brood site, whereas the wasps entering female syconia cannot produce offspring because the long-styled female flowers prevent wasp oviposition. Therefore, it is thought that the pollinator wasps must select male syconia to leave offspring, and are selected to distinguish between male and female syconia. However, selection for male and female syconia to resemble each other chemically is particularly strong for producing seeds of dioecious figs. A chemical analysis of volatiles emitted from syconia revealed that female syconia chemically mimic functionally male syconia (Soler et al. 2012). In contrast with the intersexual chemical mimicry in fig–fig wasp mutualism, the overall similarity and compositional dimorphism in floral scents between male and female flowers of *Epicephala* moth-pollinated plants in Phyllanthaceae must facilitate the delicate nocturnal pollination.

8.5 Floral Scent Profiles of *Epicephala* Moth-Pollinated and Bee/Fly-Pollinated Plants

The floral scents of 11 Phyllanthaceae plants are composed of 85 volatile compounds in total. The floral profiles of *Epicephala* moth-pollinated plants differ from those of bee/fly-pollinated plants (Fig. 8.4), but the difference is difficult to characterize. Although only one compound, 6-methyl-5-hepten-2-one, is shared by all *Epicephala*-pollinated species, it may not contribute to the attraction of *Epicephala* moths or repellence of other noneffective insect visitors because it occurs in the flowers of more than 50% of the angiosperm families investigated to date (Knudsen et al. 2006). Although 11 Phyllanthaceae species emitted linalool from flowers, (*R*)-(−)-linalool can only be detected in four *Epicephala* moth-pollinated species, *G. acuminatum*, *G. obovatum*, *G. rubrum*, and *G. lanceolatum*, and the remaining species include (*S*)-(+)-linalool in the floral scents, suggesting that the ability to synthesize (*R*)-(−)-linalool evolved after the early diversification of *Glochidion*. Therefore, there is probably no single compound that clearly distinguishes the scents of *Epicephala* moth-pollinated and bee/fly-pollinated species.

The floral scents of seven *Epicephala* moth-pollinated species are dominated by monoterpenes, (*R*)-(−)-linalool, (*S*)-(+)-linalool, and (*E*)-β-ocimene, combined with some simple aromatic compounds (e.g., benzaldehyde, phenylacetaldehyde, and eugenol) and nitrogen-bearing compounds (e.g., indole, methyl anthranilate, geranyl nitrile, and phenylacetonitrile; Okamoto et al. 2007, 2013). In general, moth-pollinated flowers produce large amounts of acyclic terpene alcohols (e.g., linalool and nerolidol), simple aromatic compounds (e.g., methyl benzoate and methyl salicylate), and nitrogen-bearing compounds (e.g., indole and methyl anthranilate; Dobson 2006). Hawk moth- and settling moth-pollinated plants differ in the volatile profile of floral scents, characterized by the dominance of methyl benzoate and scarcity of lilac compounds in the former, and by the dominance of phenylacetaldehyde or benzaldehyde and the scarcity of sesquiterpenoids and nitrogen-bearing compounds in the latter (Patt et al. 1988; Tollsten and Bergström 1993; Knudsen and Tollsten 2004; Dötterl et al. 2006). Most *Epicephala* moth-pollinated flowers have scent profiles similar to those of settling moth-pollinated flowers, although lacking lilac compounds. Furthermore, the scent profile of male flowers of *Phyllanthus reticulatus* is dominated by not only monoterpenes, (*S*)-(+)-linalool, and carene, but also a unique sulfur compound, methional (4-thiapentanal), which has not been found in the floral scents of angiosperms. It is still not known whether these unique compounds function to elicit special moth behavior on male flowers.

The floral scents of bee/fly-pollinated Phyllanthaceae plants are dominated by monoterpenoids, for example, limonene, (*E*)-β-ocimene, and (*E*)-linalool oxide furanoid, combined with a few sesquiterpenoids, for example, geranyl acetone, β-caryophyllene, humulene, elemene, and *trans*-α-farnesene. Unlike *Epicephala* moth-pollinated flowers, simple aromatic compounds tend to be infrequent in bee/fly-pollinated flowers. Such characteristics of the floral scents fall under a generalist pollination syndrome, where the plants are pollinated by a variety of nectar-seeking insects (Dobson 2006).

As discussed, the floral scents of *Epicephala* moth-pollinated and bee/fly-pollinated species cannot easily be distinguished by the presence/absence of individual components. However, the flowers of bee/fly-pollinated taxa, such as *Antidesma*, *Flueggea*, and *Phyllanthus* (subgenera *Kirganelia* and *Phyllanthodendron*), produced fewer volatile compounds (6–18) than *Epicephala* moth-pollinated flowers (17–35; Fig. 8.4; Okamoto et al. 2013), so the greater diversity of constituent chemicals may be a key to the specific attraction of *Epicephala* moths to flowers. Because diurnal pollinators such as bees and flies use mainly visual cues and olfactory cues subsidiarily to detect flowers, these bee/fly-pollinated plants may not need to produce diverse volatile compounds. The floral scents of *Epicephala* moth-pollinated plants are also less variable within species than those of bee/fly-pollinated plants (Fig. 8.7). Among the bee/fly-pollinated plant species, *Phyllanthus roseus* and *Antidesma japonicum* have unique floral scents clearly distinguished from those of other Phyllanthaceae species. Another two species, *Phyllanthus flexuosus* and *Flueggea suffruticosa*, have similar scent profiles that overlap on NMDS plots, and the intraspecific variation in the floral scent profiles is greater than those of other Phyllanthaceae plants (Fig. 8.7), suggesting that natural selection for floral scent specialization has weakened in these generalist-pollinated plants. These chemical analyses of floral scents suggest that the high host-specificity of the obligate pollination mutualism in Phyllanthaceae is underpinned by unique species-specific blends of diverse floral volatiles and the moth's high antennal sensitivity to olfactory cues. In the

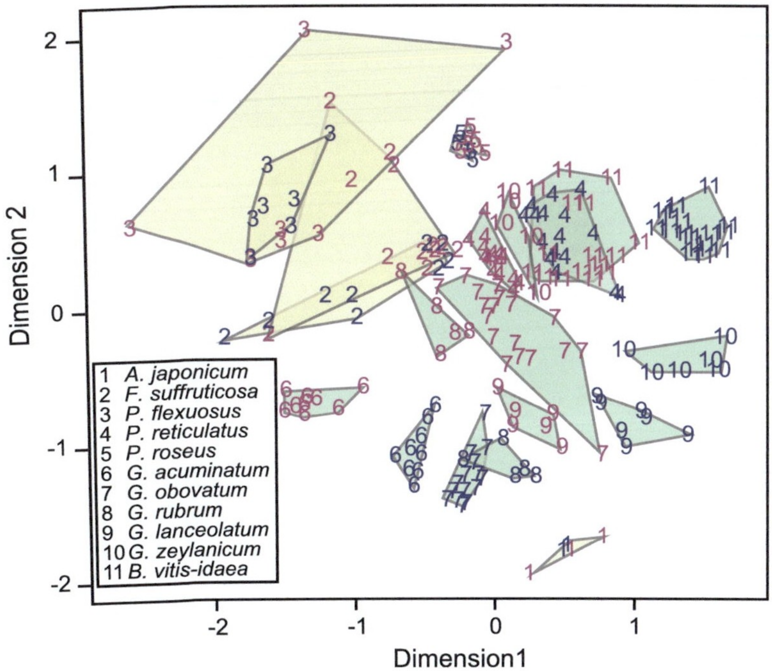

Fig. 8.7 Scatterplot resulting from the NMDS analysis based on the Bray–Curtis index of floral scents. Numbers indicate species. Male and female scent samples are indicated by blue and red coloration, respectively. Pollination system is indicated by green and yellow coloration: green for *Epicephala* moth-pollinated species and yellow for bee/fly-pollinated species

evolutionary process from generalist-pollination to specialized moth-pollination, the plants have developed species-specific blends of floral volatiles and sexual dimorphism of the floral scent to facilitate pollinator moths' missions on male and female flowers. Although the flowers of these plants are small and inconspicuous, the diversity and uniqueness of their olfactory signals are remarkable. The finely tuned flower–moth chemical communication system is an important contributor to the conspicuous diversification of moth-pollinated lineages in Phyllanthaceae.

Chapter 9
Role of Selective Flower Abortion in the Maintenance of Obligate Pollination Mutualism

Ryutaro Goto

Keywords Cost of mutualism • *Epicephala* • *Glochidion* • Mutualism stability • Oviposition • Resource limitation

9.1 Evolutionary Stability of Obligate Pollination Mutualism

Mutualism is an interaction between species, each of which gains benefits that have costs for the other. Such interactions are ubiquitous in nature and often play important roles in the maintenance of ecosystems and biodiversity. However, theoretical considerations suggest that mutualism is not evolutionarily stable because it is vulnerable to invasion by cheaters, which gain benefits without paying costs (Axelrod and Hamilton 1981; Bull and Rice 1991; Sachs et al. 2004). Thus, mechanisms preventing cheating may be essential if mutualism is to be stable (Kiers et al. 2003; Bshary and Grutter 2006; Kiers and Denison 2008; Jandér and Herre 2010; Jandér et al. 2012).

In this chapter, we focus on the evolutionary stability of obligate pollination mutualisms, in which pollinators are host-specific seed parasites. Although reproduction of the plants and their pollinators is highly interdependent, fatal conflicts of interest often develop. Active pollination by such insects features pollen collection, transfer, and deposition. As pollinators in such mutualisms are typically very small, active pollination imposes heavy costs in terms of both energy and time (Pellmyr 1997). Thus, natural selection would encourage pollinators to lay the maximum number of eggs in each flower to reduce pollination costs. However, any increase in the egg load per flower increases seed loss, because seeds are consumed by larvae. If the costs of seed predation exceed the benefits of pollination, the mutualism may simply become extinct, or plants may abandon their associations with insect

R. Goto (✉)
Department of Ecology and Evolutionary Biology, University of Michigan, 1109 Geddes Avenue, Ann Arbor, MI 48109, USA
e-mail: gotoryutaro@gmail.com

M. Kato, A. Kawakita (eds.), *Obligate Pollination Mutualism*, Ecological Research Monographs, DOI 10.1007/978-4-431-56532-1_9

partners, choosing other partners. In simple terms, the mutualism breaks down. Thus, mechanisms that prevent multiple oviposition or excessive seed predation may be necessary to maintain obligate pollination mutualism.

9.2 Selective Flower Abortion in *Yucca*–Yucca Moth Associations

Yucca is a genus of about 50 species of perennial shrubs and trees in the family Asparagaceae (formerly Agavaceae). The genus is confined to North America, particularly dry regions in subtropical to subtemperate zones. Almost all members of the genus have mutualistic pollination associations with seed-parasitic yucca moths.

Yucca filamentosa, a native of the eastern North American seaboard, is pollinated exclusively by the yucca moth *Tegeticula yuccasella*. Pellmyr and Huth (1994) compared the numbers of oviposition scars and eggs between retained and aborted flowers in a population of *Y. filamentosa* introduced into Ohio and found that the plant selectively aborted flowers with heavy egg loads. Fruit production by *Y. filamentosa* is very resource-limited; in a previous study, even when flowers were supplemented with pollen, fruit set was achieved by only ~13% of pollinated flowers (Pellmyr and Huth 1994). Thus, it is understandable that plants selectively abort low-quality flowers to manage their limited resources effectively. Under such circumstances, yucca moths must lay fewer eggs per flower to avoid progeny death upon flower abortion. Such selective flower abortion is considered to be effective for stabilizing obligate pollination mutualism.

However, subsequent studies have suggested that selective abortion of flowers with higher egg loads is not general among all *Yucca*–yucca moth associations (Addicott and Bao 1999). Some yucca moths have relatively long ovipositors, and some have relatively short ones (Davis 1967; Addicott 1996; Addicott and Bao 1999). Yuccas selectively abort flowers only when moth oviposition damages the developing ovules (Addicott and Bao 1999). Such damage is caused only by yucca moths with long ovipositors. Thus, yuccas do not selectively abort flowers with higher numbers of moth eggs if the moths have short ovipositors that superficially oviposit eggs into ovules (Addicott and Bao 1999). In addition, yucca moths often oviposit into previously parasitized flowers (Addicott and Bao 1999), indicating that selective flower abortion does not effectively encourage cooperation by the partner moths. Furthermore, factors (both biotic and abiotic) other than selective abortion also effectively prevent excessive seed predation (Segraves 2003; Bao and Addicott 1998). For example, third parties associated with mutualisms significantly reduce the costs incurred by plants; parasitoids kill yucca moth larvae within fruits (Crabb and Pellmyr 2006) and florivore beetles feed on floral styles bearing the eggs of yucca moths that oviposit superficially (Segraves 2008). Thus, any role for selective flower abortion in the maintenance of obligate pollination in a *Yucca*–yucca moth mutualism may be less important than previously thought.

9.3 Selective Flower Abortion in *Glochidion–Epicephala* Associations

Does selective flower abortion play an important role in the maintenance of other obligate pollination mutualisms? We herein focus on the obligate pollination mutualism between *Glochidion* plants and their seed-parasitic gracillariid moths. Each species of *Glochidion* is pollinated by only a host-specific moth (*Epicephala*), the larvae of which infest the seeds; each *Epicephala* larva consumes 30–50% of seeds in a fruit as the larva matures (Chap. 3). Thus, if two or more larvae are present in a fruit, almost all seeds will be eaten, reducing the benefit afforded to the plant by mutualism. If *Glochidion* trees selectively abort flowers with high egg loads, seed loss would be limited, and selection would be imposed on moths to lay fewer eggs per flower, as in *Yucca*.

The selective flower abortion hypothesis has been tested in *Glochidion acuminatum* growing on Nagakumo-toge, Amami-Oshima Island, southwestern Japan (Goto et al. 2010). The species is a monoecious tree 3–8 m in height occurring on the margins of eastern Asian subtropical forests (Govaerts et al. 2000). *G. acuminatum* produces tens to hundreds of thousands of male and female flowers once annually, in April–May. Each flower lives for about 1 week. During the flowering season, females of *Epicephala anthophilia* visit *G. acuminatum* flowers at night, attracted by the floral scent. They collect pollen on male flowers and then move to female flowers, where they deposit the pollen on the stigma and insert eggs into the stigma using their long ovipositors. Pollination/oviposition behavior is displayed on each female flower visited. Most flower abscission (including abortion) occurs by late June, soon after anthesis (Fig. 9.1). The retained flowers undergo ~3 months of dormancy and begin to develop into fruits in late September; the fruits mature by late November (Figs. 9.1 and 9.2). Each fruit usually contains six to eight seeds. *Epicephala* eggs hatch when the flower begins to develop in late September. Larvae in fruits consume the seeds within the fruits (each larva requires two or three seeds for complete maturation). After the larvae are fully grown, they escape from the fruits and become prepupae in litter under the tree. The prepupal state is maintained to the following spring.

Plants that selectively abort pollinated flowers seek to counter limited fruit production (Pellmyr and Huth 1994). To explore resource limitations in *G. acuminatum*, the fruit sets of flowers of seven hand-pollinated and naturally pollinated trees were compared. If fruit production is pollen-limited, hand-pollinated flowers should exhibit a higher fruit set than naturally pollinated flowers. If fruit production is resource-limited, hand-pollinated flowers would not exhibit as high a fruit set as naturally pollinated flowers. Outcross pollen was used to hand-

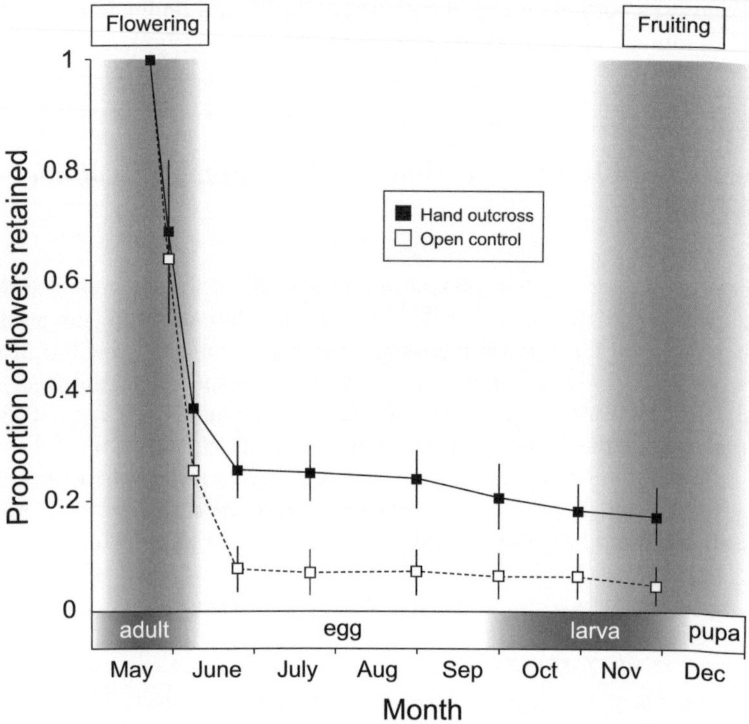

Fig. 9.1 Survivorship of *Glochidion acuminatum* flowers from anthesis to fruit maturity. *Glochidion acuminatum* produces tens to hundreds of thousands of male and female flowers from early May to middle June. Female flowers normally last ~1 week. *Epicephala* moth eggs remain dormant during the summer until pollinated flowers begin to develop in late September. *Filled and open boxes* indicate changes in the proportion of artificially cross-pollinated and control female flowers that are retained on branches respectively ($N = 7$). Corresponding life stages of the *Epicephala* moth are provided in the *bottom bar* (Reproduced from Goto et al. 2010)

pollinate 26–62 female flowers of each of seven individuals. Survivorship was compared to that of 26–67 female control flowers per plant. Survivorship was monitored every 2 weeks to late June and every month thereafter to the time of fruit maturation in November (Fig. 9.1). In late June, ~26% of hand-pollinated flowers and ~7% of naturally pollinated flowers had been retained (Fig. 9.1). This clearly shows that hand-pollinated flowers have a much higher fruit set than naturally pollinated flowers. Thus, it is likely that *G. acuminatum* is pollen-limited. Nevertheless, fruit production by *G. acuminatum* is probably resource-limited, as explained below.

At the flowering peak, 30 female flowers were randomly sampled from each of the seven experimental trees to estimate the proportions of flowers that were naturally pollinated. We found that ~76% of flowers were naturally pollinated, implying that such pollination was adequate. The average proportion of naturally pollinated flowers that developed into mature fruits was ~5.4% (Fig. 9.1); thus, the

Fig. 9.2 The development sequence from flower to fruit in *Glochidion acuminatum*: (**a**) female flowers in May; (**b**) developing fruits in October; (**c**) matured fruits in November; (**d**) dehiscent fruit in December. Main flowering season of *G. acuminatum* in Amami-Oshima Island, Japan, is Spring (around May) and the retained flowers become dormant until the Fall (**a**). Most of the retained flowers start to grow up in late October (**b**) and then develop into mature fruits within one or two months (**c**). After fruit maturation, pericarps peel off and *reddish, glossy* seeds are exposed (**d**)

plants routinely dropped ~90% of all pollinated flowers within a short period after anthesis. All hand-pollinated flowers received ample outcross pollen, whereas naturally pollinated flowers included those that were self-pollinated or that received but small amounts of pollen. Thus, the plants may have selectively aborted flowers based on pollen quantity and/or quality, as do other plant species (e.g., Huth and Pellmyr 2000). Considering that the proportion of naturally pollinated flowers was high (~76%), it seems more likely that *G. acuminatum* is resource-limited, selectively retaining only a small proportion of high-quality pollinated flowers.

Next, we explored whether selective abortion based on moth egg density occurred in *G. acuminatum*. We randomly sampled 50 retained and 50 aborted flowers from each of five plants in the flowering season of 2007. To collect aborted flowers, basket traps (1 mm diameter mesh; entrance, 600 mm × 900 mm; depth, 300 mm) were placed under the trees during the principal period of flower abortion (from May 15 to June 10). To explore whether aborted female flowers had undergone pollination, we used aniline blue to stain pollen attached to the stigma (Dafni et al. 2005). Female flowers retained on branches were sampled in late August when

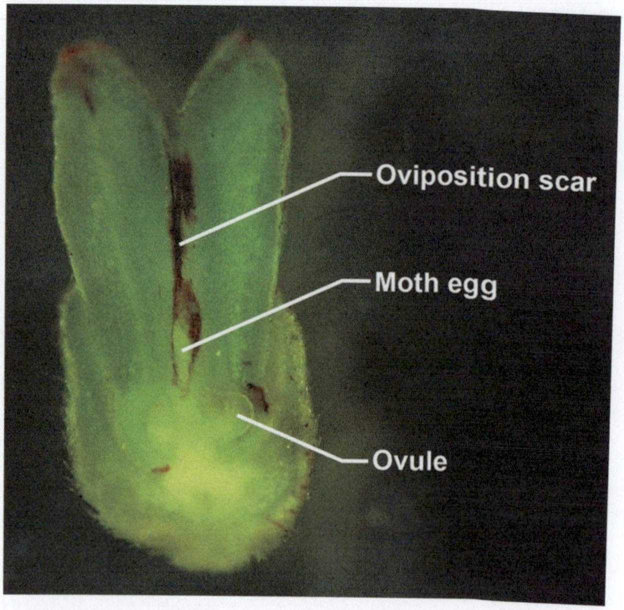

Fig. 9.3 A cross-section of a female flower of *Glochidion acuminatum*

abortion ended and flowers began to develop into fruits. The numbers of both eggs and oviposition scars per flower were counted by microscopically dissecting aborted and retained flowers (Fig. 9.3). We found a near 1:1 correspondence between the numbers of scars and eggs (2 of 574 scars were not associated with eggs). Thus, the egg number was a reliable indicator of oviposition frequency by *Epicephala*. When inserting eggs, female *Epicephala* moths sometimes pierce too deeply and penetrate the ovules. Thus, the number of ovules damaged by such penetration was also counted.

We used generalized linear mixed models (GLMMs) to explore whether the numbers of eggs and damaged ovules per flower significantly differed between aborted and retained flowers. These models allow fitting of both fixed and random factors; the random factors consider repeated measures within plants. The dependent variable was floral status (aborted or retained), the fixed variables were the numbers of eggs and damaged ovules, and a random variable was the individual plant. We found that abortion was not random among flowers (Fig. 9.4). Aborted flowers contained greater than expected proportions of two-, three-, and four-egg flowers (two: 68.0%, $n = 75$; three: 85.7%, $n = 14$; four: 100%, $n = 1$; Fig. 9.4a), whereas retained flowers contained a greater than expected proportion of zero-egg flowers (79.4%, $n = 34$; Fig. 9.4a). Similarly, the numbers of damaged ovules per flower were not randomly distributed; 85.0% ($n = 80$) and 100% ($n = 14$) of aborted flowers had one and two damaged ovules, respectively (Fig. 9.4b), suggesting that plants selectively abort flowers with higher egg loads and more damaged ovules.

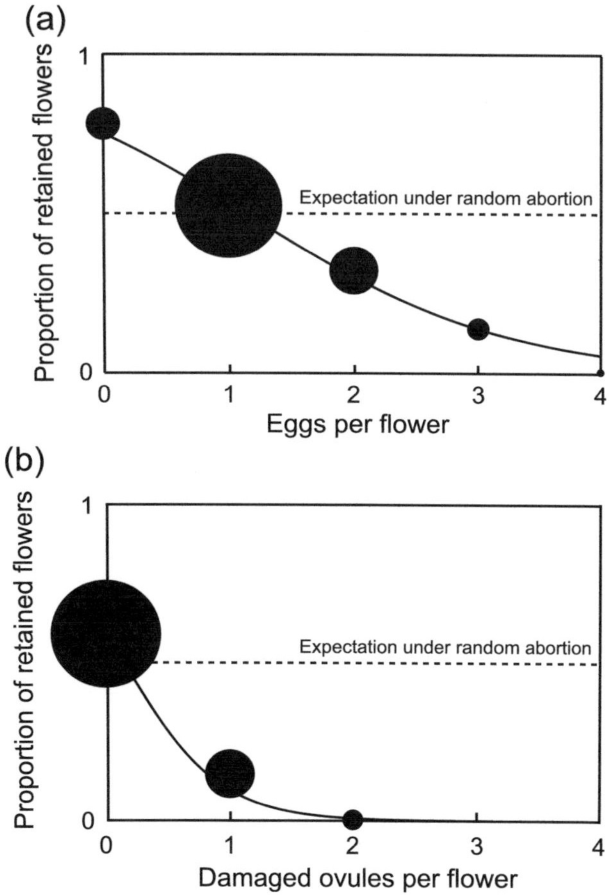

Fig. 9.4 Selective flower abortion based on number of eggs and number of damaged ovules. We sampled 50 aborted and 50 retained flowers from each of five plants and sorted the flowers by (**a**) the number of eggs carried and (**b**) the number of ovules penetrated by moth ovipositor. *Data plot* shows the relative proportion of retained flowers within the sample for each number class. If abortion is random, the regression line should approximate $y = 0.5$ (*broken line*). *Solid lines* indicate logistic regressions of flower fate (0, aborted; 1, retained) on (**a**) eggs per flower ($y = 1/(1 + e)^{-(-1.104+0.976x)}$) and (**b**) damaged ovules per flower ($y = 1/(1 + e)^{-(-0.336+2.083x)}$). Size of the *data plots* is proportional to relative frequency (Reproduced from Goto et al. 2010)

9.4 Effect of Selective Flower Abortion on Overall Seed Production

Selective abortion of flowers with high egg loads should decrease seed predation by pollinator larvae. To determine the fitness benefits afforded to host plants from selective abortion, actual seed production was compared to that expected assuming random flower abortion.

If the proportion of pollinated flowers retained to the fruiting stage (i.e., fruit set) is P, and the proportions of retained and aborted flowers with i eggs are p_i and q_i, respectively, the proportion of retained flowers with i eggs under conditions of random abortion is given by:

$$p_iP + q_i(1 - P).$$

If Q_i is the average proportion of uneaten seeds within fruits with i eggs, the proportion of uneaten seeds within fruits with i eggs under conditions of random abortion is then:

$$Q_i[p_iP + q_i(1 - P)].$$

Consequently, the overall proportion of uneaten seeds per fruit under conditions of random and selective abortion are:

$$\sum_{i=0}^{4} Q_i[p_iP + q_i(1 - P)] \quad \text{and} \quad \sum_{i=0}^{4} Q_ip_i, \quad \text{respectively.}$$

Thus, the proportion of ovules R and S, respectively, that develop into mature seeds under conditions of random and selective abortion, are:

$$R = P\sum_{i=0}^{4} Q_i[p_iP + q_i(1 - P)] \quad \text{and} \quad S = P\sum_{i=0}^{4} Q_ip_i, \quad \text{respectively.}$$

Finally, the overall increase in seed production (IS) under conditions of selective abortion is:

$$100\left(\frac{S - R}{R}\right) = 100\left\{\frac{\sum_{i=0}^{4} Q_ip_i}{\sum_{i=0}^{4} Q_i[p_iP + q_i(1 - P)]}\right\}$$
$$= \frac{8.94\,(1 - P)}{0.5262 + 0.0894P}.$$

Table 9.1 summarizes the relative proportions of flowers carrying varying numbers of eggs in random samples of aborted and retained flowers (corresponding to p_i and q_i); these numbers were obtained by dissecting 250 aborted and 250 retained flowers collected from five plants in 2008. In addition, Table 9.2 shows the proportions of uneaten seeds per fruit that had different numbers of egg capsules, evaluated using 100 fruits collected from seven plants in 2008 (corresponding to Q_i). The averaged fruit set was 5.4% (P) (Fig. 9.1). When these values are placed in the IS formula above, 61.6% (S) intact seeds remained per fruit under conditions of selective flower abortion, whereas 53.1% (R) seeds per fruit were uneaten under conditions of random abortion. Flowers with three or more eggs are extremely rare in the fruiting crop; data are thus not available for these egg numbers. Instead, the proportions of uneaten seeds in two-egg fruits were used as surrogates for the proportions of uneaten seeds in flowers with three or more eggs (Table 9.2). The increase in seed production (IS) was 15.9% when $P = 5.4$. The fruit set of pollinated female flowers varies greatly among plants (0–37.0%). However,

Table 9.1 Relative proportions of flowers carrying 0–4 eggs within random samples of retained and aborted flowers

Number of Eggs	0	1	2	3	4
Prop. among the retained flowers	0.108	0.788	0.096	0.008	0
Prop. among the aborted flowers	0.028	0.716	0.204	0.048	0.004

Data are based on 50 retained and 50 aborted flowers collected from each of five plants in May 2008

Table 9.2 Average proportions of uneaten seeds within fruits infected with 0–2 eggs

Number of Eggs	0	1	2
Number of fruits examined	6	77	17
Average prop. of uneaten seeds	1	0.614 ± 0.031	0.229 ± 0.069

Data are based on 100 fruits sampled across five plants in November 2008

the estimated fitness advantage remains largely constant across a range of fruit set (11.2–17.0%) (Fig. 9.5a).

9.5 Effects of Selective Flower Abortion on Moth Fitness

G. acuminatum aborts >90% of pollinated flowers under natural conditions (Fig. 9.1). As most pollinated flowers harbor moth eggs, flower abortion causes the death of >90% of all moth eggs. Similarly, yuccas and the senita cactus, both of which are pollinated exclusively by seed parasites, exhibit very low fruit sets of pollinated flowers (Pellmyr and Huth 1994; Addicott and Bao 1999; Holland and Fleming 1999). Theoretical considerations reveal that such a low fruit set prevents moth overpopulation and tends to stabilize the mutualism (Holland 2000). The fact that fruit set is low in *G. acuminatum* provides additional evidence supporting this hypothesis.

Under conditions of random abortion, the overall survival of moth eggs E_R equals the fruit set P. Under conditions of selective abortion, the moth egg survival E_S is:

$$Es = \frac{P\sum_{i=0}^{4} ip_i}{P\sum_{i=0}^{4} ip_i + (1-P)\sum_{i=0}^{4} iq_i}.$$

Thus, the overall decrease in moth egg survival caused by selective abortion (DS) is:

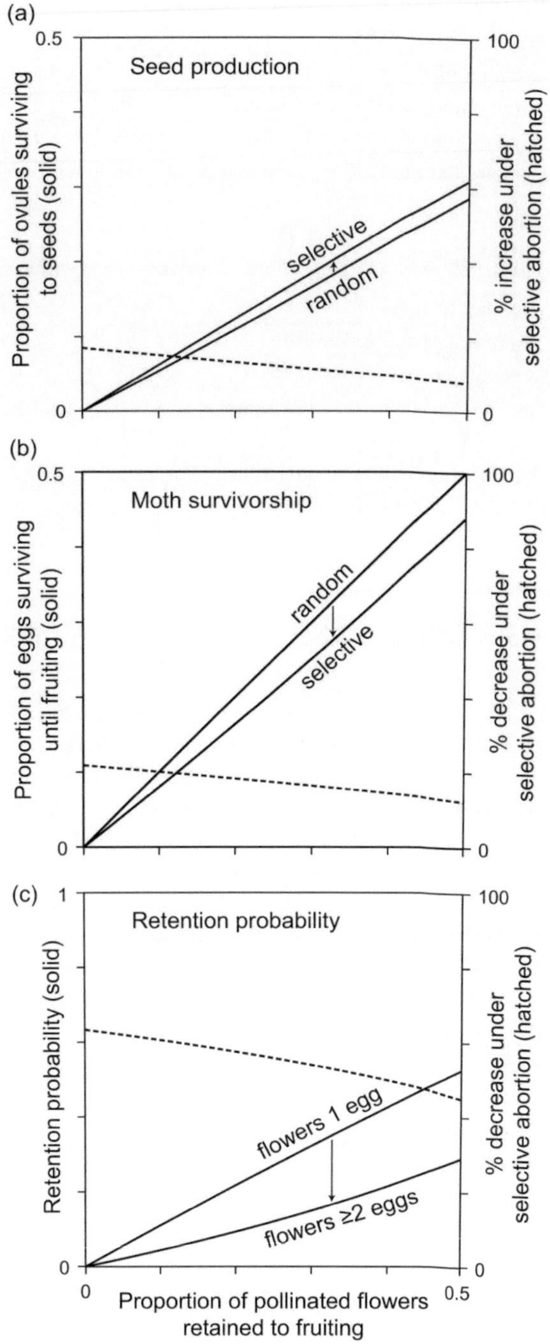

Fig. 9.5 Effects of selective flower abortion on plant and moth fitness. (**a**) *Solid lines* indicate estimated proportions of total ovules formed that survive to seeds per plant under random and selective abortion. *Hatched line* represents percentage increase in seed production under selective abortion relative to random abortion. (**b**) *Solid lines* are estimated proportions of total moth eggs laid that survive until fruiting season under random and selective abortion. *Hatched line* represents percentage decrease in population-level moth survivorship under selective abortion. (**c**) *Solid lines* indicate estimated retention probabilities of flowers with one egg and those with more than one egg under selective abortion. Percentage decrease in the probability of retention is depicted by the *hatched line* (Reproduced from Goto et al. 2010)

$$DS = 100\left(\frac{E_R - E_S}{E_R}\right) = 100\left\{1 - \frac{P\sum_{i=0}^{4} ip_i}{P\left[P\sum_{i=0}^{4} ip_i + (1-P)\sum_{i=0}^{4} iq_i\right]}\right\}$$
$$= \frac{28(1-P)}{1.284 - 0.28P}.$$

As the average fruit set equals E_R, an average of 5.4% of moth eggs laid in flowers would survive to the fruiting stage if abortion were random. However, the formula above shows that only 4.3% (E_s) of moth eggs would survive under conditions of selective abortion. This suggests that selective abortion decreases moth survivorship by 20.9% (*DS*). Natural fruit set varies from 0% to 37% among plants. Thus, the estimated decrease in moth egg survivorship caused by selective abortion would be between 16.2% and 21.8% (Fig. 9.5b).

The probability that pollinated flowers with *i* eggs are retained to the fruiting stage under conditions of selective abortion (P_i) is:

$$P_i = \frac{p_i P}{p_i P + q_i(1-P)}.$$

The percentage difference in the probability of moth eggs surviving flower abortion between one-egg flowers and those carrying more than one egg is then obtained as

$$100\left(\frac{P_1 - P_{\geq 2}}{P_1}\right) = 100\left\{1 - \frac{p_{\geq 2}P[p_1 P + q_1(1-P)]}{p_1 P\left[p_{\geq 2}P + q_{\geq 2}(1-P)\right]}\right\}$$
$$= 100 - \frac{1300P(179 + 18P)}{197P(32 - 19P)}$$
$$= \frac{106.3(1-P)}{1.684 - P}.$$

Using the average fruit set score ($P = 5.4$; Fig. 9.1), 5.9% (P_1) of the offspring of *Epicephala* females is retained to the fruiting stage when the moth oviposits in virgin flowers, whereas the figure is 2.3% (P_2) when moths oviposit in flowers that already contain egg(s). Offspring survivorship decreases 61.7% when moths oviposit in previously parasitized flowers. The natural fruit set varies from between 0% and 30.7% among plants. Thus, the estimated proportion of lost offspring attributable to multiple egg-laying ranges from 53.5% and 63.1% (Fig. 9.5c).

9.6 Do *Epicephala* Moths Avoid Multiple Oviposition in the Same Flower?

If *Epicephala* moths oviposit in flowers that already contain eggs, this substantially reduces moth fitness because of selective flower abortion. Under such circumstances, natural selection should favor moths that avoid ovipositing in previously parasitized flowers. To explore this possibility, the frequency distribution of flowers with different numbers of eggs was compared to that expected under a Poisson distribution, using the chi-square test (Goto et al. 2010). The distribution of *Epicephala* moth eggs was evaluated by counting the numbers of eggs in each of 210 flowers collected from seven plants in May 2008. The chi-square test showed that egg distribution among flowers differed significantly from random ($P < 0.01$; Fig. 9.6). We calculated Morishita's index to determine whether egg distribution was uniform or clumped. If the index is >1, egg distribution is more clumped than random (Morishita 1959). If the index is <1, egg distribution is more uniform than random (Morishita 1959). The index value was 0.44 ($F = 2.00, P < 0.01$), indicating that moth egg distribution is much more uniform than random (Fig. 9.6). Thus, moths prefer to lay eggs in virgin flowers and avoid those that already have eggs.

It is likely that moths avoid multiple ovipositing in the same flower to reduce the disadvantage created by flower abortion. However, a probable alternative hypothesis is that moths avoid larval competition in the same fruit, because only two to three larvae can grow in a single fruit (Chap. 3). Although it is difficult to separate the effects of selective flower abortion and larval competition on moth ovipositing behavior, it is at least true that selective flower abortion renders ovipositing into previously parasitized flowers an additional disadvantage.

How can *Epicephala* moths detect previously parasitized flowers? In the *Yucca*–yucca moth systems, moths lay fewer eggs in previously parasitized flowers by detecting a pheromone (Huth and Pellmyr 1999). *Epicephala* may use a similar mechanism to detect previously parasitized flowers. However, it is more probable that moths use changes in floral scents to avoid previously parasitized flowers; earlier studies have suggested that *Epicephala* moths use floral scents to locate the flowers that they visit to pollinate (Okamoto et al. 2007; Svensson et al. 2010).

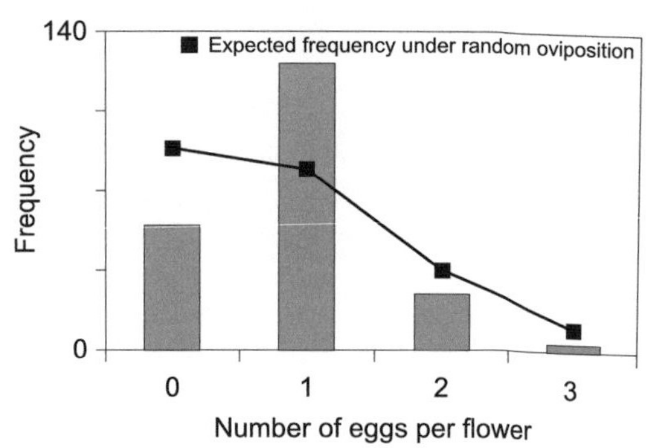

Fig. 9.6 Frequency distribution of the number of eggs per flower. *Line* indicates frequency distribution of the number of eggs per flower estimated under random oviposition, and *hatched bars* show the actual frequency distribution

9.7 Role of Selective Flower Abortion in Obligate Pollination Mutualism

G. acuminatum selectively aborted flowers with higher egg loads or higher numbers of damaged ovules. The next question is whether selective flower abortion is general among flowers of the genus *Glochidion*, which comprises more than 300 species. The number of ovules per fruit, the number of eggs per flower, and the number of seeds consumed by each larva on the path to complete development vary greatly among *Glochidion–Epicephala* systems (Chap. 3). This suggests that the criteria used to abort flowers selectively may differ among flower-abortion systems. In addition, the flowering and fruiting patterns of *G. acuminatum* are relatively unusual compared to those of other members of the family. *G. acuminatum* flowers and fruits only once annually; most *Glochidion* species produce flowers and fruits all year. Thus, resource partitioning by *G. acuminatum* to fruit and flower production probably differs from that of most other *Glochidion* species, which may cause the flower-abortion pattern to differ.

G. acuminatum is pollinated by a single species of *Epicephala* (Kawakita and Kato 2006). If a *Glochidion* were pollinated by multiple *Epicephala* species, how would selective flower abortion work to stabilize the mutualism? The same populations of *G. lanceolatum* are pollinated by two species of *Epicephala* moths on the Ryukyu Islands (Kawakita and Kato 2006). Thus, competition between the pollinator species and plant selection for the more cooperative moth would influence the evolutionary dynamics of the mutualism.

Third parties may also stabilize the cost–benefit balance of the associations between the Phyllanthaceae and *Epicephala*. Braconid wasps that parasitize *Epicephala* moths have been recorded in Phyllanthaceae species associated with obligate pollinator moths. Such braconids should reduce seed predation because they kill within-fruit moths in early larval stages. Thus, in addition to selective abortion, the activities of third parties as seeds mature would influence the evolutionary stability of the mutualism.

Part III
Evolution

Chapter 10
Cospeciation and Host Shift

Atsushi Kawakita and Makoto Kato

Keywords Coevolution • *Epicephala* • *Glochidion* • Phylogeny • Species specificity

10.1 Cospeciation of Intimately Interacting Partners

When two interacting lineages have been in intimate association during much or all of their diversification, as in the case of obligate pollination mutualisms or many host–parasite interactions, there is a probability that speciation in one group is paralleled by speciation in the other. This mode of diversification results in a pattern of shared evolutionary history between the two lineages, known as cospeciation. Cospeciation can be a nonadaptive process that occurs in the absence of selection. For example, repeated vicariance events followed by shared allopatric speciation can produce a pattern of parallel diversification (Roderick 1997). However, cospeciation can also be reinforced or directly result from an adaptive process. For example, in feather lice and their avian hosts, preening behavior of the host imposes selection on louse body size, which prevents lice from switching between hosts of different sizes (Clayton et al. 2003). In obligate pollination mutualisms, the pollinators are responsible for the fertilization among conspecific host flowers, and thus some adaptation in the plants to exclude nonlegitimate pollinators is likely present. In fact, there are several reciprocally selected traits that may reinforce plant–pollinator specialization, such as synchronized phenological patterns (Wiebes 1979; Patel and Hossaert-McKey 2000), species-specific olfactory signals (Hossaert-McKey et al. 1994; Song et al. 2001; Grison-Pigé et al. 2002, 2003; Okamoto et al. 2007; Svensson et al. 2010), and reciprocal adaptation between

A. Kawakita (✉)
Center for Ecological Research, Kyoto University, 2-509-3 Hirano, Otsu, Shiga 520-2113, Japan
e-mail: kawakita@ecology.kyoto-u.ac.jp

M. Kato
Graduate School of Human and Environmental Studies, Kyoto University, Yoshida-Nihonmatsu-cho, Sakyo, Kyoto 606-8501, Japan
e-mail: kato@zoo.zool.kyoto-u.ac.jp

M. Kato, A. Kawakita (eds.), *Obligate Pollination Mutualism*, Ecological Research Monographs, DOI 10.1007/978-4-431-56532-1_10

pollinator morphology and floral structure (Ramírez 1974; Herre 1989; van Noort and Compton 1996; Kato et al. 2003; Weiblen 2004). Thus, knowledge of the degree of cospeciation in obligate pollination mutualisms provides an essential step toward understanding the historical role of coevolution in shaping speciation and diversification in plants and pollinators.

Previous studies using the fig–fig wasp system have indicated a significant level of cospeciation at both lower and higher taxonomic levels (Herre et al. 1996; Weiblen 2000, 2001; Machado et al. 2001; Weiblen and Bush 2002). However, strict congruence of phylogenies has not been found in the fig–fig wasp association. In addition, there are several documented cases in which multiple distantly related fig wasp species associate with a single host, further indicating a lack of strict-sense cospeciation in this association (Wiebes 1979; Compton 1990; Michaloud et al. 1996; Kerdelhue et al. 1999; Lopez-Vaamonde et al. 2002; Molbo et al. 2003). Analysis of parallel cladogenesis using the yucca–yucca moth system also resulted in a similar finding; although host plant use by yucca moths is relatively conserved at higher taxomonic levels, cases of cospeciation are rare (Pellmyr and Leebens-Mack 1999; Pellmyr et al. 2007; Althoff et al. 2012).

Whereas previous studies using the fig–fig wasp and yucca–yucca moth systems provided insights into macroevolutionary patterns in these specialized interactions, the *Glochidion–Epicephala* mutualism offers a novel opportunity to corroborate and refine these earlier observations. The *Glochidion–Epicephala* system is particularly suited for such analysis, as the association is highly species-specific and diverse (Kato et al. 2003). The genus *Glochidion* comprises more than 300 species distributed in tropical Asia, Australia, and Polynesia with multiple species commonly occurring in sympatry. *Epicephala* moths associated with *Glochidion* plants are taxonomically poorly documented, but available evidence suggests that a comparable number of *Epicephala* species exist, each with a very narrow host range. This chapter provides an update to the previous analysis of cospeciation between *Glochidion* trees and *Epicephala* moths (Kawakita et al. 2004) using molecular phylogenetic analysis of 27 *Epicephala* species and 36 host *Glochidion* species (Table 10.1). Phylogenetic analyses are based on published nucleotide sequences of the internal and external transcribed spacer regions of the nuclear ribosomal DNA (ITS and ETS) for *Glochidion*, and those of the mitochondrial cytochrome oxidase subunit I gene (COI) and nuclear arginine kinase (ArgK) and elongation factor-1α (EF-1α) genes for *Epicephala* moths.

10.2 Phylogenetic Analysis and Cospeciation Test

A full list of species included in the analysis and their locality information is given in Table 10.1. The 36 *Glochidion* species sampled were collected from various localities within the distribution of the genus. Although there is little information concerning intrageneric taxonomy of *Glochidion* (Webster 1994; Govaerts et al. 2000), the samples cover a wide range of morphological variation found within this

Table 10.1 List of *Epicephala* and *Glochidion* species sampled for cophylogenetic analysis

Species	Host Plant	Locality	Note
E. anthophilia	*G. acuminatum*	Amami Island, Kagoshima, Japan	
E. bipollenella	*G. zeylanicum*	Okinawa Island, Okinawa, Japan	
E. corruptrix	*G. obovatum*, *G. rubrum*[a]	Ishigaki Island, Okinawa, Japan	
E. eriocarpa	*G. eriocarpum*	Lak Sao, Bolikhamsai, Laos	
E. lanceolatella	*G. lanceolatum*	Ishigaki Island, Okinawa, Japan	
E. obovatella	*G. obovatum*, *G. rubrum*	Cape Toi, Miyazaki, Japan	
E. perplexa	*G. lanceolatum*	Ishigaki Island, Okinawa, Japan	
E. sp. ex *G. arborescens*	*G. arborescens*	Lambir, Sarawak, Malaysia	
E. sp. ex *G. assamicum*	*G. assamicum*	Arunachal Pradesh, India	
E. sp. ex *G. benthamianum*	*G. benthamianum*	Mt. Molly, Queensland, Australia	
E. sp. ex *G. caledonicum*	*G. caledonicum*	Hienghéne, New Caledonia	
E. sp. ex *G.* cf. *zeylanicum*	*G.* cf. *zeylanicum*	Lambir, Sarawak, Malaysia	
E. sp. ex *G. collinum*	*G. collinum*, *G. cordatum*	Mt. Victoria, Fiji	
E. sp. ex *G. glomerulatum*	*G. glomerulatum*	Lambir, Sarawak, Malaysia	
E. sp. ex *G. harveyanum*	*G. harveyanum*	Mt. Windsor, Queensland, Australia	
E. sp. ex *G. littorale*	*G. littorale*	Miri, Sarawak, Malaysia	
E. sp. ex *G. lutescens*	*G. lutescens*	Lambir, Sarawak, Malaysia	
E. sp. ex *G. obscurum*	*G. obscurum*	Lambir, Sarawak, Malaysia	
E. sp. ex *G. perakense*	*G. perakense*	Kutching, Sarawak, Malaysia	
E. sp. ex *G. philippicum*	*G. philippicum*	Nanren, Pingtung, Taiwan	
E. sp. ex *G. pungens*	*G. pungens*	Mt. Lewis, Queensland, Australia	

(continued)

Table 10.1 (continued)

Species	Host Plant	Locality	Note
E. sp. ex *G. seemanii*	*G. seemanii*, *G. concolor*	Navai, Fiji	
E. sp. ex *G. sumatranum*	*G. sumatranum*	Lambir, Sarawak, Malaysia	Incorrectly labeled as *G. ferdinandii* in Kawakita et al. (2004)
E. sp. ex *G. velutinum*	*G. velutinum*	Mt. Popa, Bagan, Myanmar	
E. sp. ex *G*. sp. 1	*G*. sp. 1	Pindai, New Caledonia	
E. sp. ex *G*. sp. 2	*G*. sp. 2	Lambir, Sarawak, Malaysia	Incorrectly labeled as *G. lanceisepalum* in Kawakita et al. (2004)
E. sp. ex *G*. sp. 3	*G*. sp. 3	Phonsavan, Xiangkhoang, Laos	
—	*G. sericeum*	Lambir, Sarawak, Malaysia	
—	*G*. cf. *puberum*	Thakhek, Khammouane, Laos	
—	*G. sphaerogynum*	Ban Phon, Xiangkhoang, Laos	
—	*G*. cf. *rubrum*[a]	Thakhek, Khammouane, Laos	
—	*G*. cf. *hohenackeri*	Lak Sao, Bolikhamsai, Laos	
—	*G. lanceolarium*	Guangzhou, Guangdong, China	
—	*G. daltonii*	Lak Sao, Bolikhamsai, Laos	

[a]*G. rubrum* is used to refer to very different plants in different parts of its range. For consistency within the book, this name is used to refer to the plants called *G. rubrum* in the Japanese flora. The plants referred to as *G. rubrum* in Indochina are here labeled *G*. cf. *rubrum*, although a future taxonomic revision may reveal that the latter is the true *G. rubrum*

large genus. Representatives of *Breynia* and *Sauropus* were included as outgroups, as done previously (Kawakita et al. 2004).

Epicephala moths were obtained from 29 of the 36 above-mentioned *Glochidion* species. Based on clear morphological and molecular differences, the moths were grouped into 27 species, each using usually one but rarely two host *Glochidion* species. Species names could be reliably assigned to seven of the 27 *Epicephala* species. For outgroups, two *Epicephala* moths associated with *Breynia* and *Phyllanthus*, the close relatives of *Glochidion*, were used as in the previous analysis (Kawakita et al. 2004).

There is a large radiation of *Glochidion* and associated *Epicephala* in the French Polynesian islands (Hembry et al. 2013a), but they were not included in the present analysis because species delimitation has not been established firmly for *Epicephala*. The diversification of *Glochidion* and *Epicephala* in French Polynesia is treated in detail in Chap. 12.

Sequences used in the present analysis are all available in public databases. Alignment methods followed those described in Kawakita et al. (2004). The aligned matrices (1365 bp of ITS + ETS for *Glochidion*; 1827 bp of COI + ArgK + EF1a for *Epicephala*) were subjected to maximum likelihood (ML) phylogenetic analyses using the Treefinder software (Jobb 2011) and the substitution models chosen by the program. The robustness of the ML trees was validated by bootstrap analysis (1000 replications) using the same program. Bayesian analysis was also performed using MrBayes 3.1.2 (Ronquist and Huelsenbeck 2003) with substitution models chosen by MrModeltest 2.3 (Nylander 2004). Trees were sampled every 100 generations, and the average standard deviations of split frequencies calculated every 1000 generations. Using the stoprule option, analyses were continued until the average standard deviations of split frequencies fell below 0.01, at which point the Bayesian chains were considered to have achieved convergence. Because the average standard deviations of split frequencies were calculated based on the last 75% of all samples, the initial 25% of sampled trees were discarded as burn-in.

To assess whether species of *Glochidion* and *Epicephala* have undergone parallel diversification, we used an event-based reconciliation analysis, as implemented in the program Jane 4.01 (Conow et al. 2010). Jane uses a dynamic programming algorithm in conjunction with a genetic algorithm to find solutions (often optimal) for reconciling two phylogenies. Costs are assigned to four types of cophylogenetic events (cospeciation, duplication, host switch, and loss; for detailed terminology, see Conow et al. (2010)); optimal solutions are reconstructions with the lowest global cost. Analyses were performed with default genetic algorithm parameters. To test whether the cost of the optimal reconstruction was lower than its null expectation, plant–pollinator associations were permuted for 1000 replicates, and optimal solutions obtained for each replicate to generate a null distribution.

In addition to the Jane analysis, the extent of cospeciation was tested using the Parafit method (Legendre et al. 2002), which, rather than tree topologies, uses matrices of patristic distances (summed branch lengths along a phylogenetic tree) or phylogenetic distances calculated directly from sequence data. Whereas Jane requires fully resolved trees and thus is sensitive to selection of different phylogenetic hypotheses, Parafit is less likely to provide different results among several optimal phylogenies. In this test, distance matrices of the two groups are transformed to principal coordinates, and the trace statistic is calculated by taking plant–pollinator associations into account. The null hypothesis that the two groups are randomly associated is tested through a permutational procedure; plant–pollinator relationships are permuted to obtain a null distribution of the test statistic against which the observed value is tested. This method also allows one to test whether each plant–pollinator association contributes significantly to the global fit of the two phylogenies. This is done by calculating trace statistics with and without a given plant–pollinator link, and testing the difference between the two statistics by permutation. The analyses were done using the Copycat 1.14 program (Meier-Kolthoff et al. 2007).

For both Jane and Parafit analyses, the ML trees of *Glochidion* and *Epicephala* were used as input trees.

10.3 Cospeciation in the *Glochidion–Epicephala* Association

Figure 10.1 shows the phylogenies of *Glochidion* and *Epicephala* obtained by maximum likelihood analyses of the ITS + ETS and COI + ArgK + EF1a datasets, respectively. Although many apical branches are validated by moderate to high support values (maximum likelihood bootstrap and Bayesian posterior probability values), most higher nodes are poorly supported, highlighting the difficulty of resolving the phylogenies of *Glochidion* and associated *Epicephala*, which

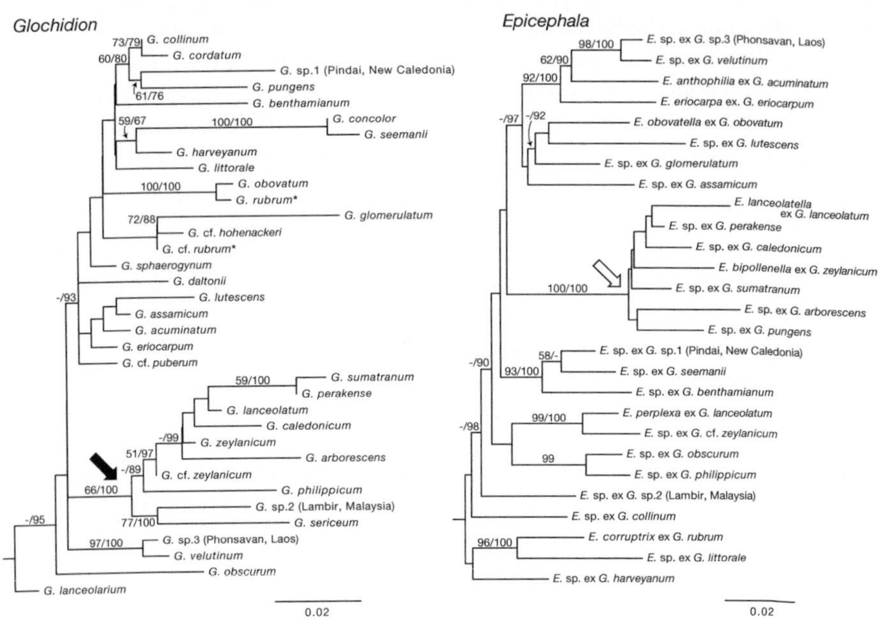

Fig. 10.1 Maximum likelihood trees for *Glochidion* and *Epicephala*. The *Glochidion* phylogeny was estimated using the combined ITS and ETS sequence data; that of *Epicephala* was based on the combined dataset of COI, ArgK, and EF-1α sequences. *Numbers above branches* are bootstrap values followed by Bayesian posterior probability (See Table 10.1 for sampling details)

probably underwent rapid initial radiations (Kawakita and Kato 2009). Thus the results of the phylogenetic reconstructions, including the positions of the root, should be taken with caution. Nevertheless, these phylogenies are useful for testing the importance of cospeciation in shaping the global phylogenetic pattern in the plants and pollinators.

Figure 10.2 shows the comparison between the *Glochidion* and *Epicephala* phylogenies with information on plant–pollinator associations. Although the two phylogenies are far from identical, there is a tendency that associated plants and pollinators occupy similar positions on the phylogeny. For example, in the *Glochidion* phylogeny, there is a relatively well-supported clade of 10 species encompassing *G. sumatranum* and *G. sericeum* (indicated by the black arrow in Fig. 10.1). The *Epicephala* species associated with many of these *Glochidion* species also form a strongly supported clade (indicated by the white arrow in Fig. 10.1), suggesting that the associated plants and pollinators underwent parallel diversification. Note that the latter *Epicephala* clade contains the pollinator of *G. pungens*. Because *G. pungens* is distantly related to species of the *G. sumatranum*–*G. sericeum* clade (Fig. 10.1), this likely represents a clear case of host switch by *Epicephala*.

Statistical analyses of cospeciation corroborate visual assessment. The cost of the optimal reconstruction using Jane, under default cost settings (cospeciation = 0; duplication = 1; host switch = 2; loss = 1), was 41, but none of the optimal solutions of randomized data (1000 replicates) had lower costs (mean ± SD, 90.5 ± 14.4), indicating that the two phylogenies are more similar than expected for a random association ($P < 0.001$). Because Jane requires fully resolved trees and thus is sensitive to phylogenetic uncertainty, the Parafit analysis, which takes tree structure into account, was performed as complementary to the Jane analysis. The Parafit analysis also found that the plant and moth phylogenies are more structurally similar than expected by chance ($P = 0.002$). Of the 31 individual plant–pollinator associations, 15 contributed significantly to the overall cophylogenetic structure (Fig. 10.2).

Studies addressing parallel diversification in plant–herbivore associations usually do not find cospeciation of the interacting lineages. Rather, host shifts are prevalent, and dramatic shifts among distantly related plant taxa are commonly observed. Therefore, the overall similarity in plant and pollinator phylogenies found in this study, together with that in the fig–fig wasp system (Herre et al. 1996; Weiblen 2000, 2001; Machado et al. 2001; Weiblen and Bush 2002), represents a special case in which plants and associated insects diversified more or less in parallel. Unfortunately, analysis of phylogenetic congruence does not identify processes that underlie the pattern of cospeciation. For example, cospeciation may arise through a number of processes including shared allopatric speciation, coevolution, and adaptation by only one group in response to the other. Thus, it is possible that plants and pollinators cospeciate as the result of shared vicariance events and that adaptive evolution may not be important in driving the overall cospeciation pattern.

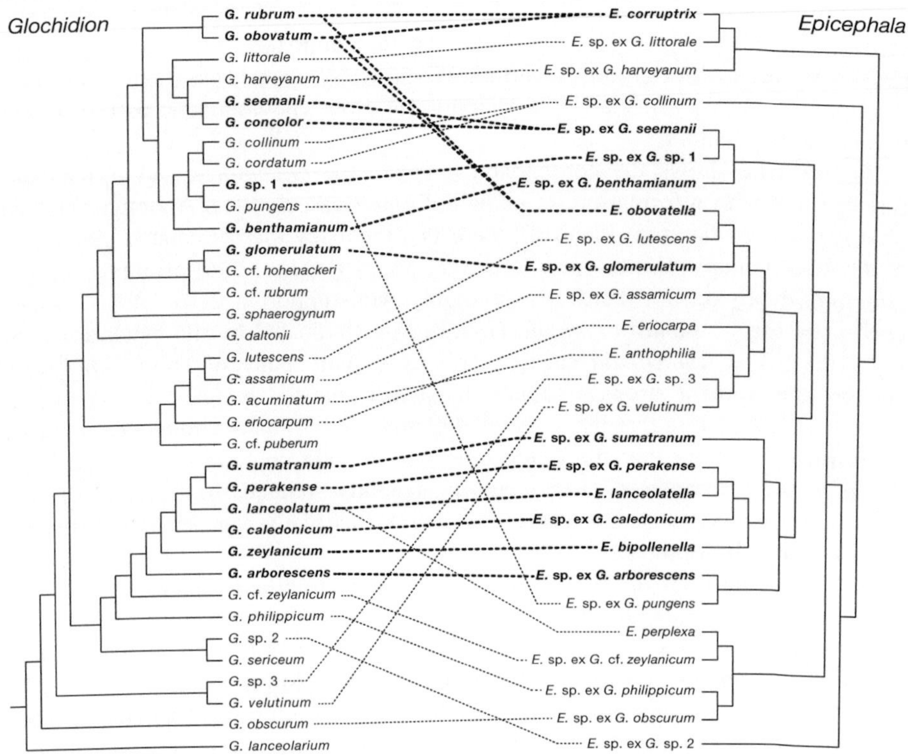

Fig. 10.2 Phylogenetic trees of *Glochidion* (*left*) and *Epicephala* (*right*) with plant–moth associations indicated using *hatched lines*. The *Glochidion* tree is the maximum likelihood topology inferred from the combined ITS + ETS sequences; the *Epicephala* tree is the maximum likelihood tree based on the combined COI, ArgK, and EF-1α sequences. Plant–pollinator links with significant contribution to overall cophylogenetic structure, as inferred by the Parafit analysis, are highlighted in *bold*

However, several observations suggest that reciprocal selection may reinforce cospeciation in obligate pollination mutualisms. For example, Weiblen and Bush (2002) demonstrated that the degree of cospeciation between *Sycomorus* figs and *Ceratosolen* pollinators is greater than that observed between the same set of host figs and nonpollinating, gall-inducing fig wasps of the genus *Apocryptophagus*. They attributed this difference to the extent of reproductive requirements by which pollinating fig wasps are constrained, such as pollen compatibility and/or reciprocal adaptation between fig wasp morphology and narrow ostiolar entrance of the host fig (Ramírez 1974; Herre 1989; van Noort and Compton 1996). Similar constraints may also have been important in preventing host shifts by *Epicephala* moths. *Glochidion* plants produce species-specific olfactory signals that attract their own *Epicephala* species, and *Epicephala* moths have distinct preferences to the floral odor of their natal hosts (Okamoto et al. 2007). Such chemical coadaptation may lock partners into strong intimacy, making host switches difficult. In *Glochidion*, the structure of the style exhibits great interspecific variation and is the principal species-diagnostic characteristic within the genus, which is in marked contrast with morphological uniformity of the male flowers (Airy Shaw 1978; Chakrabarty and Gangopadhyay 1995; Kato et al. 2003). Because *Epicephala* moths pollinate *Glochidion* flowers and oviposit in the styles using diverse and specific methods (Kato et al. 2003), this structural difference may further reinforce host plant specialization and may have played an important role in shaping the overall cospeciation pattern between *Glochidion* and *Epicephala*.

The relative importance of reciprocal adaptation in driving parallel diversification can potentially be assessed by the amount of pollinator and nonpollinator cospeciation with their host plants. For example, *Glochidion* plants are associated with leaf-mining *Diphtheroptila* moths that belong to the subfamily Ornixolinae together with *Epicephala* (Chap. 7). Although data on *Diphtheroptila* are only available for the Japanese fauna, these moths have distinctly broader host ranges than do *Epicephala* and consequently a lower level of phylogenetic similarity with their *Glochidion* hosts. For example, two *Diphtheroptila* species found in Japan both utilize *G. acuminatum* and *G. obovatum* (Chap. 7), which occupy distant positions in the *Glochidion* phylogeny (Fig. 10.1). These observations highlight the importance of coevolution in reinforcing plant–moth specialization and cospeciation in the *Glochidion–Epicephala* association.

10.4 Prevalence of Host Shift and Species-Specificity Breakdown

The intimate association between *Glochidion* and *Epicephala* is perhaps one of the most extreme cases of species-specific, plant–insect interaction known (Kato et al. 2003). It is remarkable to note that nearly all *Glochidion* species from which *Epicephala* moths were obtained had their own distinct pollinator species (Table 10.1). However, if this one-to-one rule had been maintained throughout the history of their diversification, how could the pollinators have shifted to novel hosts? If a host shift occurs successfully without violating this rule, the pollinator colonizing a new host must drive the original pollinator extinct, or the pollinator must be primarily absent on the new host. The former scenario assumes that host shift does not result in stable coexistence of two pollinator species on a single host. However, there is growing evidence showing that multiple *Epicephala* species commonly coexist on a single host. For example, *G. lanceolatum* is pollinated by *E. lanceolatella* and *E. perplexa* which can be found together on the same individual tree (Fig. 10.2; Kawakita and Kato 2006). Similarly in the fig–fig wasp system, two or more fig wasp species commonly reproduce and pollinate in a single host fig, a situation assumed to have lasted for at least a few million years (Molbo et al. 2003). Common observation of multiple pollinator yucca moth species on a single yucca host further indicates that coexistence of multiple pollinator species on a shared host do not necessarily lead to exclusion of others by a single species (Addicott 1996; Pellmyr et al. 1996b; Addicott and Bao 1999; Pellmyr 1999). Thus the observed apparent cases of host shift by *Epicephala* moths did not result solely from the colonization/exclusion process.

The alternative scenario assuming the primary absence of a pollinator does not appear to be a plausible condition in obligate pollination mutualisms, as neither the plant nor pollinator can successfully reproduce without the other. However, this process may exert where the plant colonizes a region without its original pollinator, followed by colonization of the isolated plant population by an unassociated pollinator species. For example, independent colonization of oceanic islands (e.g., French Polynesia) by unrelated plant and pollinator could lead to the formation of novel associations involving host shift. Also, some *Glochidion* species have very wide distribution ranges (e.g., *G. acuminatum* occurs from Japan to India and *G. philippicum* from Taiwan to Australia), providing a possibility that multiple, distantly related *Epicephala* species pollinate a single *Glochidion* species allopatrically across its geographic range. In plant–herbivore interactions, there are examples in which local herbivores colonize and specialize to recently introduced host plants, supporting the plausibility of this process in the organization of novel associations.

Because the known cases of species-specificity breakdown are only found in the regions where plant–pollinator association is best studied (Japan and China), the assumed one-to-one specificity is probably routinely violated in the *Glochidion–Epicephala* association. It is therefore necessary to re-examine the diversity and host specificity of *Epicephala* critically on a global scale. Importantly, *Epicephala* species co-occurring on a single *Glochidion* host are not sister species in all reported cases (Fig. 10.2); thus a host shift by either moth species is minimally needed to explain the observed pattern. This situation is analogous to those in the fig–fig wasp and yucca–yucca moth associations in which multiple distantly related pollinator species are commonly observed on a single host (Addicott 1996; Pellmyr et al. 1996b; Addicott and Bao 1999; Pellmyr 1999). The observed poor concordance between plant and pollinator phylogenies is thus likely attributable to repeated host shifts and resulting breakdown of one-to-one specificity. In light of this, it is remarkable that statistical analyses commonly find significant congruence between plant and pollinator phylogenies. Examining the relative importance of cospeciation versus host shift in generating the diversity of *Epicephala* will thus be an important topic for future research.

Chapter 11
Reversal of Mutualism

Atsushi Kawakita

Keywords Active pollination behavior • Braconidae • *Epicephala* • Gall • Parasitoid • *Phyllanthus microcarpus* • *Phyllanthus reticulatus* • Phylogeny • Proboscis

11.1 How Do Mutualisms Break Down?

A major goal in the study of mutualism is to understand how co-operation is maintained when mutualism may potentially turn into parasitism. As detailed in Chap. 8, certain mutualisms feature host sanction or partner choice mechanisms that help stabilize cooperation by decreasing the fitness of noncooperative individuals. For example, in obligate pollination mutualisms, plants selectively abscise flowers that contain high numbers of pollinator eggs relative to the amount of pollen deposited, thereby punishing individuals that impose heavy egg loads or those that do not pollinate (Pellmyr and Huth 1994; Addicott and Bao 1999; Goto et al. 2010; Jandér and Herre 2010; Jandér et al. 2012).

Although mechanisms such as host sanctions may promote the stability of mutualisms, phylogenetic analyses of mutualist lineages often indicate that nonmutualistic taxa are nested within ancestrally mutualistic lineages (Pellmyr et al. 1996a; Hibbett et al. 2000; Bidartondo and Bruns 2001; Lutzoni et al. 2001; Culley et al. 2002; Als et al. 2004), suggesting that mechanisms promoting stability over ecological timescales may be decoupled from those shaping macroevolutionary patterns. Mutualisms may break down either as the result of (1) mutualism reversal, whereby mutualists become parasites of the original mutualism, or (2) mutualism dissolution, whereby mutualists evolve alternative life histories and live independently of the original partners. Early theoretical models emphasized the likelihood of the former process, but examples of parasites that evolved from mutualists are rare. Such parasites include mycoheterotrophic plants that evolved from photosynthetic ancestors, and derived fig wasps and yucca moths

A. Kawakita (✉)
Center for Ecological Research, Kyoto University, 2-509-3 Hirano, Otsu, Shiga 520-2113, Japan
e-mail: kawakita@ecology.kyoto-u.ac.jp

M. Kato, A. Kawakita (eds.), *Obligate Pollination Mutualism*, Ecological Research Monographs, DOI 10.1007/978-4-431-56532-1_11

that consume fig/yucca seeds without pollinating the plants (Pellmyr et al. 1996a; Machado et al. 2001; Peng et al. 2008; Heraty et al. 2013). The rarity of mutualism reversal either suggests that parasite evolution is constrained in most mutualisms by mechanisms such as sanctions, or that parasites arise frequently but are evolutionarily short-lived.

Mutualism dissolution, by contrast, is much more widespread than reversals (Sachs and Simms 2006). For example, plants have repeatedly abandoned partnerships with animal pollinators, shifting to wind- or self-pollination (Culley et al. 2002). Also, various lineages of mycorrhizal fungi have reverted to saprotrophic lifestyles and live independently of plant hosts (Hibbett et al. 2000). The shift to free-living status may be common because, over evolutionary time, the cost–benefit aspect of the mutualism becomes unfavorable as partners become difficult to encounter due, for example, to decoupled biogeographical history (Pellissier et al. 2012; Espíndola et al. 2014), or the benefits gained from mutualistic partners become easily accessible in the environment, as exemplified by plants that abandon nutritional symbioses with mycorrhizal fungi or nitrogen-fixing bacteria in rich soils (Sachs and Simms 2006).

This chapter focuses on the natural history and process of mutualism reversal in the *Epicephala* moth lineage. As detailed in Chap. 6, one such reversal involves the clade of *Epicephala* moths that shifted onto herbaceous *Phyllanthus* and lost pollination behavior. Because their host plants normally attain a 100% fruit set via pollination by abundant ants that forage for nectar (Chap. 6), the benefit moths gain (seeds) is freely accessible without any need for the moths to pay the associated cost (pollination). Another case is the *Epicephala* moths on plants of the *Phyllanthus reticulatus* species complex in Taiwan. A close investigation resulted in a finding of six *Epicephala* species, of which three were mutualists and three were derived parasites. The latter did not provide benefits to the plants because they induced gall formation on flowers/buds without producing any seeds. This chapter details the natural history and evolutionary history of the six *Epicephala* species associated with plants of the *Phyllanthus reticulatus* species complex in Taiwan and suggests the possibility that the presence of a third-party partner (braconid wasps) may help explain the evolution of mutualism reversal in *Epicephala*.

11.2 *Phyllanthus reticulatus* Species Complex and Associated *Epicephala* in Taiwan

Phyllanthus reticulatus is a shrub that is common along roadsides and forest edges in the tropical regions of Asia. Although long treated as a single species, the shrub has recently been split into two species, *P. reticulatus* and *P. microcarpus*, based on several distinct morphological and ecological characteristics (Luo et al. 2011a). Both species occur throughout our study sites in Taiwan, but in many of the populations that we studied, the plants showed intermediate characteristics

Fig. 11.1 Plants of the *Phyllanthus reticulatus* species complex and associated insects. (**a**) The habit of *Phyllanthus reticulatus* along roadside in Taiwan. (**b**) Male flower of *P. reticulatus*. (**c**) Female flower of *P. reticulatus*. (**d**) Fruit of *P. reticulatus*. (**e**), Tough galls (arrows) induced by *Epicephala* sp. C. (**f**) Swollen galls induced by *E*. sp. D. (**g**) Knobbed galls induced by *E*. sp. E. (**h**) Normally developing fruits (*upward arrows*) and swollen galls (*downward arrows*) occurring on the same branch. Note the size difference. (**i**) Cross sections of a normal fruit (*upper left*), tough gall (*lower left*) and swollen gall (*right*) depicted to the same scale. Only a fraction of the seeds are destroyed by *Epicephala* larva in normal fruits, whereas galled ovules are entirely destroyed by moth larvae. Two locules are galled in the tough gall shown, one of which contains an *Epicephala* larva. All nine locules of the swollen galls had been galled. (**j**) Oviposition marks (*arrows*) left by *E*. sp. D on immature swollen galls. (**k**) Longitudinal section of a fully developed swollen gall with an irregularly developed ovule containing a larva of *E*. sp. D (*black arrow*) and undeveloped ovules (*white arrows*). Note the airspace around the infested ovule. (**l**) A braconid wasp ovipositing in a swollen gall. (**m**) An eulophid wasp resting on a tough gall produced by *E*. sp. C

suggestive of hybridization, which rendered distinctions between the two species obscure. Here it is considered that the plants belong (only) to the *Phyllanthus reticulatus* species complex; individual species names and putative hybrid status are mentioned only where appropriate.

Plants of the *P. reticulatus* species complex produce separate, small, inconspicuous male and female flowers on leaf axils (Fig. 11.1). Flowering and fruiting occur

Fig. 11.2 The six *Epicephala* moth species co-occurring on plants of the *Phyllanthus reticulatus* species complex in Taiwan. Plus (+) and minus (−) symbols indicate presence and absence, respectively. Some ovipositing females of *Epicephala* sp. E do not carry pollen on the proboscis (hence "±" under "Pollen on proboscis"). *Epicephala* sp. A, B, and F produce normal fruits containing viable seeds (drawn as filled ovules under "Flower fate"), whereas *E.* sp. C, D, and E induce galls that contain non-viable galled ovules (drawn as open ovules). Shaded areas inside galls indicate internal airspace. Fruits and galls are drawn proportional to their actual sizes. A putative hybrid between *P. reticulatus* and *P. microcarpus* is indicated as *P. ret.* × *P. mic.* under "Host species". Reproduced from Kawakita et al. (2015)

throughout the year, and *Epicephala* moths emerge many times each year. The fleshy fruits contain 12–20 ovules, of which roughly half are consumed by a single pollinator larva. Although previous chapters suggested that these plants arc pollinated by species-specific *Epicephala* moth species (e.g., Chap. 6), detailed observation of moth behavior, genital morphology, and COI sequences indicated that six distinct species are associated with the plants of the *P. reticulatus* species complex in Taiwan.

Epicephala **sp. A** is a pollinator that actively pollinates flowers and lays eggs in female flower pedicels (Fig. 11.2). Oviposition occurs only once on each flower. Pollinated flowers develop into normal fruits (Fig. 11.1), and a single larva consumes some of the developing seeds. ***Epicephala*** **sp. B**, the least common of the six species, is also a pollinator that produces normal fruits. It actively pollinates, and lays eggs superficially on the upper wall of the ovary (Fig. 11.2). This behavior is usually repeated three times on a single flower, and three eggs per flower are thus laid per visit by each moth. ***Epicephala*** **sp. C** is a parasite that induces a tough gall with a dented surface (hereafter, tough gall) on female flower buds (Fig. 11.1). None of the moths observed exhibited any pollinating behavior, and laid eggs in young female buds (Fig. 11.2), which eventually developed into tough galls. Ovipositing females do not carry pollen. ***Epicephala*** **sp. D** is also a parasite that induces a swollen gall (with an internal airspace) that is distinctly larger than a normal fruit (hereafter, swollen gall). Notably, this species has pollination behavior similar to that of the pollinator species (Fig. 11.2). The eggs are superficially laid on the upper ovary wall of the female flower. Ovipositing females carry abundant pollen on the proboscises, indicating that they had collected pollen on male flowers. Pollination and oviposition occur three times on each visit to a single flower. Infested flowers develop into swollen galls with internal airspaces (Fig. 11.1) within which ovules containing moth larvae develop into masses of endospermlike tissue that lack the features of normally produced seeds (e.g., a seed coat). Such irregular ovules are entirely consumed by the larvae of *E.* sp. D; thus, although whether the irregularly developed ovules retain the germination ability has not been tested, the presence of *E.* sp. D is clearly not beneficial to the plant. Ovules that are not attacked by moth larvae inside the swollen gall remain undeveloped (Fig. 11.1). ***Epicephala*** **sp. E** is (similarly) a parasite that induces a gall with a weakly knobbed surface that is similar in size to a normal fruit (hereafter, knobbed gall). Adults also display the pollination behavior (Fig. 11.2). However, ovipositing moths sometimes do not carry pollen, suggesting that pollen collection may be occasionally omitted in this species. Oviposition occurs once per visit, and infested flowers develop into galls containing irregularly developed ovules typical of those induced by *E.* sp. D, but lacking the internal airspace. ***Epicephala*** **sp. F** is a pollinator that lays eggs into the apical stigmatic pits of female flowers (Fig. 11.2). Only one egg is laid per visit.

Analysis of COI sequences identified six distinct clades (Fig. 11.3) that corresponded perfectly with the observed differences in adult behavior, genital morphology, and the characteristics of the fruits/galls from which adults were reared. In mainland China, there are three additional *Epicephala* species associated with plants of the *P. reticulatus* species complex (Li and Yang 2015), although the

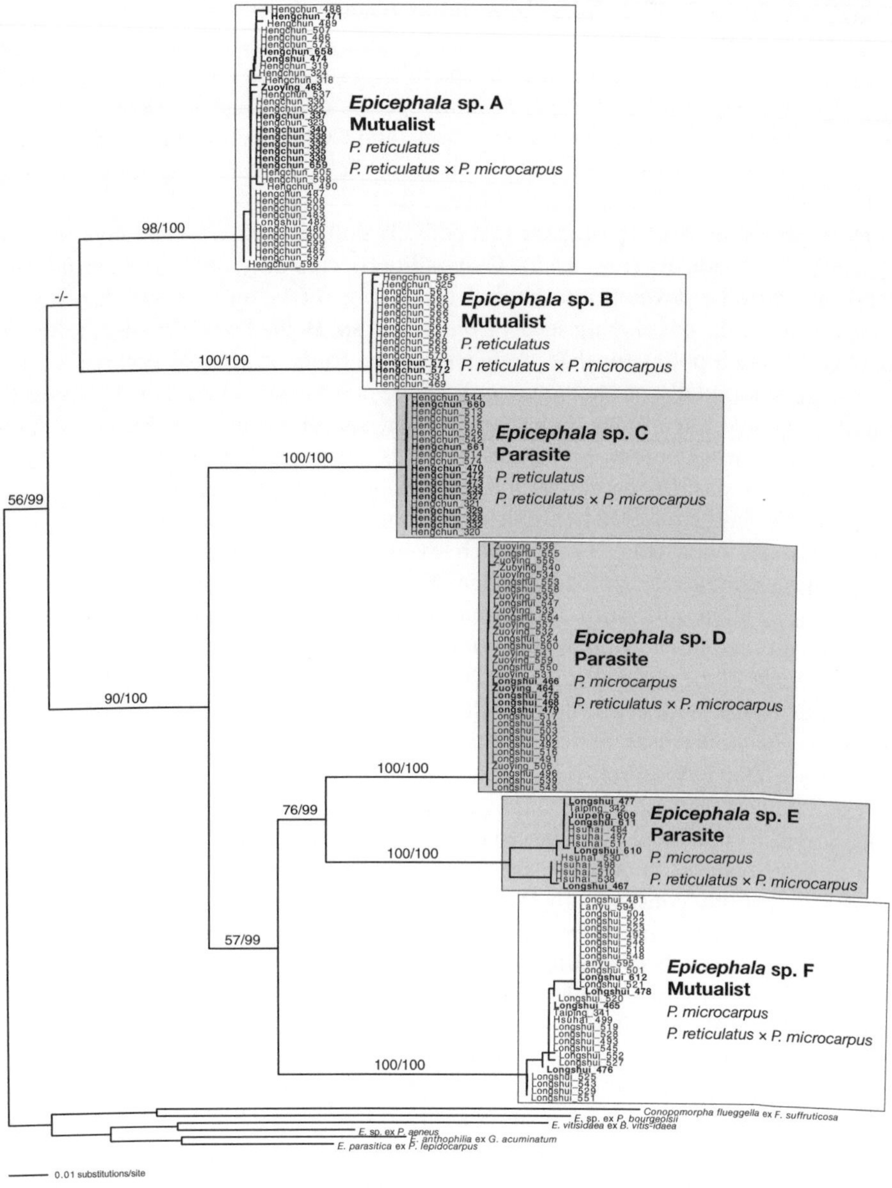

Fig. 11.3 Phylogeny of 174 *Epicephala* moths collected from plants of the *Phyllanthus reticulatus* species complex based on COI gene sequences. The numbers above branches are maximum-likelihood bootstrap values followed by the Bayesian posterior probabilities. Individual moths are labelled with locality name followed by unique numbers. Individuals for which oviposition behavior was observed in the field are indicated in bold. Parasitic species are shaded in grey. The taxonomy of the host with which each species is associated is shown

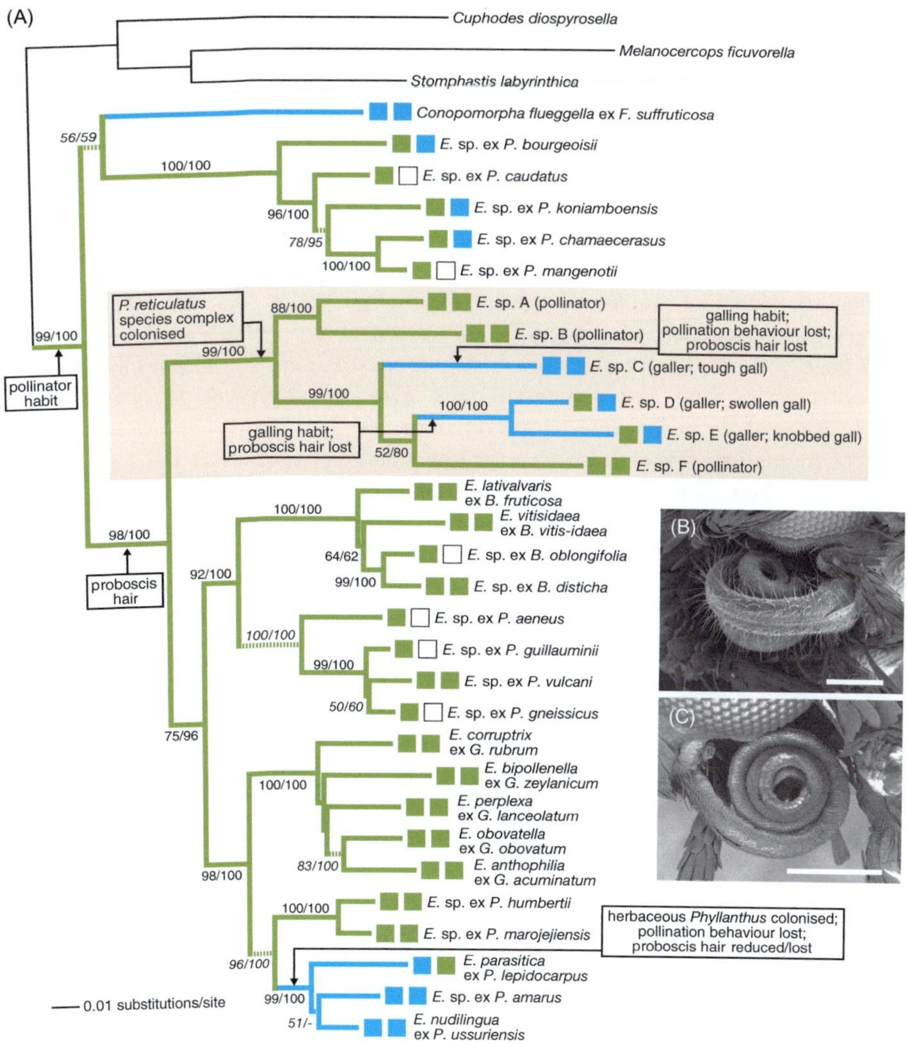

Fig. 11.4 Reversal of mutualism in *Epicephala*. (**a**) Maximum-likelihood phylogeny of the genus *Epicephala* based on combined data from the COI, ArgK, and EF1α genes. Numbers at the branches are the maximum-likelihood bootstrap values followed by the Bayesian posterior probabilities. Nodes unsupported upon individual gene analysis are indicated as hatched lines, with support values in italics. The monophyletic clade containing the six species associated with plants of the *Phyllanthus reticulatus* species complex is highlighted in the shaded box. Mutualistic and parasitic lineages are coloured green and blue, respectively. Boxes located to the left of the terminal taxon names indicate the presence/absence of pollination behaviour (*left*; present when *green*) and hairs on the proboscis (*right*; present when *green*; *open boxes* indicate missing data). Major evolutionary events are indicated in boxed notes. Genus abbreviations are: *E.*, *Epicephala*; *F.*, *Flueggea*; *P.*, *Phyllanthus*; *B.*, *Breynia*; and *G.*, *Glochidion*. (**b**) Scanning electron micrograph (SEM) of female proboscises of *Epicephala* sp. A (pollinator). (**c**) SEM of female proboscises of and *E.* sp. D (galler). Reproduced from Kawakita et al. (2015)

larval ecology of these species has not yet been studied in detail. Because plants of the *P. reticulatus* species complex is distributed throughout tropical Asia and Africa, it is likely that numerous more species with varying adult and larval ecology will be found throughout the range.

Phylogenetic analysis of *Epicephala* and *Conopomorpha flueggella* associated with a broad diversity of Phyllantheae plants indicated that the six species associated with the *P. reticulatus* species complex were monophyletic, and that the parasitic species were derived from pollinating ancestors (Fig. 11.4a). It is unclear whether mutualism reversal occurred only once, or many times, within the clade, because the level of statistical support at internal nodes was low (Fig. 11.4a). Reversion to parasitism also occurred in a clade of *Epicephala* moths associated with the weed *Phyllanthus*, as previously reported (Chap. 6). The nonpollinating *Conopomorpha flueggella* associated with *Flueggea suffruticosa* is embedded within *Epicephala* in this tree. However, it is not clear if this represents another case of mutualism reversal, because of low statistical support at the basal nodes (Fig. 11.4a). Female proboscis sensilla that are characteristic of the core *Epicephala* clade (Chap. 5) were absent in the three gall-forming *Epicephala* species found in the present study (Fig. 11.4b, c), indicating that the sensilla were lost as the galling habit evolved.

11.3 Third-Party Partner of the Mutualism

Because variation in gall traits, or in the galling habit in general, may be caused by selection imposed by natural enemies (Stone and Schönrogge 2003; Bailey et al. 2009), we hypothesized that galling in *Epicephala* evolved as a defense against a specialist parasitoid, rather than as a response to a shift in the cost–benefit balance

Table 11.1 Intensity of parasitism by braconid and eulophid wasps on *Epicephala* sp. A (pollinator) and E. sp. C (gall maker)

	Fruit (*E.* sp. A)	Gall (*E.* sp. C)	Significance[a]
Selectivity of wasp oviposition			
Natural abundance	282	258	
Braconid ovipositions	27	1[b]	$P < 0.001$
Eulophid ovipositions	1	25	$P < 0.001$
Parasitism rate			
Number examined	176	189	
Fruit/gall with braconid	31	2	$P < 0.001$
Fruit/gall with eulophid	1	95	$P < 0.001$

Data are based on fruit/gall samples collected from six *Phyllanthus reticulatus* individuals at the Hengchun population, Taiwan

[a]Significance based on Fisher's exact test

[b]A single wasp alighted on the gall and attempted oviposition but failed to pierce ovipositors through gall wall

of the interaction of the moth with plants. This hypothesis was tested using two *Epicephala* species: those that induce tough galls and those that induce swollen galls. These species were chosen because both gall types were abundant at the study sites, and because the defensive functions of these gall traits were relatively straightforward in both species.

The most prevalent natural enemies of *Epicephala* larvae are specialized *Bracon* wasps associated with most *Epicephala* species studied to date. The wasps insert their ovipositors into fruits/galls containing later-instar *Epicephala* larvae and lay single eggs on the body surface of a moth larva. The wasp larvae develop by consuming the moth larvae externally, and emerge as adults from the fruits/galls. Gall toughness may render it impossible for the piercing wasp ovipositor to penetrate the gall wall, whereas the internal airspace of the swollen gall may increase the distance between the gall surface and *Epicephala* larvae inside galled ovules, rendering the wasp ovipositor too short for effective oviposition.

Field observation of wasp oviposition on tough galls produced by *E.* sp. C, and subsequent dissection of the galls, indicated that the braconid wasp seldom attacked the larvae of *E.* sp. C (Table 11.1). In only one instance did we observe the braconid attempting to oviposit on the tough gall, but the wasp failed to pierce the gall wall. Rather, *E.* sp. C moths were regularly infested by a eulophid, *Aprostocetus* sp. (Fig. 11.1), which was far less abundant on *E.* sp. A that produces normal fruits. *Aprostocetus* wasps thus far have not been found from *Epicephala* species associated with plants other than those of the *P. reticulatus* species complex. Braconid ovipositions into tough galls occurred significantly less frequently, and those by eulophids occurred significantly more frequently than expected by consideration of the natural abundances of fruits and galls (Table 11.1). The same pattern was observed when actual parasitism levels (by braconids and eulophids) inside galls and fruits were evaluated (Table 11.1).

In contrast, *E.* sp. D, which induces a swollen gall, was susceptible to braconid attack (Fig. 11.1). However, this species was more likely to escape braconid parasitism when larger galls were infected. Overall gall size increased as more ovules were infested in each flower (Fig. 11.5), indicating that gall size was dependent on the number of moth larvae infesting the gall. In turn, the number of ovules infested per gall did not affect the size of infested ovules per se, indicating that the increase in overall gall size was not due to increase in the size of infested ovules but was solely attributable to an increase in the volume of the internal airspace. In addition, galls containing parasitized moth larvae were smaller than those with intact moth larvae only (Fig. 11.5), suggesting that larval induction of gall development ceases as moth larvae become parasitized. Thus, the number of infested ovules per gall is a better proxy of the gall size at which braconid ovipositions occur. Logistic regression of the parasitism rate (the proportion of moth larvae parasitized per gall) on the number of infested ovules (a proxy for gall size) showed that the parasitism level decreased as more larvae infested the gall, and hence, as galls became larger (Fig. 11.5).

Phylogenetic analysis of the *Bracon* COI sequences suggested that the wasps reared from fruits and galls of plants in the *P. reticulatus* species complex group

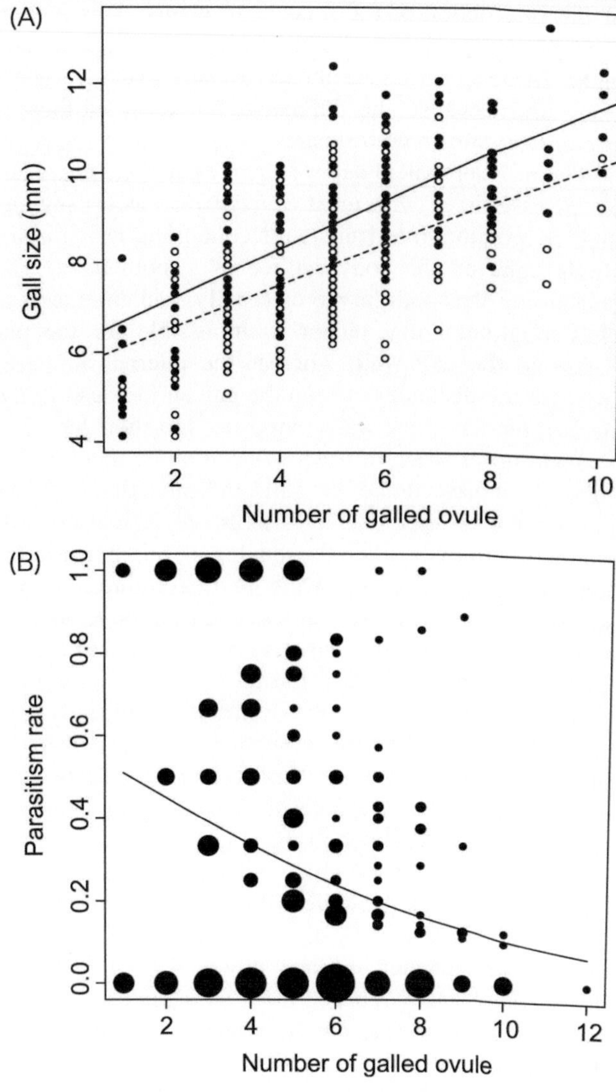

Fig. 11.5 Variation in the size of swollen galls and braconid oviposition success. (**a**) Relationship between the number of infested ovules per gall and overall gall size (N = 348). White and black circles indicate galls with and without parasitized moth larvae. (**b**) Logistic regression of the proportions of parasitized moth larvae on infested ovule numbers per gall (N = 380). The area of each circle is proportional to the sample size. Reproduced from Kawakita et al. (2015)

into two separate clades (Fig. 11.6). There was a clear difference in the pattern of dorsal markings on the thorax and abdomen of male wasps between the two clades (Fig. 11.6), suggesting that each clade represents distinct species. The first species contained wasps reared from tough galls produced by *E.* sp. C and fruits of *P. microcarpus* produced by *E.* sp. F. The second species contained wasps reared from fruits of *P. reticulatus* produced by *E.* sp. A and swollen galls produced by *E.* sp. C (Fig. 11.6). There was no support from the COI phylogeny that these two *Bracon* species are sister to each other (Fig. 11.6).

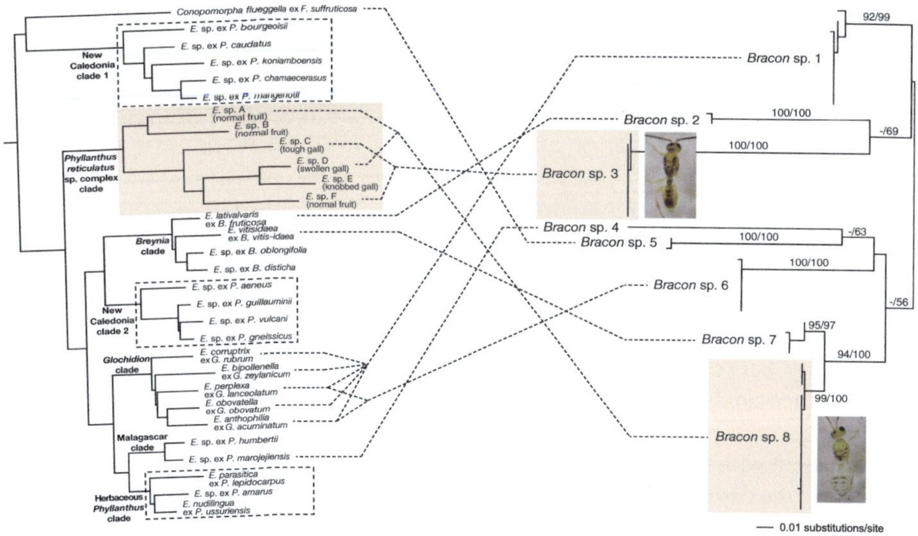

Fig. 11.6 Phylogeny of 49 *Bracon* wasps (*right*) based on COI gene sequences and their associations with *Epicephala* moth hosts. The wasps were divided into eight distinct clades, which are considered as species. The numbers above the branches are the maximum-likelihood bootstrap values followed by the Bayesian posterior probabilities (given only for nodes above species level). Hatched lines connect each wasp species with its host *Epicephala* species. The two *Bracon* species that attack *Epicephala* associated with plants of the *Phyllanthus reticulatus* species complex (sp. 3 and sp. 8) can be distinguished based on the markings on the dorsal surface of thorax and abdomen in the males: the former possesses dark markings at the centre of first to sixth metasomal terga (T1–T6), both sides of the propodeum and areas around the wing base, whereas the latter possesses only light markings on T3–T6 that are intermitted centrally by non-pigmented area and no mesosomal markings as in sp. 3 (photos). Although *Bracon* wasps are prevalent among *Epicephala* moths, they are absent from *Epicephala* that colonized New Caledonia and those associated with herbaceous *Phyllanthus* (indicated by hatched boxes). Islands and herbaceous host may have provided *Epicephala* moths with enemy-free space. Reproduced from Kawakita et al. (2015)

11.4 Third-Party Partner as a Driver of Mutualism Reversal

Examples of parasites evolving from mutualistic ancestors offer opportunities to study the circumstances under which such evolution occurs, but documented examples of mutualism reversals are still limited (Pellmyr et al. 1996a; Machado et al. 2001; Peng et al. 2008). As detailed in this chapter, the Phyllantheae–*Epicephala* association offers an exciting opportunity to study mechanisms underlying the shift to parasitism because parasitic *Epicephala* arose multiple times in the history of the mutualism (Fig. 11.4).

Nevertheless, identifying the cause of the mutualism reversal found in the three galler *Epicephala* species is not straightforward. One possibility is that galling afforded a more cost-effective means of gaining resources than pollinating.

However, the parasitic *E.* sp. D and sp. E both retain pollination behavior in addition to investing in the chemical substances that induce the galls, indicating that the cost paid by gallers is not substantially lower, if it is lower at all, than that paid by the mutualistic ancestors. Alternatively, galling may improve the nutritional value of larval food, or allow the moths to circumvent selective flower abscission, and thereby increase larval survival. However, nutritional improvement or flower retention does not comprehensively explain some features of the galls, such as toughness or presence of airspace inside the gall.

We hypothesize that the mutualism reversal evident in the present study is a by-product of an adaptation made by *Epicephala* moths to escape braconid parasitism. Proof that galling is an effective defense mechanism requires a comparison of mortality between gallers and nongalling ancestors at the time when galling evolved; such analysis is impossible using long-established extant galler lineages. Nevertheless, the patterns of parasitism experienced by the galler species studied are consistent with the possibility that galling initially evolved as a defense, as often assumed in the case of galls produced by many other insects (Stone and Schönrogge 2003; Bailey et al. 2009). For example, *E.* sp. C, bearing tough galls, is virtually free of braconid parasitism, whereas *E.* sp. D, with swollen galls, experiences higher survival when the galls have a larger internal airspace. Presently, neither species necessarily experiences lower parasitism than closely related mutualist species (overall parasitism rates are 41.7% ($n = 240$) and 27.7% ($n = 1724$) for the parasitic *E.* sp. C and D, respectively, and 22.5% ($n = 151$) for the mutualist *E.* sp. A, calculated based on the original data used for Table 11.1 and Fig. 11.5), probably because enough time has elapsed for the parasitoid community to adapt to gallers. For example, *E.* sp. C is presently attacked by a eulophid wasp, and *E.* sp. D continues to suffer high-level braconid parasitism, possibly because the wasp ovipositor coevolved (became longer) with increasing gall size. Escape from parasitoid attack by gall induction may be a common evolutionary trajectory in *Epicephala* because similar galls are produced by *Epicephala* moths associated with *Glochidion obovatum* in Japan (Chap. 5) and *Phyllanthus* in Madagascar, both distantly related to the *Epicephala* species associated with plants of the *P. reticulatus* species complex (Chap. 5).

It is still unclear why *E.* sp. D and sp. E retain pollination behavior despite development of a galling ability. One possibility is that although pollination is unnecessary, the behavior cannot be easily lost because it is tightly integrated into the sequence of *Epicephala* oviposition. This was suggested by the observation that some individuals of *E.* sp. E did not have pollen on the proboscis, that they sometimes oviposited in buds, and that proboscis hairs were absent in both species. These findings indicate that selection toward pollen transport was relaxed. Unlike pollination behavior, however, proboscis hair may be a labile character that can be quickly lost after galling evolves. In any case, the evolution of galling did not eliminate the cost of pollination behavior. Whether proboscis hairs are associated with substantial costs (e.g., impeding of feeding) remains to be determined. The proboscis constitutes less than 0.3% of moth body mass, and the microscopic structures evident on the surface thereof are unlikely to be costly, at least energetically (also see Pellmyr 1997 for a discussion of the similarly small structural cost of the pollen-manipulating appendages of yucca moths).

Some mutualism theories suggest that long-term persistence of mutualisms is facilitated by mechanisms that stabilize the cost–benefit ratios of the interacting partners (Sachs et al. 2004). However, this study leads to a hypothesis that factors extrinsic to the mutualism can have large effects on the evolutionary fate of mutualisms, regardless of whether the pairwise interaction continues to favor cooperation. In the present study system, derived parasitic *Epicephala* species are able to coexist stably with related mutualistic *Epicephala* species on a single host. However, if parasitic species limit the persistence of mutualist populations via processes such as resource competition or reproductive interference, it is possible that mutualism collapses solely via a process unrelated to the cost–benefit balance of the interaction. Our study thus highlights the need to explore how factors extrinsic to a mutualism may shape the macroevolutionary dynamics of that mutualism; such work will improve our understanding of mutualism stability.

Chapter 12
Phyllantheae–*Epicephala* Mutualistic Interactions on Oceanic Islands in the Pacific

David H. Hembry

Keywords Dispersal • *Epicephala* • *Glochidion* • Host shift • Leafflower • Oceanic islands • Pacific • *Phyllanthus*

12.1 Introduction

Oceanic islands, and the organisms that live on them, have long served as models for the study of evolution. Formed de novo by volcanism or by the uplift of previously submerged rock, oceanic islands have never been connected to continents. They are colonized by a limited number of founding lineages that arrive via long-distance dispersal. The resulting discrete and isolated nature of these communities and the organisms within them have led oceanic islands to be used as "natural laboratories" since the time of Darwin by researchers interested in speciation (Darwin 1859; Coyne and Orr 2004; Goodman et al. 2012), adaptive radiation (Lack 1947; Carlquist 1974; Chiba 2004; Grant and Grant 2008), and community assembly (MacArthur and Wilson 1967; Gillespie 2004; Casquet et al. 2015). The utility of oceanic island biotas as models for evolutionary insights is indicated by their ongoing adoption in new areas of evolutionary ecology, such as diversification dynamics (Economo and Sarnat 2012; Bennett and O'Grady 2013), host–microbe interactions (Ort et al. 2012, O'Connor et al. 2014), and coevolutionary biology (Hembry et al. 2013a).

The diversity of Phyllantheae (leafflower plants) in tropical continental regions is striking (Govaerts et al. 2000; Kathriarachchi et al. 2006; Kawakita and Kato 2004a, b, 2009; this volume). Phyllantheae have, however, not only remained on continents. They have also successfully colonized remote oceanic islands throughout the Pacific Ocean, including some of the most isolated archipelagos in the world (Fig. 12.1). In many cases, their pollinating *Epicephala* moths have also colonized these islands with them. Finally, Phyllantheae have diversified in many of these island groups, undergoing endemic radiations in some of the larger archipelagos.

D.H. Hembry (✉)
Department of Ecology and Evolutionary Biology, University of Arizona, Tucson, AZ 85721, USA
e-mail: hembry@post.harvard.edu

M. Kato, A. Kawakita (eds.), *Obligate Pollination Mutualism*, Ecological Research Monographs, DOI 10.1007/978-4-431-56532-1_12

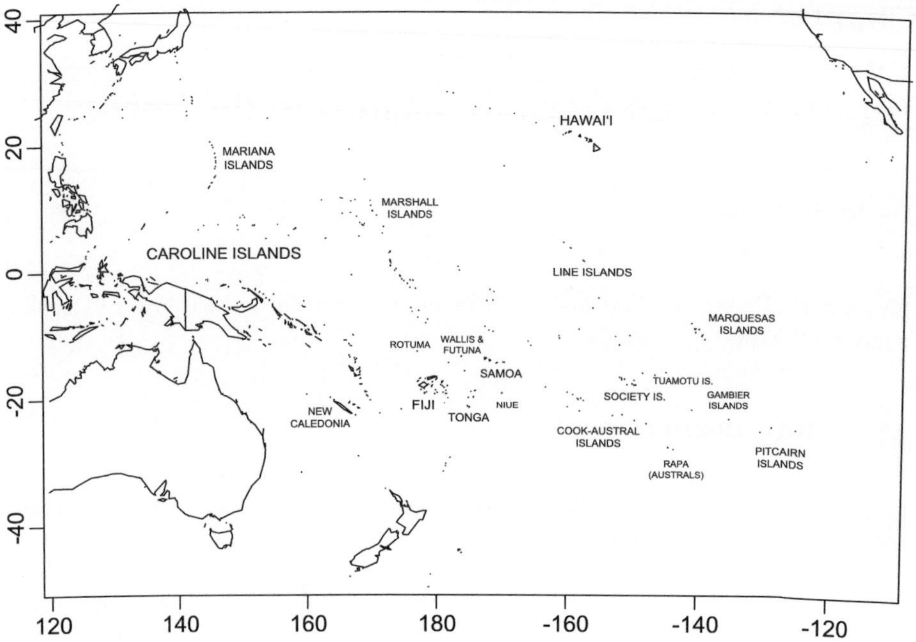

Fig. 12.1 Map of Pacific Basin with archipelagos mentioned in text indicated

These insular Phyllantheae thus have potential to serve as "natural experiments" for the study of diversification and coevolution.

This chapter summarizes what is known about the biogeography of Phyllantheae and their interactions with *Epicephala* moths, with particular attention to the clade *Glochidion* (Figs. 12.2, 12.3, 12.4), across the tropical Pacific Basin. The area considered spans over 11,000 km from Palau in the west to the Pitcairn Islands in the east, and contains regions traditionally referred to as Micronesia (Mariana Islands, Palau, Federated States of Micronesia) and Polynesia (independent Samoa, American Samoa, Tonga, Wallis and Futuna, Niue, Cook Islands, French Polynesia, Pitcairn Islands, and Hawaii), as well as Fiji (often treated biologically as part of Polynesia). Strictly speaking, not all of these islands are "oceanic", particularly a number of those in the western part of Micronesia and the Fijian archipelago. However, they share a remoteness from continental regions, relative geologic youth (most formed in the Miocene or more recently), and a common biota of primarily Paleotropical origin.

These islands may crudely be classified as either atolls or high islands. Atolls are ring-shaped coral reefs found atop submerged volcanoes; they reach a maximum elevation of only a few meters above sea level, and in many places the reef is continuously awash. Atolls support a limited terrestrial biota of plants and animals that tolerate this high-salt, ephemeral, and low-water environment. They are inhabitable by people, particularly with the aid of the specialized agriculture and technology invented by Polynesians, but are vulnerable to anthropogenic sea level rise.

Fig. 12.2 *Glochidion* plants native to Pacific islands. (**a**) *Phyllanthus longfieldiae* (syn. *Glochidion longfieldiae*) is a shrub or tree endemic to Rapa, Austral Islands. (**b**) *P. nadeaudii* (syn. *G. nadeaudii*) is a tree with distinctive drooping branches endemic to Moorea, Society Islands. (**c**) *P. emarginatus* (syn. *G. emarginatum*) is a prostrate shrub endemic to Raiatea, Society Islands. (**d**) *P. orohenense* (syn. *G. orohenense*) is a small tree of cloud forests endemic to Tahiti, Society Islands. (**e**) *P. marchionicus* (syn. *G. marchionicum*) is a tree found in a wide variety of habitats, endemic to the Marquesas Islands. Here it grows in dry, perturbed vegetation on Nuku Hiva. (**f**) *P. florencei* (syn. *G. societatis*) is found on both basaltic and calcareous substrates in the

Good examples of atolls are Rangiroa and Fakarava in the Tuamotu Islands (French Polynesia) or the Marshall Islands. With very few exceptions, these atolls do not host Phyllantheae, unless they are uplifted (see below).

Very generally, high islands are all other islands, whether tens or thousands of meters in elevation above sea level. Many of these high islands, particularly in eastern Micronesia and Polynesia, are formed by volcanoes. These high islands may form linear chains of increasing age as a result of the movement of the Pacific Plate over fixed hotspots in the Earth's mantle; good examples include the Society, Marquesas, Samoan, and Hawaiian archipelagos. Others form chains without age progression near subduction zones, such as is the case for parts of the Mariana Islands and Tonga. A few of these islands are still geologically active and have ongoing eruptions; these include the island of Hawaii (the "Big Island"), Savai'i in independent Samoa, Niuafo'ou in Tonga, and Mehetia in the Society Islands (French Polynesia). Soils on these islands are of basaltic origin. Other high islands are makatea islands, in which a heavily eroded volcanic island surrounded by a coral reef was subsequently uplifted, leading to a new high island with an outer limestone ring and inner basaltic core; good examples include several of the southern Cook Islands such as Mangaia. Finally, yet other high islands, such as Palau, Guam, much of Fiji, and some of the Austral Islands (French Polynesia) have had repeated periods of volcanism and uplift, and do not fit as neatly into the subcategories of high islands as presented above. In general, high islands support a more diverse biota than do atolls; the largest islands are known for extensive endemism and even adaptive radiation in their biota, and this is the case for Phyllantheae as well.

A grey area are those atolls that have been uplifted a few meters by lithospheric flexure from adjacent younger volcanoes; examples in the Pacific include Niau and Anaa in the Tuamotu Islands (French Polynesia). Other atolls have been uplifted by tens of meters or more; examples include (confusingly) Makatea in the Tuamotu Islands and Henderson in the Pitcairn Islands. Even a minimal amount of uplift allows these former atolls to support elements of a high island biota, including Phyllantheae.

12.2 Note on Taxonomy

Phyllanthus s. l. and *Glochidion* in particular are in a state of taxonomic flux in the Pacific. Most species of *Phyllanthus* s. l. from this part of the world are part of the clade *Glochidion* and have traditionally been assigned names in that genus. Because many recent studies have resolved a monophyletic *Glochidion* nested within a

---

Fig. 12.2 (continued) Leeward Society Islands (Maupiti, Tahaa, Raiatea, Huahine). Here it grows on calcareous substrate on Motu Auira, an offshore islet of Maupiti

Fig. 12.3 Female flowers of Pacific island *Glochidion*. Photos not to scale relative to one another. (**a**) *Phyllanthus samoanus*, Tutuila, American Samoa. (**b**) *P. cuspidatus* (syn. *Glochidion cuspidatum*), Tutuila. (**c**) *P.* sp., Mangaia, Cook Islands. (**d**) *P.* sp., Ātiu, Cook Islands. (**e**) *P. raivavense* (syn. *G. raivavense*), Rurutu, Austral Islands. (**f**) *P. raivavense*, Raivavae, Austral Islands. (**g**) *P. temehaniensis* (syn. *G. temehaniense*), Raiatea, Society Islands. (**h**) *P. emarginatus* (syn. *G. emarginatum*), Raiatea. (**i**) *P. st-johnii* (syn. *G. myrtifolium*), Raiatea. (**j**) *P. huahineense* (syn. *G. huahineense*), Huahine, Society Islands. (**k**) *P. florencei* (syn. *G. societatis*), Huahine. (**l**) *P. temehaniensis*, Huahine. (**m**) *P. nadeaudii* (syn. *G. nadeaudii*), Moorea, Society Islands. (**n**) *P. manono* (syn. *G. manono*), Moorea. (**o**) *P. taitensis* (syn. *G. taitense*), Tahiti, Society Islands. (**p**) *P. orohenense* (syn. *G. orohenense*), Tahiti. (**q**) *P. tuamotuensis* (syn. *G. tuamotuense*), Niau, Tuamotu Islands. (**r**) *P. marchionicus* (syn. *G. marchionicum*), Nuku Hiva, Marquesas Islands. (**s**) *P. longfieldiae* (syn. *G. longfieldiae*), Rapa, Austral Islands. (**t**) *P. rapaense* (syn. *G. rapaense*), Rapa

Fig. 12.4 Foliage and fruits of Pacific island *Glochidion*. (**a**) *Phyllanthus samoanus*, Tutuila, American Samoa. (**b**) *P. wilderi* (syn. *Glochidion wilderi*), Mangareva, Gambier Islands. (**c**) *P. emarginatus* (syn. *G. emarginatum*), Raiatea, Society Islands. (**d**) *P. nadeaudii* (syn. *G. nadeaudii*), Moorea, Society Islands (**e**) *P. taitensis* (syn. *G. taitense*), Tahiti, Society Islands. (**f**) *P. papenooense* (syn. *G. papenooense*), Tahiti, Society Islands. (**g**) *P. huahineense* (syn.

grade of *Phyllanthus* s. l. (Kathriarachchi et al. 2006; Kawakita and Kato 2009; Luo et al. 2011a, b; this volume), Wagner and Lorence (2011), as part of the Flora of the Marquesas project, chose to assign valid names in *Phyllanthus* to all described species in *Glochidion* reported from Micronesia, Fiji, and Polynesia (as recommended by Hoffman et al. 2006), and assign new epithets in cases of homonymy. They also performed some useful revisions facilitating future work on this group, combining *Glochidion tooviianum* into *Phyllanthus marchionicus* (syn. *G. marchionicum*) and assigning a new epithet to *P. samoanus*, previously erroneously assigned to *Glochidion ramiflorum*. Most (but not all) of these taxa in *Phyllanthus* retain valid synonyms in *Glochidion*.

For consistency with the local biodiversity and conservation literature in the South Pacific (much of which now uses the new combinations in *Phyllanthus*), with the ongoing Flora of Micronesia project (Wagner et al. 2012), with peer-reviewed literature on Pacific terrestrial ecology (Kahn et al. 2015), and with recent peer-reviewed literature on Pacific *Glochidion* (Hembry 2013, Hembry et al. 2013a, b), this chapter uses the names in *Phyllanthus*, but at first mention, states the synonyms in *Glochidion* where they exist. The name "*Glochidion*" continues to be used in this chapter to designate the clade *Glochidion* as resolved by many recent phylogenetic studies, identical to *Glochidion* sensu lato, for example, the genus *Glochidion* ignoring Wagner and Lorence's (2011) revisions. The reader should bear in mind that nearly all species of *Phyllanthus* s. l. in the remote Pacific are part of the clade *Glochidion*.

Unrelated to nomenclatural issues, there is some unrecognized diversity of *Glochidion* in the remote Pacific. Taxa from Tonga, Wallis and Futuna, Rotuma, and the Cook Islands have not been revised, although in a few cases they have been assigned to species found elsewhere. One taxon from Rapa was not described by Florence (1997a) but informally assigned a name (*Glochidion* "*fosbergii*") which is present on some herbarium labels. Morphological examination of fresh material indicates that several unrecognized taxa are likely present in the Cook and northern Austral Islands (Fig. 12.3c–f). Finally, widespread species from the Leeward Society Islands (particularly Bora Bora, Tahaa, Raiatea, and Huahine), such as *P. temehaniensis* (syn. *G. temehaniense*; Fig. 12.3g, l) and *P. st-johnii* (syn. *G. myrtifolium*), differ in floral characters among islands and are in need of more taxonomic attention. To date, molecular phylogenetic analysis using Sanger-sequenced nuclear ribosomal and chloroplast data has been unsuccessful at resolving relationships among most southeastern Polynesian *Glochidion* (Hembry et al. 2013a; Fig. 12.5).

Fig. 12.4 (continued) G. huahineense), Huahine, Society Islands. (**h**) *P. longfieldiae* (syn. *G. longfieldiae*), Rapa, Austral Islands

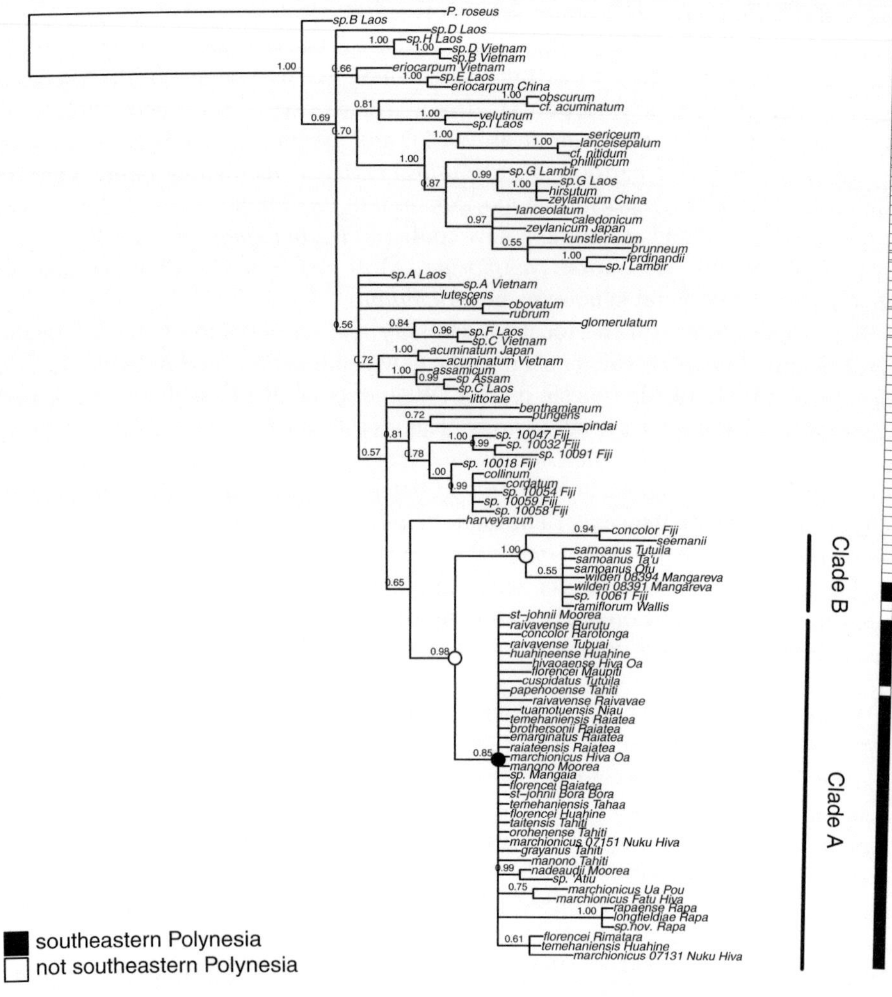

Fig. 12.5 Bayesian consensus tree for continental and Pacific island *Glochidion* from Hembry et al. (2013a). Node labels represent posterior probabilities. Tip labels refer to specific epithets of *Glochidion* species. Colored *boxes* and *circles* represent present-day distributions and ancestral state reconstructions of biogeography at selected nodes (*black*: southeastern Polynesia; *white*: not southeastern Polynesia; reprinted with permission of the Royal Society)

12.3 Diversity of *Glochidion* in the Pacific

Glochidion are found on most high island chains in the Pacific, with a few notable exceptions. With 54 described species (not including undescribed taxa), this represents a major fraction of global diversity in this clade. All but one of these species (*Phyllanthus littorale*, syn. *Glochidion littorale*) are endemic to this region, and nearly all are single-archipelago endemics. Those species reported from more than one archipelago have not been recently revised based on morphological or molecular data. *Glochidion* have notably radiated in the Fijian archipelago (18 spp.) and the Society Islands (14 spp.); based on current taxonomy, other archipelagos with notable diversity include the Caroline Islands (9 spp.) and the Cook-Austral Islands (≥6 spp.). The relatively large (by surface area) Samoan (3 spp.) and Marquesas (2 spp.) archipelagos show less diversity than the Societies or Fiji.

12.3.1 Micronesia: Caroline Islands (9 spp.)

The Caroline Islands, comprising volcanic and limestone islands as well as atolls, are spread across almost 3000 km of the west Pacific Ocean, east of the Philippines and north of New Guinea. They are politically divided between the Republic of Palau (Palau; also spelled Belau) and the Federated States of Micronesia (often abbreviated FSM). The FSM comprise four main island groups: Yap, Chuuk, Pohnpei, and Kosrae.

The islands making up Palau have a complex geology, including atolls, volcanic islands, and uplifted limestone islands. They are the closest to Asia and Australasia of any of the island groups considered here, being only 900 km from Mindanao (Philippines), Maluku (Indonesia), and New Guinea. The largest island in this archipelago is Babeldaob, with an area of 331 km^2 and reaching a maximum elevation of 224 m. Two *Glochidion* species, *P. kanehirae* (syn. *G. kanehirae*; also found in the FSM) and *P. otobedii* (syn. *G. palauense*; restricted to Palau) are found on multiple islands (including Babeldaob) in this archipelago (Wagner et al. 2012). The single-island endemic *P. macrosepalus* (syn. *G. macrosepalum*) is found only on Babeldaob (Wagner et al. 2012). In addition, *P. littoralis* (syn. *G. littorale*), a distinctive coastal species with coriaceous leaves and enlarged, bright red fruit, is known from the island of Oreor (Koror; Wagner et al. 2012). This latter species is also found in coastal (including mangrove) and riparian areas in South and Southeast Asia (Govaerts et al. 2000, Nguyen 2007, van Welzen and Chayamarit 2015, Hembry, unpublished data).

Pohnpei (formerly Ponape), located 2600 km east of Palau and 1500 km north of the Solomon Islands, in the FSM, is the tallest island in Micronesia (772 m elevation) as well as one of the largest (344 km^2). Five different species of *Glochidion* are known from Pohnpei. These are the four Pohnpei-endemic species *P. cleistanthoides* (syn. *G. cleistanthoides*), *P. hosokawae* (syn. *G. hosokawae*), *P. ponapense* (syn. *G. ponapense*), and *P. websteri* (syn. *G. websteri*), and the more widely distributed *P. senyavinianus* (syn. *G. puberulum*, *G. senyavinianum*; also found in the Chuuk and Hall Islands; Hosokawa 1935, Wagner et al. 2012). However, *P. hosokawae* and *P. websteri* are only known from the type collections (Wagner and Lorence 2011). Pohnpei is the only volcanic high island in the Senyavin Island group, the remainder of which are atolls and lack *Glochidion* (Wagner et al. 2012).

Chuuk (formerly Truk) State consists of several island groups in the central FSM, 1900 km east of Palau and 700 km west of Pohnpei. Economically, Chuuk State is dominated by the Chuuk Islands, a group of volcanic islands enclosed within a single atoll lagoon (Chuuk Lagoon), geologically analogous to the Gambier Islands in French Polynesia (below). These islands have the nonendemic *Glochidion* species *P. kanehirae* and *P. senyavinianus* (Wagner et al. 2012). *P. senyavinianus* is also reported from Nomwin Island, an atoll 82 km north of Chuuk in the Hall Islands (also part of Chuuk State; Wagner et al. 2012). Finally, *P. kanehirae* is the only *Glochidion* reported from Yap, a group of islands made of uplifted continental crust 450 km northeast of Palau (Wagner et al. 2012). Interestingly, Kosrae (formerly Kusaie), despite being an apparently suitable volcanic island (110 km^2, 634 m elevation) and no more remote than any other major island in Micronesia, has no reported *Glochidion* (Wagner et al. 2012).

12.3.2 Micronesia: Mariana Islands (1 sp.)

The Mariana Islands are a linear archipelago 700 km in length with a north–south orientation at the boundary of the Pacific and Philippine Plates, 850 km northeast of Yap, 1000 km northwest of Chuuk, and 550 km south of the Volcano Islands (e.g., Iwo Jima, Japan). The southern islands are primarily uplifted limestone islands and the northern ones primarily volcanic. The southernmost island, Guam, has both volcanic and limestone components to its geology and politically comprises the US territory of Guam. The islands to the north of Guam form a different US territory, the Commonwealth of the Northern Mariana Islands. The only *Glochidion* reported from the Marianas is *Phyllanthus mariannensis* (syn. *Glochidion marianum*), and it is only known from Guam (Wagner et al. 2012), where it grows commonly on both limestone and basaltic substrates, including apparently as an early successional shrub or small tree in old fields (Stone 1970). Interestingly, no *Glochidion* are reported from the Northern Mariana Islands (Fosberg et al. 1975, Wagner et al. 2012).

12.3.3 Fiji (18 spp.)

Fiji, politically the Republic of Fiji, is the largest (in terms of total area), most populous, and oldest of the island groups under consideration here. Its geology is considerably more complex than that of these other island groups. Fiji is dominated by two large islands, Viti Levu (10,388 km^2, maximum elevation 1324 m) and Vanua Levu (5587 km^2, maximum elevation 1111 m). These islands are surrounded by hundreds of other islands. The oldest dated rock formations in Fiji are from the late Eocene to early Oligocene on Viti Levu. Although many rock formations in Fiji are of volcanic origin, the geology of the archipelago is far more complex than that of any other island group considered here, and the reader is referred to the reviews by Rodda (1994) and Colley (2009).

Fiji has 18 species of *Glochidion*, of which apparently 17 are endemic (Smith 1981). This is a greater diversity of *Glochidion* than on any oceanic archipelago in Micronesia or Polynesia, and would be expected given the large area and great age of this archipelago. Most of these species are found on the largest islands of Viti Levu, Vanua Levu, and Taveuni, as well on as the surrounding islands. Although some of these species are extremely common and widespread on the main islands (e.g., *Phyllanthus cordatus* syn. *G. cordatum*, and *P. concolor* syn. *G. concolor*; Smith 1981, Hembry, unpublished data), three are only known from the type collections (*P. euryoides* syn. *G. euryoides*, *P. podocarpus* syn. *G. podocarpum*, and *P. inusitatus* syn. *G. inusitatum*). In contrast to the many species found on the main and surrounding islands, *P. calciphilus* (syn. *G. calciphilum*), a shrub or gnarled tree reaching 1 m in height, appears to be endemic to the Lau Group in southeastern Fiji (270 km southeast of Viti Levu), where it is known from two limestone islands (Fulaga and Kabara). The only *Glochidion* species native but not endemic to Fiji is *P. concolor*, which Smith (1981), after examining a large number of specimens from elsewhere in the Pacific, considered also to be present in Tonga; Wagner and Lorence (2011) ascribe *Glochidion* from Rarotonga and Niue to this species, although molecular data indicate that which has been ascribed to *P. concolor* aff. in Rarotonga is distantly related to *P. concolor* in Fiji and should not be considered the same taxon (Hembry et al. 2013a).

Molecular phylogenetic analysis (Kawakita et al. 2004, Hembry et al. 2013a) indicates that *Glochidion* in Fiji fall into at least two separate clades (Fig. 12.5). One of these clades may be closely related to species from Australia and New Caledonia; the other is closely related to *P. samoanus* from American Samoa, *P. wilderi* (syn. *G. wilderi*) from Mangareva (Gambier Islands), and undescribed *Glochidion* from Wallis. Whether this indicates two separate colonizations of Fiji by *Glochidion*, or a role for Fiji as a source area for *Glochidion* found further east in Polynesia, is unclear.

12.3.4 Western Polynesia: Rotuma (1 sp.)

Rotuma, 450 km north of the main Fijian archipelago, consists of about 10 islands of volcanic origin that are politically part of the Republic of Fiji. The largest is Rotuma, 44 km^2 in area and reaching a maximum elevation of 256 m. McClatchey et al. (2000) report *Glochidion* from these islands, which they assign to *Glochidion ramiflorum*. This plant may be closely related to *P. samoanus* from Samoa or *P. concolor* from Fiji.

12.3.5 Western Polynesia: Wallis and Futuna (1 sp.)

Wallis and Futuna is a French overseas collectivity consisting of the volcanic islands of Wallis (‘Uvea; 300 km northeast of Vanua Levu), and 260 km distant, the islands of Futuna and Alofi, 370 km west of Savai‘i. St. John and Smith (1971), Morat and Veillon (1985), and Meyer (2007b) report *Glochidion* (ascribed, presumably incorrectly, to *G. ramiflorum*) from several sites in primary and secondary inland and coastal forest, as well as in *Dicranopteris* savanna (*toafa* in Wallesian), on Wallis and Futuna. Molecular phylogenetic analysis indicates *Glochidion* from Wallis are closely related to *P. samoanus* from American Samoa and *P. wilderi* from Mangareva (Hembry et al. 2013a; Fig. 12.5). Wagner and Lorence (2011) list no species of any *Phyllanthus* s. l. from these islands.

12.3.6 Western Polynesia: Samoa (3 spp.)

The Samoan archipelago is politically divided between the Independent State of Sāmoa (also known as Western Samoa; hereafter, “Independent Samoa”) and the US territory of American Sāmoa. Independent Samoa is dominated by the largest (and oldest) islands in the archipelago, ‘Upolu and Savai‘i, whereas American Sāmoa contains the smaller and younger islands of Tutuila, ‘Aunu‘u, Ofu, Olosega, and Ta‘ū (the last three comprise, collectively, the Manu‘a Islands). ‘Upolu has been dated to 2.3 Ma, whereas Tutuila has been dated to 1.2 Ma, and the Manu‘a Islands to 0 Ma (Clouard and Bonneville 2005). Given the age progression, Savai‘i is presumably the oldest island, but is also geologically active, with contemporaneous eruptions and lava flows.

The Samoan archipelago has three described species of *Glochidion*, all of which are endemic to the archipelago. These are *Phyllanthus samoanus* (no valid name in *Glochidion* exists, but these have traditionally and incorrectly been referred to *Glochidion ramiflorum*; Figs. 12.3a and 12.4a), known from primary and secondary forest and pastures from sea level to 1400 m elevation on all the main Samoan islands; *P. cuspidatus* (syn. *G. cuspidatum*; Fig. 12.3b), known from secondary forest, ridges, and clearings at 300–750 m elevation on Savai‘i, ‘Upolu, and Tutuila; and *P. christophersenii* (syn. *G. christophersenii*), known only from cloud forests at 1000–1550 m on Savai‘i (Whistler 2004; Wagner and Lorence 2011). *P. samoanus* are reported from early succession vegetation from lava flows on Savai‘i (Whistler 2002). *P. cuspidatus* and *P. samoanus* can be found in close proximity along the ridgeline of Mount ‘Alava on Tutuila, within the National Park of American Sāmoa. Compared to most coexisting species pairs of *Glochidion* in the Pacific, *P. samoanus* and *P. cuspidatus* are quite distantly related (Fig. 12.5). *P. samoanus* is more closely related to *Glochidion* from Fiji, Wallis, and the Gambier Islands, whereas *P. cuspidatus* is more closely related to the radiation of Southeastern Polynesian *Glochidion* (Hembry et al. 2013a).

12.3.7 Western Polynesia: Tonga (≥1 sp.)

Tonga is an archipelago spread across 650 km of ocean, east of Fiji and south of Samoa, consisting of a mixture of limestone and volcanic islands located along the subduction zone at the boundary of the Indo-Australian and Pacific plates. *Glochidion* in Tonga have not been revised. Those on ‘Eua (Drake et al. 1996), Vava‘u and Tongatapu (Burkill and Crosby 1901), Niuatoputapu (St. John 1977), and Niuafo‘ou (Uhe 1974) have been ascribed to *G. ramiflorum* or its synonyms. Smith (1981) was of the opinion that several species of *Glochidion* were present on Tonga, and ascribed some specimens from ‘Eua to *P. concolor* (syn. *G. concolor*, also known from Fiji). *P. concolor* is the only *Glochidion* listed for Tonga by Govaerts et al. (2000).

12.3.8 Western Polynesia: Niue (1 sp.)

Niue is a flat, makatea (limestone karst) island located 270 km from Tonga and 1100 km from Rarotonga (Cook Islands), making it the easternmost island in Western Polynesia. Politically, it is a self-governing state in free association with New Zealand. There is only one species of *Glochidion* reported from Niue, which Sykes (1970) assigned to *G. ramiflorum*. Wagner and Lorence (2011) suggest that these may actually be *P. concolor*. Sykes (1970) reported that this *Glochidion* was common in secondary forest and on the Upper Terrace (or Mutalau Reef) of the island.

12.3.9 Southeastern Polynesia: Cook and Austral Islands (≥6 spp.)

The southern Cook and Austral Islands are two political units considered geologically to form part of the same complex archipelago spanning 2000 km. Politically, the Cook Islands are an independent country in free association with New Zealand (similar to Niue), whereas the Austral Islands are part of the French overseas territory (“overseas country,” *pays d’outre-mer*) of French Polynesia. Geologically, the Cook and Austral islands were formed as a result of several hotspots under the Pacific Plate. How many hotspots were responsible, and which islands were formed by which hotspots, is contentious and not reviewed here (Clouard and Bonneville 2005; Maury et al. 2014).

Independently of their geologic history, the Cooks and Australs are best thought of as comprising three units. The first, the southern Cook Islands, consists of a group of uplifted makatea (karst) islands (Mangaia, Ātiu, Ma‘uke, and Miti‘āro), one almost-atoll (‘Aitutaki), and one reef-platform island (Takūtea) within 300 km of the geologically young (1.4 Ma; Clouard and Bonneville 2005) volcanic island of Rarotonga. The Southern Cooks also include two atolls (Palmerston and Manuae). *Glochidion* are found on Rarotonga, Mangaia, Ātiu, Ma'uke, and Miti'āro, but are interestingly apparently absent from 'Aitutaki (McCormack 2007). The *Glochidion* from Rarotonga have been tentatively assigned to *Phyllanthus concolor* (syn. *G. concolor*), a species from Fiji and Tonga (McCormack 2007; Wagner and Lorence 2011), but molecular phylogenetic analysis suggests that these two taxa are distantly related and these Rarotonga *Glochidion* merit a different name (Hembry et al. 2013a; Fig. 12.5). The *Glochidion* from the Ngapūtoru group (Ātiu, Ma'uke, and Miti'āro) and Mangaia have been tentatively assigned to *P. taitensis* (syn. *G. taitense*), a species from the Society Islands (McCormack 2007; Fig. 12.3c–d). There has been some uncertainty as to whether there are two coexisting species of *Glochidion* on Rarotonga and/or Mangaia, which remains unresolved (W. Sykes, G. McCormack, pers. comm., 2008). Differences in female flower morphology between trees from Mangaia (Fig. 12.3c), Ātiu (Fig. 12.3d), and *P. taitensis* from the Society Islands (Fig. 12.3o; Hembry, unpublished) suggests that the diversity of Cook Islands *Glochidion* has been underestimated. Rarotonga has some of the most intact native forests in Polynesia, and *Glochidion* is infrequent in wet forests on this island (Hembry, unpublished notes). *Glochidion* is locally common on makatea substrate on both Ātiu and Mangaia (Hembry, unpublished notes). Finally, *Glochidion* is also present on volcanic-origin soils in the center of Mangaia, growing in *Dicranopteris* fernlands that have been replanted in the late twentieth century by exotic trees such as *Pinus* and *Acacia* (Hembry, unpublished notes).

The second unit of the Cook–Austral chain is the northern Austral Islands, consisting, from west to east, of Maria, Rimatara, Rurutu, Tubuai, and Raivavae. Maria is an atoll that lacks *Glochidion* (Meyer 2013). Rimatara and Rurutu are both high islands that have recently experienced some secondary uplift (Maury et al. 2014), and thus, like Mangaia, possess both makatea substrate and soils of volcanic origin. Both Tubuai and Raivavae are old (9.1 and 6.3 Ma, respectively) and highly eroded high islands (Maury et al. 2014). Florence (1997a) ascribed Rimatara *Glochidion* to *P. florencei* (syn. *G. societatis*), a species common in the Leeward Society Islands; however, morphological differences in the female flowers are evident (Hembry, unpublished) suggesting that the Rimatara population may constitute a distinct taxon. On Rimatara this taxon is extremely rare, but grows on both makatea and volcanic-origin soils. The single species of *Glochidion* described from Rurutu, Tubuai, and Raivavae is the archipelago-endemic *P. raivavense* (*G. raivavense*; Fig. 12.3e–f). On all three islands—whose vegetation is considerably degraded (Meyer et al. 2014)—*P. raivavense* is found on volcanic soils; it is locally common on Rurutu in partially degraded, secondary vegetation, less common in secondary forests on Raivavae, and extremely rare on Tubuai. Considerable differences in stylar column morphology between the Raivavae specimens (Fig. 12.3f) and those from Rurutu and Tubuai (Fig. 12.3e; Hembry, unpublished) suggest the latter may merit recognition as a distinct taxon.

The third unit consists of the remote islands of Rapa (Rapa Iti) and Marotiri, the southernmost islands in Southeastern Polynesia. *Glochidion* are only known from Rapa, a circular volcanic island with near-temperate climate 650 m in elevation, 40 km^2 in area, and 4.6 million years old (Clouard and Bonneville 2005). (Marotiri is a cluster of islets and rocks, with a highly impoverished flora; J.-Y. Meyer, pers. comm., 2008.) Two *Glochidion* are described from Rapa: *P. longfieldiae* (*G. longfieldiae*; Figs. 12.2a, 12.3s and 12.4h), which is common in *Dicranopteris* savanna, open areas, and native forest from near sea level up to at least 500 m elevation (D. Hembry, unpublished notes), and *P. rapaense* (*G. rapaense*; Fig. 12.3t), which is more restricted in its distribution and found primarily at low elevations. A third taxon has been recognized in herbarium collections, which Jacques Florence omitted from his *Flore de la Polynésie française* (1997a) with the intention that the late Grady Webster describe it as "*Glochidion fosbergii*", the name by which it is labeled on some herbarium sheets (J. Florence, pers. comm., 2006). This taxon remains undescribed. Molecular phylogenetic analysis indicates that these three taxa, which are endemic to the island, form a monophyletic group to the exclusion of other *Glochidion* (Hembry et al. 2013a; Fig. 12.5). Fieldwork on Rapa in 2008 and examination of herbarium specimen labels at PAP and BISH suggest that these species are usually found singly, but one locality where all three can be found within 500 m of each other is known (D. Hembry, unpublished data).

12.3.10 Southeastern Polynesia: Society Islands (14 spp.)

The Society Islands are a volcanic hotspot archipelago 680 km east of the Cook Islands and 550 km north of the northern Australs. The high islands of the archipelago comprise, in descending age order, Maupiti, Bora Bora (Porapora), Tahaa (Taha'a), Raiatea (Ra'iatea), Huahine, Maiao (Mai'ao), Moorea (Mo'orea), Tahiti, and Mehetia (Me'eti'a). (Names in parentheses indicate Tahitian orthography, where they differ from English.) Maupiti is the oldest at 4.6 Ma, and Mehetia is the only geologically active subaerial volcano in the chain (Clouard and Bonneville 2005). The islands vary considerably among each other in maximum elevation, size, and topographic complexity; Tahiti, at 1045 km^2 area and 2242 m maximum elevation (Mt. Orohena) is by far the largest island in Southeastern Polynesia, and the tallest island anywhere in the South Pacific outside of New Zealand and New Guinea.

The Society Islands have 14 species of described *Glochidion* (Figs. 12.2b–d, 12.2f, 12.3g–p and 12.4c–g). Under the current taxonomy, all are endemic to the archipelago, with the exception of *Phyllanthus florencei* (syn. *G. societatis*), to which *Glochidion* from Rimatara in the northern Australs are assigned by Florence (1997a; see *Cook and Austral Islands*, above). Maupiti and Bora Bora are both small old islands with a single species of *Glochidion* each, *P. florencei* and *P. st-johnii* (syn. *G. myrtifolium*), respectively. Neither of these species are single-island endemics and both are relatively common in secondary forest vegetation. The larger islands of Tahaa, Raiatea, Huahine, Moorea, and Tahiti each have multiple species of *Glochidion*, with some habitat diversity seen among species within the same island (see Fig. 12.4).

12.3.11 Southeastern Polynesia: Tuamotu Islands (2 spp.)

The Tuamotus are a chain of atolls located in the center of Southeastern Polynesia. The volcanic cores on which these atolls rest probably date back tens of millions of years (see dates in Clouard and Bonneville 2005) and may represent the remnants of an ancient vast archipelago in the central Pacific that played a role in late Tertiary Pacific biogeography (Hembry and Balukjian 2016). Presently nearly all of the atolls are too ephemeral and too low in elevation (1–4 m above sea level) to support a high island biota, including *Glochidion*. Several of the western Tuamotus have been secondarily uplifted by lithospheric flexure from Tahiti (McNutt and Menard 1978), and consequently are able to support elements of a high island flora, including *Glochidion*. These are Makatea (*G. wilderi*; syn. *Phyllanthus wilderi*), Niau (*P. tuamotuensis*; syn. *G. tuamotuense*; Fig. 12.3q), and Anaa (*P. wilderi*). The taxonomy of these species is unusual; both are also reported from the Gambier Islands, ≥1200 km distant (Florence 1997a), and these relationships have yet to be tested with molecular data (Hembry et al. 2013a). On Niau, which has some of the most ecologically intact forests anywhere in southeastern Polynesia, 7–8 m above sea level, *P. tuamotuensis* is a common forest canopy tree and also a small tree of forest margins, on makatea substrate (D. Hembry, unpublished notes). *P. wilderi* is found in primary and degraded forest on the plateau of Makatea (Florence 1997a) and was also reported recently from several calcareous islets (*motu* in Tahitian) on Anaa, 4–5 m above sea level (Butaud and Jacq 2009).

12.3.12 Southeastern Polynesia: Gambier Islands (2 spp.)

The Gambier Islands are a cluster of islets and one central island, Mangareva, in the same lagoon, at the southeastern end of the Tuamotu Archipelago. They thus should perhaps be thought of as the last high islands of the ancient Tuamotu chain. They form an alignment with some of the southern Tuamotus (Hereheretue and Moruroa) as well as with the Pitcairn Islands to the east (Clouard and Bonneville 2005). The Gambiers have suffered extensive human impacts; the islands are covered today in vegetation consisting almost entirely of forest plantations and invasive species, and what remains of the native flora consists of isolated shrubs in a sea of invasives (Kirch 2004, Butaud 2009). In the twentieth century, *Glochidion* was reported from both Mangareva and the adjacent islet of Taravai (Florence 1997a). These are ascribed to *Phyllanthus tuamotuensis* (syn. *G. tuamotuense*) from Taravai (Florence 1997a) and possibly also one site on Mangareva (W. Teamotuaitau, pers. comm.,

2008), and *P. wilderi* (syn. *G. wilderi*) from Mangareva (Florence 1997a). As stated above, these species are also reported from the Tuamotus (Florence 1997a), and Hembry et al. (2013a) were unable to obtain samples of *P. wilderi* from the Tuamotus or *P. tuamotuensis* from the Gambiers to test these taxonomic hypotheses with molecular data. As of 2008, *P. wilderi* was extremely rare as saplings in forests dominated by invasive species near where it had been previously reported by Florence (1997a) and was found as a small tree on cliffs (Hembry 2013; J.-F. Butaud, pers. comm., 2008; Fig. 12.4b). To the best of my knowledge, no one has been able to confirm the continued presence of *Glochidion* on Taravai since Florence's collections in 1994 (Florence 1997a). The Gambier *P. wilderi* represents a separate colonization of Southeastern Polynesia from the rest of the *Glochidion* in this region, and is more closely related to *Glochidion* from Samoa, Wallis, and Fiji (Hembry et al. 2013a; Fig. 12.5). It thus represents an evolutionarily unique lineage from a conservation perspective.

12.3.13 Southeastern Polynesia: Pitcairn Islands (2 spp.)

The Pitcairn Islands are two high islands (Pitcairn and Henderson) and two atolls (Oeno and Dulcie) that are part of the same alignment as some of the southern Tuamotus (Hereheretue, Moruroa) and the Gambiers (Clouard and Bonneville 2005). Located 400 km east of the Gambiers, the Pitcairns are politically not part of French Polynesia, but instead an overseas territory of the United Kingdom. Pitcairn is a volcanic island, dated to 0.6 Ma and reaching a maximum elevation of 347 m (Clouard and Bonneville 2005). Pitcairn was inhabited and then abandoned by Polynesians before European arrival and is inhabited today by people of British and Polynesian heritage; its native vegetation is confined to small patches of forest and zones of *Dicranopteris* and *Nephrolepis* fernland (Waldren et al. 1995). Henderson is in contrast an uplifted atoll 30 m high, uninhabited by people today (although it was inhabited and then abandoned by Polynesians before European contact; Weisler 1995), and covered in far more intact native forest (Waldren et al. 1995). Two species of *Glochidion, Phyllanthus pitcairnense* (syn. *G. pitcairnense*) and *P. comitus* (syn. *G. comitum*) are known from Pitcairn, where both are found in grass/fern vegetation and forest edges (Florence et al. 1995; Waldren et al. 1995; Florence 1997b; Wagner and Lorence 2011). *P. pitcairnense* is also found on Henderson, where it is locally abundant in native forests dominated by *Pisonia grandis* (Nyctaginaceae; Florence et al. 1995, Waldren et al. 1995). Both species are endemic to the Pitcairn Islands.

12.3.14 Southeastern Polynesia: Marquesas Islands (2 spp.)

The Marquesas Islands are the most remote archipelago from continents in the world, 4800 km from Mexico and ~7000 km from Australia. They are volcanic high islands with basaltic soils and extremely limited coral reef formation. They form a linear age progression similar to the Austral or Society Islands, ranging over a period from 5.5 Ma (Eiao) to 1.5 Ma (Fatu Hiva; Clouard and Bonneville 2005). The habitat diversity of the Marquesas is of particular interest. In addition to tropical wet and mesic forests like much of the rest of Southeastern Polynesia, the Marquesas are tall enough (>800 m elev.) to have cloud forest (Meyer 2010). Unlike much of the rest of the tropical Pacific considered here, the Marquesas also have remnants of true dry forest, dominated by *Sapindus* and *Xylosma*, on low-elevation, leeward slopes of the islands (referred to as *fenua ataha* in Marquesan or *terres désertes* in French; Florence and Lorence 1997).

Phyllanthus marchionicus (syn. *G. marchionicum*) is a widely distributed, ecologically labile species found from 50 to 1130 m above sea level, in wet, mesic, cloud, and some areas of dry forest on Nuku Hiva, Ua Pou, Ua Huka, Hiva Oa, Tahuata, and Fatu Hiva (Florence 1997a, Butaud et al. 2008, Wagner and Lorence 2011, Hembry, unpublished observations; Figs. 12.2e and 12.3r). The taxon *Glochidion tooviianum*, endemic to the Toovii Plateau on Nuku Hiva and distinguished from *P. marchionicus* only by foliar and floral pubescence, was combined with the latter taxon by Wagner and Lorence (2011). The only other described species of *Glochidion* from the Marquesas is *P. hivaoaense* (*G. hivaoaense*), endemic to cloud forests between 700 and 1200 m elevation on Hiva Oa and Tahuata (Florence 1997a, Wagner and Lorence 2011). This species is morphologically very distinct from *P. marchionicus*, but is very similar to *P. taitensis* from Tahiti.

12.3.15 Places with no *Glochidion*

The widespread distribution of *Glochidion* in remote regions of the tropical Pacific is remarkable, given its apparent reliance on specialized pollination by *Epicephala* moths (Kato et al. 2003, Hembry et al. 2012) and the many plant and insect lineages that have failed to naturally colonize the eastern Pacific (particularly east of the Andesite Line; Gressitt 1956; Stoddart 1992; Weston and Crisp 1996). As stated above, *Glochidion* are absent from atolls that have not been uplifted, likely because these atolls have been too ephemeral during the Holocene, or because *Glochidion* may not be sufficiently salt-tolerant to grow there. Nevertheless, there are many high islands that do not appear to have ever had native *Glochidion*. These include some small high islands in southeastern Polynesia (Eiao, Hatutaa, and Mohotani in the Marquesas; Mehetia and Maiao in the Societies; Wagner and Lorence 2011, Meyer 2007a, Meyer et al. 2009), Kosrae (Wagner et al. 2012), the Northern

Mariana Islands (Wagner et al. 2012), the Hawaiian Islands (Wagner et al. 1990), Kermadec Islands (Sykes and Campbell 1977, Sykes and West 1996), and Easter Island (Rapa Nui; Etienne et al. 1982). The pollen record of Easter Island has received extensive attention and includes many Paleotropical elements, but no evidence of *Glochidion* has been reported (e.g., Flenley et al. 1991, Mann et al. 2008). Meyer (2007a) expressed surprise at not finding *Glochidion* on Maiao, only 78 km from Moorea and otherwise ecologically suitable and supposed that *Glochidion* was once present but was extirpated (along with much of the rest of the native flora) due to human activities.

12.4 Ecological Diversity and the Radiation of Southeastern Polynesian *Glochidion*

With 23 described and several undescribed taxa, *Glochidion* are actually one of the largest endemic plant radiations in Southeastern Polynesia, less species-rich than only *Myrsine* (Primulaceae; 27 spp.), *Cyrtandra* (Gesneriaceae; 27 spp.), and *Psychotria* (Rubiaceae; 24 spp.) (Meyer 2004). As might be expected as a result of this diversity, *Glochidion* show a range of variation in their habitats. Some taxa, including *Phyllanthus florencei* (syn. *Glochidion societatis*), *P. huahineense* (syn. *G. huahineense*), *P. tuamotuensis*, and *P. wilderi*, are found growing on calcareous substrate near sea level, including on calcareous islets (*motu* or cognates in Tahitian and many Polynesian languages) on Huahine and Maupiti (Figs. 12.2f and 12.4 g) and on makatea substrate on uplifted islands (Niau, Makatea, Anaa, Ātiu, Mangaia). The other species are found on substrates of basaltic origin; whether this apparent dichotomy is a result of strict habitat requirements or different presence of calcareous and basaltic substrates on different islands is unclear. *P. florencei* and *P. huahineense* are found on both calcareous and basaltic sites on the same islands (Hembry, unpublished data).

Within basaltic substrates, there is a range of both habitat and elevational variation among species. In common with much of the native flora, on most islands, *Glochidion* are more common at middle and high elevations (e.g., >300 m) than at low elevations. This is best ascribed to human-induced disturbance at low elevations, rather than strict habitat requirements. A few species appear primarily restricted to low elevations, primarily *P. florencei* in the Leeward Society Islands (Maupiti, Raiatea, and Huahine) and *P. rapaense* and *P.* sp. undescribed ("*fosbergii*") on Rapa. A few taxa appear to be restricted to very high elevations and/or specialized montane vegetation communities; these include *P. hivaoaense* (restricted to cloud forests 700–1200 m elevation on Hiva Oa and Tahuata), *P. orohenense* (syn. *G. orohenense*; Fig. 12.2d; restricted to cloud forests 900–1750 m on Tahiti), and *P. grantii* (syn. *G. grantii*; restricted to montane plateaus and ridge forests 435–730 m on Raiatea and Tahaa; Florence 1997a, Wagner and Lorence 2011). Other species are extremely localized in distribution and habitat requirements: *P. emarginatus* (syn. *G. emarginatum*; Figs. 12.2c and 12.4c) is endemic to the windswept scrublands of Te Mehani Rahi Plateau and Mt. Toomaru on Raiatea (580–930 m elev.), *P. raiateaensis* (syn. *G. moorei*) is endemic to Te Mehani Plateau on Raiatea, *P. nadeaudii* (syn. *G. nadeaudii*; Figs. 12.2b and 12.4d) is endemic to particular mountains on Moorea (420–900 m elev.), and *P. papenooense* (syn. *G. papenooense*; Fig. 12.4f) is restricted to wetlands and riparian habitats at a few localities on Tahiti (Florence 1997a, Hembry, unpublished data, J.-Y. Meyer, pers. comm., 2007). In contrast, a few species are clearly habitat generalists: *P. taitensis* (syn. *G. taitense*; Fig. 12.4e) is widespread on Tahiti from 50–1500 m elevation in a range of mesic, wet, and cloud forests in various stages of perturbation; *P. florencei* is found on both basaltic and calcareous substrates on Maupiti and Huahine; and *P. marchionicus* (Fig. 12.2e) is widely distributed in wet, mesic, cloud, and dry forest environments in the Marquesas, particularly Nuku Hiva (Florence 1997a, Hembry, unpublished notes).

Nearly all of these species can be found either as shrubs or trees and these growth forms are probably best thought of as plastic in these species, but *P. emarginatus*, in its windswept habitat on the Te Mehani Rahi Plateau on Raiatea, grows as a prostrate shrub often <1 m high and with creeping branches up to >1 m long (Figs. 12.2c and 12.4c; Hembry, unpublished). Florence (1997a) reports that *P. manono* (syn. *G. manono*) and *P. marchionicus* can reach heights up to ≥ 10 m; on Moorea and Niau, respectively, *P. nadeaudii* and *P. tuamotuensis* of similar heights are important components of the native forest overstory (Hembry, unpublished notes). Smith (1981) reports that in Fiji *P. seemannii* and *P. concolor* may rarely reach 20 and 25 m in height, respectively. Many of these *Glochidion* can be reproductive at <1 m tall (Hembry, unpublished notes), but these individuals may actually be very old ones that have resprouted repeatedly from older rootstock.

Within some of the larger Society Islands, some of the *Glochidion* species appear to have differing habitat requirements, although zones of coexistence are known to occur. For instance, on Tahiti, *P. taitensis* (Figs. 12.3o and 12.4e) is ecologically widespread, *P. manono* is found primarily in drier (e.g., mesic) environments below 1000 m elevation, *P. grayanus* (syn. *G. grayanum*) is found primarily in wet environments between 60 and 1040 m elevation, *P. orohenense* (Figs. 12.2d and 12.3p) is primarily restricted to high-elevation cloud forest, and *P. papenooense* (Fig. 12.4f) is always found in association with wetlands or flowing water (Florence 1997a, Hembry, unpublished notes). On neighboring Moorea, *P. manono* (Fig. 12.3n) is widespread below 600 m, *P. nadeaudii* (Figs. 12.2b, 12.3m, and 12.4d) is found on certain mountains between 420 and 900 m (contact zones exist), and *P. st-johnii* (syn. *G. myrtifolium*) is restricted to a few sites in mesic or wet forest (Florence 1997a, Hembry, unpublished notes). On Huahine, *P. florencei* and *P. huahineense* (Figs. 12.2f and 12.3j–k) are found below 420 m, and *P. temehaniensis* (syn. *G. temehaniense*; Fig. 12.3l) above 300 m (Hembry, unpublished). On Raiatea, *P. florencei* and *P. st-johnii* (Fig. 12.3i) are primarily found below 400 m, *P. emarginatus* (Figs. 12.2c, 12.3h, and 12.4c), *P. grantii*, *P. raiateaensis*, and *P. temehaniensis* (Fig. 12.3g) are restricted to the two Te Mehani plateaus and other montane environments between 470 and 930 m, and

P. brothersonii (syn. *G. brothersonii*) is known only from two low-elevation sites (Florence 1997a, Hembry, unpublished). Published specimen locality data indicate that similar patterns may exist on Tahaa and Savai‘i (Florence 1997a, Wagner and Lorence 2011). This habitat diversity indicates that the diversification of *Glochidion* in the Society Islands (and possibly elsewhere in the Pacific) has likely occurred at least in part along ecological axes. *Glochidion* are interestingly absent from elevations above 900 m on Moorea (they are not known from the summit of Mt. Tohiea, 1207 m elevation; Hembry, unpublished, J.-Y. Meyer, pers. comm., 2014) and above 1750 m on Tahiti (Florence 1997a), indicating that the radiation of *Glochidion* has not been able to establish in all environments on these islands.

Some *Glochidion* show evidence of resilience to disturbance, as is known elsewhere in the world (e.g., Asia; Kawakita 2010). *Glochidion* commonly grow on lava flows 100–250 years in age on Savai‘i (Whistler 2002) and 20–54 years in age on Niuafo‘ou in Tonga (Uhe 1974), and resprout in areas burned by wildfire on Moorea (Hembry, unpublished). *Glochidion* thus fit with the island biology syndrome of early-successional or "weedy" plants being successful colonizers of oceanic islands (Carlquist 1974). How human-induced disturbances following Polynesian colonization may have influenced *Glochidion* is unclear, but they do appear in the prehuman (e.g., >1000 year. BP) Holocene pollen records on Rimatara (Prebble and Wilmshurst 2009), Rapa (Kennett et al. 2006, Prebble 2014), and Moorea (Kahn et al. 2015), and in the posthuman record on Rimatara (Prebble and Wilmshurst 2009). On Moorea, before human arrival, they may have been a dominant component of coastal and swamp vegetation along with *Ficus*, *Pandanus*, *Hibiscus tiliaceus*, and other strand trees (Kahn et al. 2015), perhaps analogous to coastal wetland vegetation on Huahine today.

12.5 Diversity of Other Phyllantheae in the Pacific

Other Phyllantheae in addition to *Glochidion* are not particularly diverse on oceanic islands in the Pacific, in contrast to the spectacular radiation of *Gomphidium* (*Phyllanthus*) in Melanesia, particularly New Caledonia (Webster 1986, Schmid 1991). Palau has two species, *Phyllanthus palauensis* on volcanic soils in savannas and wet forests, and *P. rupiinsularis* on limestone soils near the coast (Wagner and Lorence 2011). On many islands in the Marianas, the shrubs *P. marianus* and *P. saffordii* grow on limestone outcroppings and in savannas, respectively; *P. marianus* is also found on Ulithi Atoll near Yap in the FSM (Stone 1970; Wagner and Lorence 2011). Fiji has four endemic shrubby *Phyllanthus* (*P. pergracilis*, *P. heterodoxus*, *P. wilkesianus*, and *P. smithianus*; Smith 1981, Webster 1986). Tonga has the endemic shrub *P. amicorum* on ‘Eua (Webster 1986). *P. societatis* has a disjunct distribution on coralline substrates on Makatea, the Cooks (‘Aitutaki, Ātiu, Ma‘uke, and Miti‘āro), and Nauru in Micronesia (Florence 1997a; Wagner and Lorence 2011). The subshrub *P. pinaiensis* (Florence 1997a refers to this species as *P. urceolatus*) is endemic to the Society Islands, apparently extirpated on Tahiti but present at a few mesic forest sites on Moorea and Raiatea (Florence 1997a). *P. aoraiensis* is an extinct Tahiti endemic shrub, last collected in 1857 (Florence 1997a). *P. pacificus* is a relatively common shrub in the Marquesas and has a montane prostrate form on Fatu Hiva that is considered conspecific (Florence 1997a, Wagner and Lorence 2011). The Hawaiian Islands have two native (and endemic) Phyllantheae, each found on several of the major islands: the endangered dry/mesic forest tree *Flueggea neowawraea*, and the scandent shrub *P. distichus* (Wagner et al. 1990).

The pantropical weeds *P. amarus*, *P. debilis*, *P. tennellus*, *P. urinaria*, and *P. niruri* are introduced and common at low elevations in the Pacific (Smith 1981, Florence 1997a). Smith (1981) and Florence (1997a) considered *P. virgatus* to be an Austronesian (e.g., Polynesian) introduction to the remote Pacific for medicinal use. *Flueggea virosa* is a tree widely distributed from the Philippines and Moluccas into Melanesia, Fiji, and western Polynesia; it is considered by many (Smith 1981, Whistler 2004) to be an Austronesian introduction to Fiji, Rotuma, Tonga, and/or Samoa because of its traditional uses and frequent occurrence near human settlement. *Breynia* is not native anywhere in Micronesia, Fiji, or Polynesia, but *Breynia disticha* cv. "Roseo-picta" is cultivated widely in Fiji (Smith 1981) and French Polynesia (J.-F. Butaud, pers. comm., 2015).

12.6 Diversity of *Epicephala* in the Pacific

Before recent work (Kawakita et al. 2004, Hembry et al. 2012, 2013a), *Epicephala* had been reported from very few islands in the Pacific. Meyrick described *E. acrocarpa* from 'Upolu in Samoa (Meyrick 1927) and also ascribed to Australian *E. colymbetella* moths that were collected in the Marquesas Islands (Meyrick 1928). Clarke (1986) collected more specimens of *Epicephala* from Nuku Hiva and Fatu Hiva and described these, along with Meyrick's 1928 material, as *E. spinula*, endemic to the Marquesas. Clarke also collected (but did not describe) *Epicephala* from Pohnpei, which are held in the Smithsonian Institution. Several undescribed *Epicephala* specimens from the Society Islands are held in the Bernice P. Bishop Museum in Honolulu.

Recent rearing surveys targeting *Epicephala* have revealed them to be present on nearly all species of *Glochidion* for which they have been examined in Southeastern Polynesia (the few exceptions are cases where the *Glochidion* species are rare and only 0–2 fruit were collected in the field; Hembry et al. 2012; Fig. 12.6). These surveys have also yielded *Epicephala* associated with multiple species of *Glochidion* in Fiji (Kawakita et al. 2004, Hembry et al. 2013a) and *P. samoanus* and *P. cuspidatus* in American Samoa (Hembry et al. 2013a). No *Epicephala* have been found associated with several species of *Phyllanthus* not in *Glochidion* (*P. pinaiensis* on Moorea, *P. pacificus* in the Marquesas, or *P. societatis* in the Cooks; Hembry et al. 2013b). Morphological (Hembry et al. 2012) and molecular phylogenetic (Hembry et al. 2013a) analysis indicates that in southeastern

Fig. 12.6 *Glochidion–Epicephala* interactions on Pacific islands. (**a**) *Epicephala* moth visiting male *P. grayanus* (syn. *G. grayanum*) flower, Tahiti, Society Islands. (**b**) *Epicephala* inserting proboscis into female *P. grayanus* flower, Tahiti. (**c**) *Epicephala* inserting ovipositor into female *P. grayanus* flower, Tahiti. (**d**) *Epicephala* inserting ovipositor into female *P. temehaniensis* (syn. *G. temehaniense*) flower, Raiatea, Society Islands. (**e**) Pollen on the proboscis of an *Epicephala* captured after being observed pollinating and ovipositing into flowers of *P. temehaniensis*, Raiatea. (**f**) Pollen on the proboscis of a female *Epicephala spinula* collected in 1968 on Nuku Hiva, Marquesas Islands by J. F. G. Clarke (1986). (**g**) Pollen on the proboscis of a female *Epicephala* collected in 1953 on Pohnpei, Caroline Islands by J. F. G. Clarke. (**h**) *Epicephala* larva emerged from *P. grayanus* fruit and spinning a cocoon, Tahiti. (**i**) Dehisced fruit of *P. manono* (syn. *G. manono*) showing intact seeds with red arils (*red*, *left*) and seeds consumed by *Epicephala* larva (*right*, *brown*), Moorea, Society Islands (Reprinted with permission of the Royal Society from Hembry et al. 2012. Photos (**d**) and (**h**) courtesy of Tomoko Okamoto (Gifu University))

Polynesia, these *Epicephala* constitute a number of undescribed species, most of which are endemic to single islands or single archipelagos.

In southeastern Polynesia, *Epicephala* have been observed visiting male flowers, actively pollinating female flowers, and ovipositing into female flowers (Hembry et al. 2012; Fig. 12.6a–d) in the same manner as has previously been reported from Asia (Kato et al. 2003, Kawakita and Kato 2006). Female individuals of all minimally monophyletic *Epicephala* clades (tentative species) bear pollen-carrying hairs on the proboscis (Hembry et al. 2013a), like pollinating Asian *Epicephala* but unlike nonpollinating *Epicephala* (Kawakita and Kato 2009). Museum specimens of *Epicephala* collected by Clarke in the Marquesas and on Pohnpei also bear pollen on the proboscis, indicating flower-visiting behavior (Hembry et al. 2012; Fig. 12.6f–g).

Molecular phylogenetic analysis (Hembry et al. 2013a; Fig. 12.7) indicates that these Southeastern Polynesian *Epicephala* are biphyletic, resulting from at least two separate colonizations. Samoan *Epicephala* (from Tutuila) are at least biphyletic, as are Fijian *Epicephala*. What is presumed (based on divergence time estimation) to be the older colonization (Clade Y of Hembry et al. 2013a) has since diversified into a number of locally endemic subclades with distinct male genitalia; tentatively, these are distributed in the Marquesas (*E. spinula*), Windward Society Islands (Tahiti and Moorea), Leeward Society Islands (Raiatea, Huahine, Tahaa), Rapa, and possibly also the Cook Islands (at least Ātiu) and the uplifted Tuamotu Islands (at least Niau). There may be a "back-colonization" of Fiji and Samoa nested within this clade. The younger clade (Clade Z of Hembry et al. 2013a) is a single, widely distributed morphospecies of *Epicephala* found on at least 12 different host *Glochidion* species on 13 islands in the Cooks, Australs, and Societies. This clade is either a single widespread generalist, or a recently diversifying species complex. In either case, however, it indicates rapid shifts onto many new hosts, as well as rapid range expansion, following its arrival in Southeastern Polynesia. Both *Epicephala* clades coexist on the larger Society Islands (Tahiti, Moorea, Huahine, Raiatea). Whether both clades coexist on any of the northern Austral or Cook Islands remains unclear. The remaining Fijian and Samoan taxa sampled by Hembry et al. (2013a) form a third clade, not necessarily closely related to the other two, in which an *Epicephala* from *P. cuspidatus* is sister to the remaining Fijian accessions.

12.7 Hypotheses of Dispersal and Interaction Reassembly

Inasmuch as available evidence indicates that the mutualistic interaction between *Glochidion* and *Epicephala* in the remote Pacific is similar to that in Asia, how did these plants and insects disperse to these remote islands? *Glochidion* and *Epicephala* are unlikely to have colonized Southeastern Polynesia in a single dispersal event because of the differences in generation times between these organisms. If a fruit with both *Epicephala* larvae and viable *Glochidion* seeds

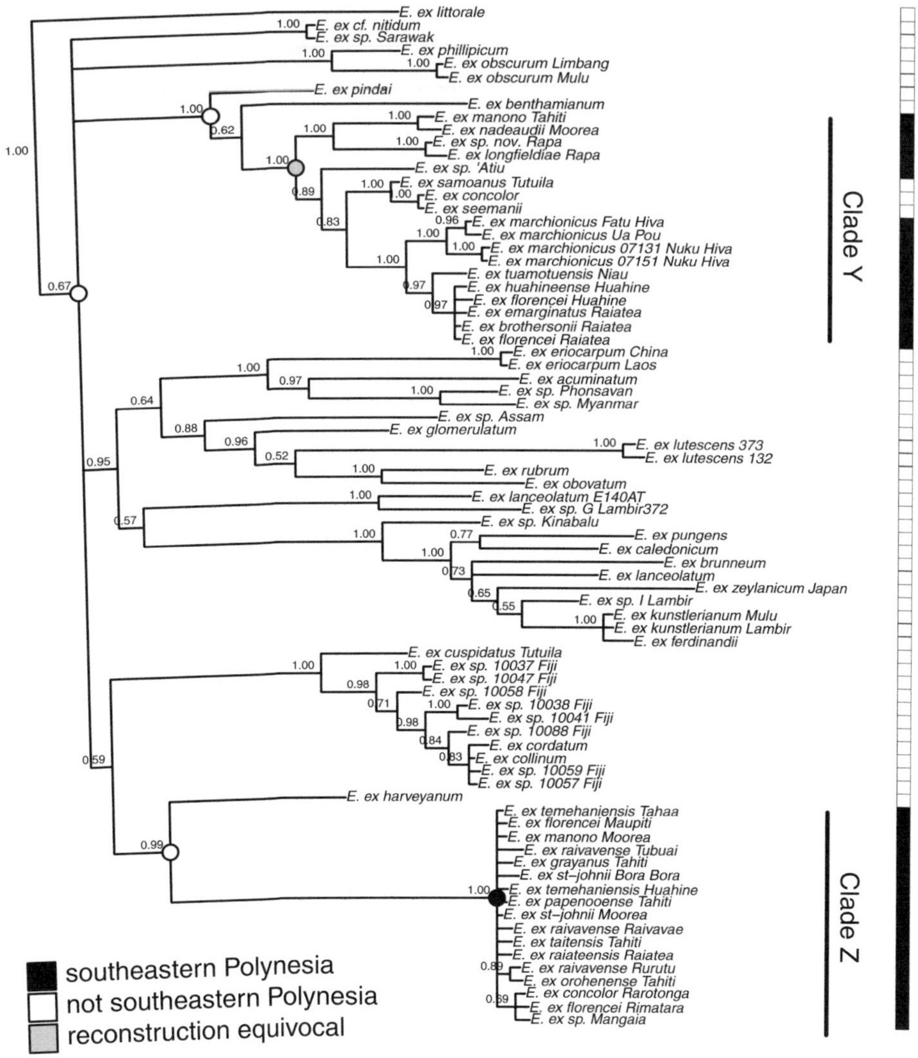

Fig. 12.7 Bayesian consensus tree for the *Epicephala* clade associated with *Glochidion* globally, from Hembry et al. (2013a). Node labels represent posterior probabilities. Tip labels refer to the specific epithets of host *Glochidion* species. Colored *boxes* and *circles* represent present-day distributions and ancestral state reconstructions of biogeography at selected nodes (*black*: southeastern Polynesia; white: not southeastern Polynesia; *grey*: biogeographic reconstruction equivocal). Southeastern Polynesian *Epicephala* fall into two distantly related clades; Samoan and Fijian *Epicephala* are also polyphyletic (Reprinted with permission of the Royal Society)

landed on an island, the *Epicephala* would likely eclose in a few weeks, whereas the seed would require at least a few years to grow into a reproductively mature plant that could produce developing fruit to feed the *Epicephala*'s offspring. The same constraint would likely apply to a vegetation raft containing *Epicephala* larvae or pupae and *Glochidion* roots, shoots, or seeds. Most likely, *Glochidion* and *Epicephala* colonized these islands separately. It is difficult to distinguish among specific hypotheses of dispersal in this mutualism. However, because *Epicephala* lack known alternate hosts, it is likely that *Glochidion* arrived first and persisted for some period of time (possibly reproducing vegetatively, or by selfing, or even being visited by some rare alternate pollinator), after which *Epicephala* colonized the island and resumed pollination and seed-predation behavior. Because the oceanic islands of Southeastern Polynesia have never been connected to each other, this process must have repeated itself on each island where these taxa are present today.

Glochidion may have initially colonized via either vegetation rafts or seeds dispersed by birds. Polynesian *Glochidion* show strong ability to resprout from rootstock following cutting (Hembry, unpublished notes) and so it is possible that they could grow from fragments of branches or roots. The fruit of both continental and insular *Glochidion* dehisce to reveal seeds with red arils that may be attractive to frugivorous birds (Fig. 12.5i); perhaps unsurprisingly, starlings in Samoa (Whistler 2004), Henderson Fruit Doves (*Ptilinopus insularis*) on Henderson (Brooke and Jones 1995), and Ultramarine Lorikeets (*Vini ultramarina*) on Ua Huka (J.-F. Butaud and F. Jacq, pers. comm., 2015) have all been reported as feeding on *Glochidion* fruit. Before human colonization, different species of fruit doves and lorikeets were widespread throughout Polynesia (Steadman 2006). *Epicephala* may have colonized as larvae inside seeds dispersed by birds, or by vegetation rafts, or as adult moths dispersed via wind. Small moths are heavily represented in the Lepidoptera fauna of oceanic islands (Clarke 1971) and the fact that *Epicephala* colonized Southeastern Polynesia (and some individual islands) more than once (Hembry et al. 2013a), suggest that *Epicephala* may actually be very good dispersers. Regardless of the mechanisms, *Glochidion* and *Epicephala* are exceptions to the widely invoked "rule" in island biology that island plants and pollinators tend to be generalists, and that taxa with specialized biotic interactions are unable or unlikely to colonize oceanic islands (Hembry et al. 2012). This also provides insights into the global biogeography of Phyllantheae (Kawakita 2012).

12.8 What Do *Glochidion* and *Epicephala* on Oceanic Islands Reveal About How They Coevolve?

The most striking aspect of the phylogenetic history of *Glochidion* and *Epicephala* on oceanic islands is that the monophyletic (except for *P. wilderi*) radiation of southeastern Polynesian *Glochidion* has been independently colonized twice by *Epicephala* (Fig. 12.8). Although repeated independent colonizations of the same archipelago may seem remarkable, it has been reported with increasing frequency in the Pacific biogeography literature in such organisms as snails (Holland and Cowie 2009), spiders (Casquet et al. 2015), birds (Cibois et al. 2011), and plants (Clark et al. 2009; Baldwin and Wagner 2010; Cantley et al. 2014).

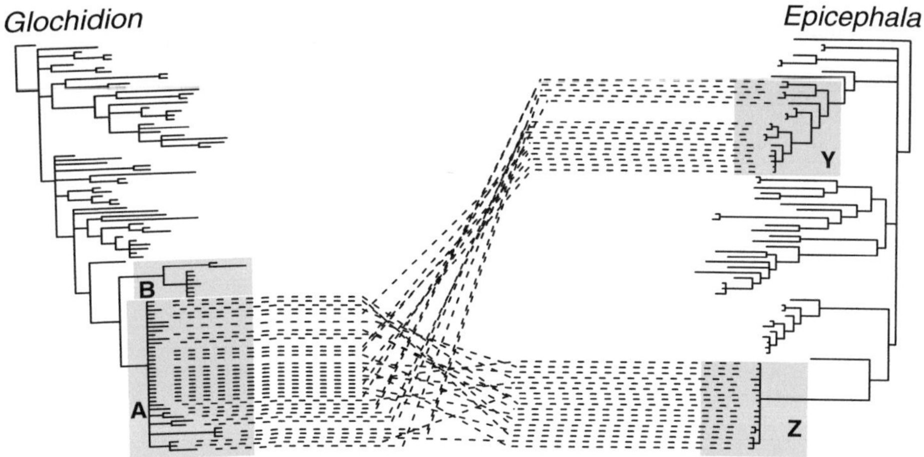

Fig. 12.8 Cophylogenetic patterns between southeastern Polynesian *Glochidion* and *Epicephala*, from Hembry et al. (2013a). *Letters* refer to clades containing Southeastern Polynesian taxa, as described in text. *Dashed lines* represent associations between southeastern Polynesian taxa; associations between other taxa omitted for clarity. Two distantly related clades of Epicephala, representing separate colonizations (*Y* and *Z*), are associated with a single clade (*A*) that contains all but one species of Southeastern Polynesian *Glochidion*. *B* represents a clade containing several *Glochidion* taxa from the Gambier Islands, Samoa, Fiji, and Wallis (Reprinted with permission of the Royal Society)

In addition to being just an additional example of repeated, long-distance dispersal, this finding reveals two very interesting things about *Epicephala* and its association with Phyllantheae plants. First, it shows that although available data indicate *Glochidion*-associated *Epicephala* are extremely specialized in ecological time, at least in continental regions (Kawakita and Kato 2006, Zhang et al. 2012a), in at least some cases, *Epicephala* lineages have the capacity to recognize and shift onto novel hosts with which they have no recent coevolutionary history. When the ancestor of the younger *Epicephala* colonization (Clade Z) first arrived in Southeastern Polynesia, it presumably survived and left offspring because it was able to recognize and use *Glochidion* species that were not those it or its ancestors had recently encountered. Second, associations between *Epicephala* and *Glochidion* can be extremely dynamic over relatively short periods of evolutionary time. Not only did the ancestor of the younger *Epicephala* colonization leave offspring by using a novel host, but its descendants then shifted onto another 12 islands and many other novel *Glochidion* hosts on those islands. This rapid and widespread host-shift occurred despite all the factors—the over-water barriers to dispersal between these islands, the unusual stimuli emitted by these other *Glochidion*, and the presence of potential competitors in the form of the older *Epicephala* lineage that was already there—that might have prevented it. The arrival of this new *Epicephala* lineage presumably completely reorganized the pattern of interactions between *Glochidion* and *Epicephala* on each of the islands it colonized. Whether this kind of rapid and widespread host-shift is possible between *Glochidion* and *Epicephala* on less isolated islands, or in continental regions, is unclear and awaits further study. However, it is interesting in the context of some of the host-shifts that have been inferred by other phylogenetic studies of *Epicephala* on *Glochidion* and *Breynia* in Asia (Kawakita et al. 2004, Kawakita and Kato 2006, Zhang et al. 2012b, Hembry et al. 2013a).

12.9 Conclusions

Leafflower–leafflower moth (Phyllantheae–*Epicephala*) interactions are widely distributed across oceanic islands in the tropical Pacific. In particular, *Glochidion* are found on many high islands in the tropical Pacific as far east as the Marquesas and Pitcairn Islands at the eastern end of the Indo-Malayan biogeographic region. They have radiated notably in Fiji (18 spp.) and the Society Islands (French Polynesia; 14 spp.), and are one of the largest endemic plant radiations in Southeastern Polynesia (Cook Islands, French Polynesia, and Pitcairn Islands). On many of the larger Pacific islands, co-occurring *Glochidion* differ in their habitat and altitudinal range, indicating that the diversification of *Glochidion* in this region has occurred at least in part along ecological axes. Pacific Island *Glochidion* are pollinated by *Epicephala* moths in the same manner as their congeners are in Asia and Australasia. Molecular phylogenetic analysis indicates that *Glochidion* and *Epicephala* have colonized the remote Pacific multiple times; in particular, Southeastern Polynesian *Glochidion* have been independently colonized twice by *Epicephala*. The more recent of these colonizing *Epicephala* lineages has rapidly spread across at least 12 new host *Glochidion* species on 13 islands, indicating that rapid and widespread host-shifts are possible in this specialized mutualism. Other Phyllantheae are present in Micronesia, Fiji, and Polynesia, but none of them are as diverse as the radiations of *Glochidion* in this region or *Gomphidium* (genus *Phyllanthus*) in New Caledonia.

Acknowledgements I thank Gerald McCormack (Cook Islands Natural Heritage), Jean-François Butaud (environmental consultant, Tahiti), and Atsushi Kawakita (Kyoto University) for their comments and corrections on this manuscript, and David Lorence (National Tropical Botanical Garden), Warren Wagner (Smithsonian Institution), and Jean-Yves Meyer (Délégation à la Recherche, Government of French Polynesia) for discussion and providing copies of relevant literature.

Chapter 13
Evolution and Diversity of Obligate Pollination Mutualisms

Atsushi Kawakita and Makoto Kato

Keywords Coevolution • Evolutionary stability • Fig–fig wasp mutualism • Host specificity • Leafflower–leafflower moth mutualism • Reciprocal diversification • Yucca–yucca moth mutualism

13.1 Diversity of Obligate Pollination Mutualisms

It is an exciting time for the study of obligate pollination mutualisms. New mutualisms continue to be discovered, and information on individual systems is rapidly growing. Presently, there are at least seven plant lineages apart from Phyllanthaceae that contain plants pollinated by seed-parasitic insects (Fig. 13.1, Table 13.1). There is little doubt that more such lineages will be discovered, inasmuch as one or two new mutualisms continue to be uncovered each decade. The abundance and heterogeneity of documented cases of obligate pollination mutualism offer an unprecedented opportunity to examine key topics of broad ecological and evolutionary relevance. In this chapter, we review the basic natural history of the seven mutualisms known outside of Phyllanthaceae (Fig. 13.1, Table 13.1) and address the following questions that are critical to our understanding of obligate pollination mutualisms.

1. Why do plants specialize to seed-parasitic pollinators despite the high cost imposed by the seed-feeding pollinator larvae?
2. How are mutualisms maintained despite the potential for selfish partners to disrupt the interaction?
3. Is pollinator specificity reinforced, and if so, why?

A. Kawakita (✉)
Center for Ecological Research, Kyoto University, 2-509-3 Hirano, Otsu, Shiga 520-2113, Japan
e-mail: kawakita@ecology.kyoto-u.ac.jp

M. Kato
Graduate School of Human and Environmental Studies, Kyoto University, Yoshida-Nihonmatsu-cho, Sakyo, Kyoto 606-8501, Japan
e-mail: kato@zoo.zool.kyoto-u.ac.jp

M. Kato, A. Kawakita (eds.), *Obligate Pollination Mutualism*, Ecological Research Monographs, DOI 10.1007/978-4-431-56532-1_13

Fig. 13.1 Obligate pollination mutualisms. (**a**) The leafflower–leafflower moth mutualism. A female *Epicephala anthophilia* actively depositing pollen on *Glochidion acuminatum* flower in Amami-Oshima Island, Japan. (**b**) The fig–fig wasp mutualism. A female of an unidentified pollinating fig wasp species entering the syconium of *Ficus auriculata* in Phonsavan, Laos. (**c**) The yucca–yucca moth mutualism. A female *Tegeticula yuccasella* actively depositing pollen on *Yucca filamentosa* in New York, USA. (**d**) The senita cactus–senita moth mutualism. A female *Upiga virescens* actively pollinating the stigma of *Pachycereus schottii* with its abdomen. Photo credit: J. Nathaniel Holland. (**e**) The *Lithophragma*–*Greya mutualism*. *Greya politella* (Prodoxidae) ovipositing into a flower of *Lithophragma cymbalaria* in California, USA

4. Do obligate pollination mutualisms drive the reciprocal diversification of plants and pollinators?

We use the term obligate pollination mutualism to refer only to associations involving pollinators that are seed-parasitic. Although there are examples of specialized nursery pollination mutualisms in which the plants provide decaying floral parts as brood sites of pollinator larvae (Sakai 2002), these associations do not raise the type of questions listed above because decaying floral parts are not costly for the plant and because pollinator fitness is unlinked to pollination success. The above challenges posed by obligate seed-parasitic pollination mutualisms offer valuable opportunities to test and refine existing theories in ecology and evolution and thus are the focus of this chapter.

13.1.1 Fig–Fig Wasp Mutualism

Among the documented examples of obligate pollination mutualisms, those of figs and yuccas are arguably the best known (Janzen 1979; Herre et al. 2008; Pellmyr 2003). All of the >750 species in the genus *Ficus* (Moraceae), distributed throughout the tropical and subtropical areas of the world are pollinated by the fig wasps currently classified in the subfamilies Tetrapusiinae, Kradibiinae, and Agaoninae of the family Agaonidae (Heraty et al. 2013; Fig. 13.1b). That fig wasps are involved in the maturation of the cultivated fig, *Ficus carica*, was already known by Aristotle and Theophrastus as early as 350 BC (Condit 1947; Goor 1965). Ramírez (1969) was the first to recognize that many fig wasps pollinate actively; the females collect pollen into thoracic pollen pockets shortly after emergence, and later place the pollen on the stigmas of flowers in the figs where they oviposit, an observation later substantiated by Galil and Eisikowitch (1969). A smaller number of fig wasps pollinate passively, with dehiscent anthers depositing pollen on the surface of adult female wasps as they emerge from the syconia (Ramírez 1969; Kjellberg et al. 2001). Each fig wasp species is usually associated with only one *Ficus* species, although the long-presumed one-to-one relationship often breaks down, as many *Ficus* species host two or more fig wasp species sympatrically (Molbo et al. 2003; Machado et al. 2005; Haine et al. 2006; Su et al. 2008; Sun et al. 2011; Cornille et al. 2012). Two fig wasp species have lost the pollinating habit and became

Fig. 13.1 (continued) (Saxifragaceae). Photo credit: John N. Thompson. (**f**) The *Silene–Hadena* mutualism. A female *Hadena bicruris* ovipositing into a flower of *Silene latifolia*. Photo credit: Anne-Marie Labouche. (**g**) The globeflower–globeflower fly mutualism. *Chistocheta* flies at the entrance of a *Trollius europaeus* flower. Photo credit: Station Alpine Joseph Fourier. (**h, i**) The *Rheum nobile–Bradysia* mutualism. A female *Bradysia* sp. fungus gnat visiting the flowers of *Rheum nobile* (**h**) whose inflorescence axis is entirely covered with non-green bracts (**i**). Photo credit: Bo Song

Table 13.1 Comparison of the eight known obligate pollination mutualisms

Trait	Reciprocal obligate pollination mutualism			Asymmetrical Obligate Pollination Mutualism				
	Leafflower–leafflower moth	Fig–fig wasp	Yucca–yucca moth	Senita cactus–senita moth	Saxifragaceae–*Greya*	*Silene–Hadena*/*Perizoma*	Globefower–globeflower fly	*Rheum nobile–Bradysia*
Plant								
Family	Phyllanthaceae	Moraceae	Agavaceae	Cactaceae	Saxifragaceae	Caryophyllaceae	Ranunculaceae	Polygonaceae
Genera	*Glochidion, Breynia, Phyllanthus*	*Ficus*	*Yucca, Hesperoyucca*	*Pachycereus*	*Lithophragma, Heuchera*	*Silene*	*Trollius*	*Rheum nobile*
No. of species	>500	>750	ca. 40	1	2	ca. 10	1	1
Geographic range	Pantropical	Pantropical	North and Central America	North and Central America	North America	Holarctic	Europe	Himalayas
Habit	Tree or shrub	Tree or shrub	Woody succulent	Woody succulent	Perennial	Perennial	Perennial	Perennial
Sexual system	Monoecious	Monoecious, functionally dioecious	Hermaphrodite	Hermaphrodite	Hermaphrodite	Gynodioecious, dioecious	Hermaphrodite	Hermaphrodite
No. ovules per flower	6–20	1	200–300	ca. 200	150–400	ca. 40	ca. 50	1
Flowering season	Continuous or seasonal	Continuous or seasonal	June–July	April–June	April–June	Spring–autumn	June–August	June
Self compatibility	Compatible	Unknown?	Compatible	Incompatible	Incompatible	Compatible	Incompatible	Compatible
Nectar production	None or little	None	None or little	None or little	Present	Present	Present	None
Pollinator attractant	**Scent**	**Scent**	**Scent**	Color and scent?	Color and scent	Color and scent	Color and scent	Scent, color?

Limitation of fruit production	Variable	Unknown	Resource-limited	Resource-limited	Variable	Variable	Pollen-limited	Resource-limited
Fruit type	Capsule, berry	Fig	Dry or fleshy capsule	Berry	Capsule	Capsule	Seedhead	Samara
Floral specialization to pollination	Nocturnal flowering, reduced nectar, style fusion, stamen fusion	Urn-shaped inflorescence	Nocturnal flowering, reduced nectar	Small flower size, reduced nectar	Multivariate floral traits	None	Globular flower	Showy bracts?
Selective abortion of flowers with high egg load	Present (not all species)	Present (strength varies among species)	Present (not in all plant–pollinator combinations)	Attacked fruits invariably abscise	Absent	Absent	Absent	Infested flowers more likely retained
Defensive trait against seed parasite	Elongated gynophore, enlarged tepals	Long styles	Constricted ovary	None	None	Long corolla	Adonivernith production	None
Pollinator								
Order	Lepidoptera	Hymenoptera	Lepidoptera	Lepidoptera	Lepidoptera	Lepidoptera	Diptera	Diptera
Family	Gracillariidae	Agaonidae	Prodoxidae	Pyralidae	Prodoxidae	Noctuidae, Geometridae	Anthomyiidae	Sciaridae
Genus	*Epicephala*	Various genera	*Tegeticula*, *Parategeticula*	*Upiga*	*Greya*	*Hadena*, *Perizoma*	*Chiastocheta*	*Bradysia*
No. of species	>500?	>750	ca. 20	1	2	ca. 10	6	1
No. of host per species	1–2	1–2	1–6	1	1–6	1–4	1	1

(continued)

Table 13.1 (continued)

Trait	Reciprocal obligate pollination mutualism			Asymmetrical Obligate Pollination Mutualism				
	Leafflower–leafflower moth	Fig–fig wasp	Yucca–yucca moth	Senita cactus–senita moth	Saxifragaceae–*Greya*	*Silene–Hadena/Perizoma*	Globefower–globeflower fly	*Rheum nobile–Bradysia*
Active pollination	**Present**	**Present**	**Present**	**Present**	Absent	Absent	Absent	Absent
Specialized pollen collecting structures	Proboscis sensilla	Pollen pockets, coxal comb	Maxillary tentacles	Abdominal scales	None	None	None	None
Co-pollinators	**None**	**None**	**None**	Halictid bees	Bees, flies	Various Lepidoptera	Bees, beetles	Autonomous self-pollination results in seed set
Oviposition site	Inside/outside of ovary	Ovule	Ovary tissue	Petal	Ovary	Ovary, petal	Carpel	Ovary
No. of eggs per flower	1–3	1–20 (per syconium)	one to several	1	1	1	one to several	1
No. of larvae per fruit	≤ 3	Many (per syconium)	≤ 6	1	1	1	≤ 10	1
Percent seed destruction per fruit	Variable: 0–100%	Variable: 0–50%	Variable: 0–30%	100% (attacked fruits abscise)	Variable: 15–30%	100%	Variable: 0–50%	100%
Generations per year	Multivoltine or univoltine	Multivoltine	1	4–5	1	1?	1	1
Pupation site	On litter	In seed	In soil	In cactus branch	In soil	In soil	In soil	In seed
Mating site	On host plant	In mature fig	On host	On cactus spines	On host	Unknown	In flower	On host

Parasitoid of pollinator larva	*Bracon* (Braconidae)	Non-pollinating fig wasps (Pteromalidae)	*Digonogastra* (Braconidae)	Unknown	Unknown	Unknown	Unknown	Unknown
Inquiline	*Bracon* (Braconidae); seed-parasitic moths (Tortricidae, Pyralidae, Carposinidae)	Non-pollinating fig wasps (Agaonidae, Pteromalidae, Ormyridae, Eurytomidae, Torymidae); fruit flies (Tephritidae, Lonchaenidae)	*Prodoxus* (Prodoxidae); florivore bug (Coreidae) and beetle (Tenebrionidae)	Unknown	Unknown	Unknown	Unknown	Unknown
Association								
Plant dependence on seed-parasitic pollinator	**Obligate**	**Obligate**	**Obligate**	Facultative	Facultative	Facultative	Facultative	Facultative
Pollinator dependence on plant	**Obligate**	**Obligate**	**Obligate**	**Obligate**	**Obligate**	**Obligate**	**Obligate**	Obligate?
Synergistic diversification	**Yes**	**Yes**	**Yes**	No	No	No	No	No

Information is based on various sources

nonmutualistic: *Ceratosolen galili* on *Ficus sycomorus* in Africa (Galil and Eisikowitch 1969; Compton et al. 1991) and an undescribed *Eupristima* species on *F. altissima* in China (Peng et al. 2008). The close relatives of pollinating fig wasps are the so-called nonpollinating fig wasps that join the association by either inducing galls on fig ovules or exploiting the galls induced by other fig wasps (Weiblen 2002). They offer an excellent comparative model to test hypotheses on cospeciation and the evolution of host specificity (e.g., Weiblen and Bush 2002; Marussich and Machado 2007).

13.1.2 Yucca–Yucca Moth Mutualism

Engelmann (1872) was the first to observe yucca moths in the flowers of yuccas, and suggest that they are essential for yucca pollination (Pellmyr 2003). However, it was Charles Riley who unfolded the natural history of the mutualism in great detail, including the moth's ability to pollinate actively using the remarkably developed maxillary tentacles (Riley 1872, 1880, 1881, 1892; Fig. 13.1c). Yuccas, native to North and Central America, are plants of the genus *Yucca* and the monotypic *Hesperoyucca* (Agavaceae), of which the latter was formerly considered a section within *Yucca*. They are iconic components of the arid vegetation of North and Central America, although a few species occur in the rainforests of southernmost Mexico. Molecular phylogenetic analysis clearly shows that *Hesperoyucca* is distantly related to *Yucca* (McKain et al. 2016). A close relative of *Hesperoyucca* is *Hesperoe*, which has no association with yucca moths and relies instead on hummingbirds for pollination (Pellmyr and Augenstein 1997). Thus, the obligate mutualism with yucca moths is believed to have evolved independently twice: in *Hesperoyucca* and the common ancestor of *Yucca* (Bogler et al. 1995; McKain et al. 2016). Pollinating yucca moths belong to the genera *Tegeticula* and *Parategeticula*, which form a well-supported clade within the Prodoxidae family (Brown et al. 1994; Pellmyr and Leebens-Mack 1999). Two derived *Tegeticula* species, *T. intermedia* and *T. corruptrix*, independently lost the pollinating behavior and oviposit in young fruits produced by other pollinating yucca moths; they are "cheaters" in this mutualistic system (Addicott 1996; Pellmyr et al. 1996b; Pellmyr 1999). Sister to the clade of *Tegeticula* and *Parategeticula* is the genus *Prodoxus*, or bogus yucca moths, which consists of species that feed on the nonreproductive organs of yuccas and other Agavaceae (leaves, inflorescence stalks, or fruit walls; Pellmyr et al. 2006). As with nonpollinating fig wasps, they provide important comparative models for understanding the process of speciation in yucca moths (e.g., Althoff et al. 2001, 2012).

13.1.3 Senita Cactus–Senita Moth Mutualism

In contrast to the fig and yucca mutualisms, which have been known for more than 100 years, other examples of mutualisms between plants and pollinating seed parasites were discovered relatively recently. Fleming and Holland (1998) studied the interaction between the senita cactus (*Pachycereus schottii*; Cactaceae) and senita moth (*Upiga virescens*; Pyralidae) in the Sonoran Desert, and made the remarkable finding that the moth is an active pollinator; the female collects pollen on specialized abdominal scales, actively deposits pollen on stigmas (Fig. 13.1d), and lays a single egg on a flower petal. Fruits attacked by moth larvae abscise and produce no viable seeds, but only a fraction of the moth eggs actually survive to become seed predators, therefore the interaction is mutualistic overall (Holland and Fleming 1999). The cactus is also visited by halictid bees during the daytime, but its nocturnal flowering and temperature-dependent flower closing limit the overall contribution by bees to the fruit set (Holland and Fleming 2002). The basal lineages of *Pachycereus* produce bat-pollinated flowers that open at night (Fleming et al. 1996, 2001), therefore nocturnal flowering may not be a trait that has evolved as a result of specialization. Rather, small flowers and reduced nectar production may be adaptations associated with moth pollination (Hartmann et al. 2002).

13.1.4 **Lithophragma–Greya** *Mutualism*

The family Prodoxidae, to which yucca moths belong, is a basal clade of Lepidoptera characterized by a piercing ovipositor, and contains species that feed parasitically on flowers and other plant parts in the Rosaceae, Grossulariaceae, and Saxifragaceae (Pellmyr et al. 1996b). Of these, *Greya politella* and *G. enchrysa*, occurring in North America, are seed parasites of *Lithophragma* and *Heuchera* (Saxifragaceae). The *Greya* females pollinate their host plants passively as they oviposit in the flowers (Pellmyr and Thompson 1992; Pellmyr et al. 1996b; Fig. 13.1e). *G. politella* has been recorded from several *Lithophragma* and *Heuchera* species, whereas *G. enchrysa* is known only from *H. cylindrica*. It is important to note that *Lithophragma* and *Heuchera* plants used by *Greya* are also pollinated by bumblebees, solitary bees, or bombyliid flies, therefore the mutualistic effect of pollination by *Greya* is often swamped by these copollinators (Thompson and Pellmyr 1992). Nevertheless, in populations of *Lithophragma* where copollinators are rare or absent, *Greya* is the main pollinator (Thompson and Cunningham 2002; Thompson et al. 2010), and traits that enhance moth pollination have likely evolved (Thompson et al. 2013; Friberg et al. 2014).

13.1.5 Silene–Hadena/Perizoma *Mutualism*

As new mutualisms involving pollinating seed parasites were being discovered, increased attention was paid to whether moth seed parasites of *Silene* (Caryophyllaceae) contribute to the pollination of their hosts (e.g., Brantjes 1976a,b; Pettersson 1991a,b; Westerbergh 2004; reviewed in Kephart et al. 2006). Plants of *Silene* and several allied genera distributed widely in the Holarctic are attacked by seed-feeding moths in *Hadena* (Noctuidae; Fig. 13.1f) and *Perizoma* (Geometridae). Both male and female moths visit flowers nocturnally to consume nectar, and in doing so cause pollination (Labouche and Bernasconi 2010). However, because *Silene* flowers are visited by a variety of other nocturnal moths and diurnal bees and flies, *Hadena* and *Perizoma* are rarely major pollinators, and usually have only parasitic effects on their hosts. However, in populations where copollinators are rare or in years when copollinators are scarce, these seed parasites may act as mutualists of their host plants (Westerbergh 2004). In *Silene latifolia*, long corolla tubes act to constrain oviposition by *Hadena bicruris* to sites of high egg mortality (Labouche and Bernasconi 2013). If such traits shift the cost–benefit balance in favor of plants, there is the potential for this interaction to evolve toward mutualism.

13.1.6 Globeflower–Globeflower Fly Mutualism

Most of the above examples of obligate pollination mutualism involve moths as pollinators (a notable exception is the figs pollinated by fig wasps), but the pollinators of the last two examples are dipteran insects. Pellmyr (1989) studied the pollination ecology of *Trollius europaeus* in great detail and found that, of the four species of *Chiastocheta* flies (Anthomyiidae) that consume the seeds of *T. europaeus* in Sweden, three are likely mutualistic pollinators that deliver a net benefit to the plant (Fig. 13.1g). Studies of *T. europaeus* in other parts of Europe show that the plant is pollinated by *Chiastocheta* flies throughout its range, although the species composition of *Chiastocheta* varies geographically (Jaeger and Desprès 1998; Desprès et al. 2007). The flies do not pollinate actively; pollination occurs as the flies mate and eat pollen and nectar in the flowers. Notably, the tepals of *T. europaeus* form a globe with a narrow opening at the apex (Fig. 13.1g), which serves as the entrance and exit for *Chiastocheta* flies. Because related *Trollius* species pollinated by bumblebees have disc-shaped flowers (Pellmyr 1992), the globular flower of *T. europaeus* likely evolved to exclude visitors other than *Chiastocheta*. Regardless of such a specialization, *Trollius europaeus* has copollinators in many populations, and the mutualism is asymmetrically obligate wherein plants do not always entirely depend on the fly (Suchan et al. 2015).

13.1.7 *The* Rheum nobile–Bradysia *Mutualism*

A symbolic plant of the high Himalayas, *Rheum nobile*, attracts strong botanical interest owing to its showy nongreen bracts that conceal the entire stout, conical inflorescence axis (Fig. 13.1h). The pollinator of this remarkable plant has long remained uninvestigated, but recently, Song et al. (2014) found that the seed-feeding fungus gnat, *Bradysia* sp. (Sciaridae), is its sole pollinator; however, it is still unknown whether the plant is the only host of the insect. Fungus gnats pollinate passively as the female insects crawl on the inflorescence inside the bracts in search of suitable oviposition sites (Fig. 13.1i). Because the flowers are uniovulate, infested fruits will not have viable seeds; however, the fungus gnats pollinate more flowers than are used for oviposition, resulting in a net mutualistic outcome. During anthesis, the flowers produce an unusual floral volatile, 2-methyl butyric acid methyl ester, which is attractive to female fungus gnats (Song et al. 2014). Because the plants can set seeds by autonomous self-pollination, the degree of reciprocal dependence is also not high in this system as compared to other highly obligate interactions.

13.2 Reciprocal and Asymmetrical Obligate Pollination Mutualism

Among the eight nursery pollination mutualisms involving seed-parasitic pollinators, only three are reciprocally obligate pollination mutualisms, whereas the remaining five are asymmetrically obligate such that the plants are pollinated not only by the seed-parasitic insect partners but also by copollinators. Intriguingly enough, the three reciprocally obligate pollination mutualisms, that is, the leafflower–leafflower moth, fig–fig wasp, and yucca–yucca moth mutualisms, share the following properties (Table 13.1): (1) adult females actively pollinate the flowers; (2) the attractant released by the flowers is floral scent, which can be finely tuned into private communication signals owing to the countless combinations of volatile compounds; (3) the plants are unable to set seeds in the absence of the seed-parasitic insects; and (4) the plants and the pollinators have undergone reciprocal synergetic diversification. Thus, reciprocal obligate pollination mutualisms likely derived from asymmetrical obligate pollination mutualisms by the following three innovations: pollinator's adoption of active pollination behavior, development of volatile chemical communication between plants and pollinators, and elimination of copollinators by the plants through plant–pollinator coevolution. The resulting high reciprocal dependence and exceedingly high host-specificity have canalized gene flows and driven the reciprocal diversification of the plants and the seed-parasitic pollinators. We discuss these processes in detail in the following sections.

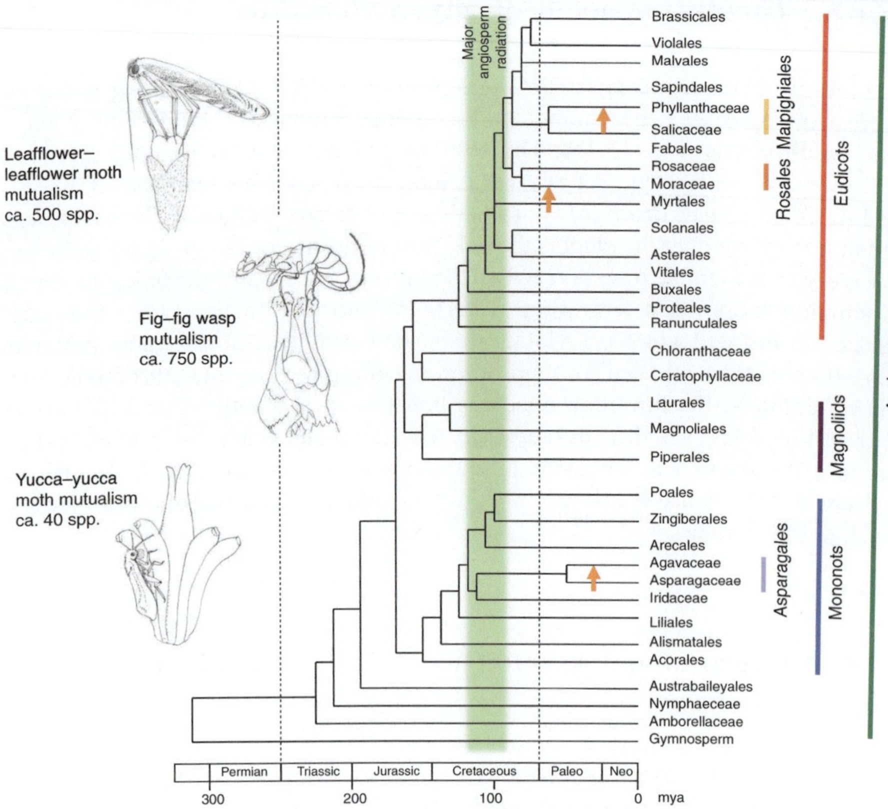

Fig. 13.2 A chronogram of angiosperm diversification indicating the origins of the three reciprocally obligate pollination mutualisms. The chronogram is based on Zeng et al. (2014)

The three reciprocally obligate pollination mutualisms are ancient and diverse (Fig. 13.2). The origin of the fig–fig wasp mutualism dates back to 60 mya, and subsequent codiversification has produced an impressive diversity of plant and pollinator species (there are more than 750 *Ficus* species; Rønsted et al. 2005). Reliable dating of the origins of the yucca–yucca moth and leafflower–leafflower moths is hampered by the paucity of lepidopteran fossils, but the current best estimates of the ages of active pollinators are 32–40 mya for the yucca moths (Pellmyr and Leebens-Mack 1999; Gaunt and Miles 2002) and 25–30 mya for the leafflower moths (Kawakita and Kato 2009). Within Agavaceae, specialization to pollinating yucca moths evolved twice in *Hesperoyucca* and *Yucca*, the latter of which has diversified into approximately 40 spp. In Phyllanthaceae, the mutualism arose at least five times independently, and concurrent radiations in these plant lineages resulted in a total of approximately 500 plant species that are obligately pollinated by leafflower moths (Chapter 6). Intriguingly, once specialized, no fig, yucca, or leafflower species has reverted to pollination by other agents, which strongly indicates that pollination by seed parasites is a highly successful and stable strategy for the plants.

13.3 Origin of Obligate Pollination Mutualisms

Historically, figs and yuccas have been the only groups of plants that are known to be pollinated by seed-parasitic insects. Because of their rarity among angiosperms, obligate pollination mutualisms were thought to evolve only under exceptional circumstances where plants have no means of achieving successful pollination other than specializing to seed parasites. Because seed destruction reduces plant reproductive output, copollinators, if present, are thought to provide better pollination services than seed parasites and limit plant specialization to seed-parasitic pollinators. In fact, the exclusion of copollinators has not occurred in some of the above examples, including *Lithophragma* plants pollinated by *Greya* moths (Thompson and Pellmyr 1992) and *Silene* plants pollinated by *Hadena* and *Perizoma* moths (Kephart et al. 2006). However, many others exhibit a moderate to considerable degree of specialization (Table 13.1), calling into question the view that specialization occurs under exceptional circumstances. Furthermore, detailed phylogenetic analysis of the yucca and leafflower lineages indicates that specialization to yucca moths and leafflower moths, respectively, occurred twice in yuccas and five times in leafflowers (Bogler et al. 1995; Kawakita and Kato 2009; McKain et al. 2016). Increasingly, it seems that, under some conditions, seed parasites provide superior pollination services and are preferred by plants over potential copollinators.

Seed-parasitic pollinators may be effective at transferring pollen because they usually have narrow diets, and hence a strong affinity to their host plants compared with generalist floral visitors that use various other plants as food sources. Actively pollinating seed parasites, such as leafflower moths, fig wasps, yucca moths, and senita moths, may be particularly effective pollinators that waste far less pollen than do generalist visitors. Thus, seed parasites may exert major positive effects on the male reproductive success of the plant, perhaps large enough to offset the impact of seed destruction on a plant's female reproductive success, thereby allowing plants to specialize to seed-parasitic pollinators. Increased use of seed-parasitic pollinators always decreases a plant's fitness through the female function; therefore, specialization, if it occurs at all, must occur solely via fitness gain through the male function.

Comparisons of pollen-to-ovule ratios between Phyllanthaceae plants specialized to *Epicephala* and those pollinated by other insects provide evidence of high pollen transfer efficiency by active pollinators (Mochizuki and Kawakita, unpublished data). If the pollinator of a given plant species wastes a large proportion of pollen picked up from the anthers, then plants should produce excess pollen to supplement the loss, leading to high pollen-to-ovule ratios, and vice versa. Low pollen-to-ovule ratios among *Epicephala*-pollinated species are best illustrated by differences in the relative abundance of male and female flowers (Fig. 13.3).

Fig. 13.3 Male-to-female flower ratios in Phyllanthaceae plants. (**a**) A flowering branch of *Epicephala pollinated Glochidion zeylanicum*, showing few male flowers (*upward arrows*) and several female flower clusters (*downward arrows*), each bearing ca. 10 flowers. (**b**) A flowering branch of dipteran-pollinated *Phyllanthus flexuosus*, showing few female flowers (*downward arrows*) and many male flowers (not indicated by arrows)

Plants specialized to *Epicephala* usually have excess female flowers relative to male flowers, whereas those pollinated by other insects have more male than female flowers. The pollen-to-ovule ratio is also much lower in figs that are pollinated by actively pollinating fig wasps than in those pollinated by passively pollinating fig wasps (Kjellberg et al. 2001; Jousselin et al. 2003), and a similarly low pollen-to-ovule ratio is suggested for yuccas (Pellmyr 2012). Although these comparisons are limited to active versus passive pollinators, active pollinators, at least, are probably remarkably effective at transferring pollen.

13.4 Evolutionary Stability of Obligate Pollination Mutualisms

Mutualisms may collapse if mutualists' costs and benefits are not balanced. As seen in Chapter 9, host sanction is one mechanism whereby hosts selectively allocate resources to the more cooperative partners, consequently reducing the fitness of selfish partners. In the case of obligate pollination mutualisms, sanction takes the form of selective abortion of flowers (or syconia, in the case of figs) with heavy egg loads. Since its first discovery in *Yucca filamentosa* (Pellmyr and Huth 1994), selective abortion has been confirmed in other yucca species (Richter and Weis 1995; Addicott and Bao 1999), in figs (Jandér and Herre 2010; Jandér et al. 2012; Wang et al. 2014), and in one *Glochidion* species (Goto et al. 2010). Host sanction has also been demonstrated in other mutualisms such as those involving plant–microbial relationships (Kiers et al. 2003, 2011), and this phenomenon provides a general explanation for the evolutionary persistence of mutualisms.

Host sanction contributes to the stability of the mutualisms in which it occurs, but it is not a universal mechanism that allows mutualisms to persist. For example, later studies in yuccas showed that selective abortion does not occur when yucca moths with short ovipositors infested the flowers, probably because yuccas have no means of detecting moth oviposition (Addicott and Bao 1999). Moreover, although host sanction can prevent the spread of an overexploiting phenotype, it does not necessarily guarantee the benefit to a host. In the case of the leafflower–moth mutualism, for example, plants do not produce any offspring, even with selective flower abscission, if moth populations increase to the point where their eggs routinely exceed numbers that cause total seed destruction. Consequently, factors that limit the costs of mutualisms, regardless of the presence of host sanctions, are also important (Segraves 2003, 2008; Althoff et al. 2005, 2013; Crabb and Pellmyr 2006; Dunn et al. 2008).

The cost of mutualism can be alleviated by the presence of a third-party partner, such as parasitoids or predators of the pollinators (Segraves 2008; Althoff et al. 2013). Dunn et al. (2008) demonstrated that in *Ficus rubiginosa*, parasitic fig wasps that oviposit from outside the syconium are more likely to attack pollinator larvae in ovules that are closer to the syconium wall. The resulting gradient in offspring viability likely poses selection on pollinating fig wasps to avoid outer ovules, thus limiting the overall intensity of seed destruction by pollinator larvae. Parasitoids also have a positive effect on seed production in the leafflower–moth mutualism; in *Breynia vitis-idaea*, the larvae of the pollinator attacked by braconid parasitoids consume fewer seeds than do those that develop normally (Kawakita and Kato 2004b). In contrast, the effect of parasitoids is obscure in the yucca–moth mutualism (Crabb and Pellmyr 2006). However, florivorous beetles sometimes consume moth eggs while feeding on style tissue, and this may have a positive effect on yucca seed production (Segraves 2008), although florivore effects on plants can vary from positive to negative depending on their annual abundance (Althoff et al. 2005, 2013). Overall, there is ample evidence showing that third-party partners

Fig. 13.4 Defensive plants traits in the Phyllanthaceae–*Epicephala* mutualism. (**a**, **b**) Elongated gynophores (stalk-like structures on fruits) of *Breynia vitis-idaea*. There is large within-population variation in gynophore length, and individuals with longer gynophores suffer less damage by seed-feeding *Epicephala* larvae. (**c**) Female flowers of New Caledonian *Phyllanthus nothsii* are covered with exaggerated tepals that create distance between the ovary and ovipositing *Epicephala* moth. Flowers indicated by arrows are female flowers, whereas those not indicated by arrows are male flowers. Note the size difference between female and male flowers. (**d**) An *Epicephala* egg laid on the ovary through the tepals. The associated *Epicephala* species has an elongate abdomen that likely coevolved with flower tepal size

contribute to limiting the costs of mutualism for plants in obligate pollination mutualisms.

Another mechanism that alleviates costs, but has thus far attracted little attention, is the evolution of traits that make host resources less exploitable (i.e., defense traits; Oliver et al. 2009). For example, in the *Silene latifolia–Hadena bicruris* interaction, long corolla tubes constrain moth oviposition to sites where eggs suffer high mortality without negatively affecting pollination efficiency, thus possibly preventing moth overexploitation (Labouche and Bernasconi 2013). Some individuals of *Breynia vitis-idaea* have remarkable gynophores in their fruits (Fig. 13.4a,b), whose function has been unknown. However, recent research shows that this is also a form of defense; fruits with longer gynophores suffer less seed predation by moth larvae because hatched larvae must bore through the elongating gynophore to reach seeds, and often die in the process (Furukawa and Kawakita, unpublished data). Several *Phyllanthus* species in New Caledonia have well-developed tepals that create distance between the ovary and ovipositing moths (Fig. 13.4c,d), and associated *Epicephala* species have remarkably elongated abdomens that have likely coevolved with tepal size. Constricted ovaries in *Yucca baccata*, which reduces the number of surviving moth larvae by limiting access to ovules (Bao and Addicott 1998), are another clear example of defense. Defense traits are not always morphologically obvious; *Trollius* plants produce a C-glycosyl-flavone, adonivernith, in the carpel wall in response to infestation by *Chiastocheta* larvae, which reduces larval seed feeding ability (Ibanez et al. 2009). Note that host defense can be viewed alternatively as plant cheating (Bao and Addicott 1998), and thus itself is a source of mutualism instability; mutualisms will not persist if such a defense completely prevents the pollinator from gaining any benefit. Regardless, the prevalence of plant defenses across many mutualisms strongly indicates that the evolution of such traits is a major process by which plants limit the heavy costs imposed by seed-feeding pollinator larvae.

Interestingly, whereas the cost of mutualism for the plants (seeds) is substantially high in obligate pollination mutualisms, the cost incurred by the pollinators (pollination) is probably very small (Pellmyr 1997, 2012). As a consequence, there are no clear examples of pollinator adaptations that result directly from selection to reduce the cost of pollination. For example, two derived yucca moth species do not possess the pollinating behavior, but the loss of pollinating behavior is more likely a by-product of a phenological shift to an unexploited seed resource, after which pollination behavior became redundant (Pellmyr 2003). Loss of pollination behavior in *Epicephala* also did not happen through selection for cheating. One lineage colonized herbaceous *Phyllanthus*, which regularly attain full fruit sets through pollination by ants (Kawakita and Kato 2009), and others evolved gall-inducing ability and no longer rely on pollination (Kawakita et al. 2015). There are also two derived nonpollinating species among fig wasps, but the evolutionary forces that gave rise to these species remain unknown. It is possible that active pollination is costly in fig wasps (Jandér and Herre 2010); if so, the loss of pollination behavior may have evolved solely to eliminate the cost of pollination. Interestingly, the nonpollinating *Eupristina* sp. that coexists with the pollinator *Eupristina altissima* on *Ficus altissima* has smaller egg loads and suffers higher larval mortality than do pollinator species; the apparent rarity of derived nonpollinators among the fig wasps may thus be explained by a competitive disadvantage of nonpollinators relative to mutualists (Zhao et al. 2014). A notable pollinator adaptation is found in the *Bradysia* fungus gnat that pollinates *Rheum nobile*; this gnat appears to manipulate its host chemically so that flowers containing *Bradysia* eggs are more likely to be retained than those without eggs (Song et al. 2016).

Overall, the pollinators in obligate pollination mutualisms impose heavy costs on plants through seed destruction, and even develop strategies to exploit their hosts selfishly (as seen in the above examples of derived nonpollinators or host-manipulating *Bradysia*). Heavy costs and selfish partners are clearly threats to the persistence of mutualisms, but multiple factors contribute to reducing their impact, such as host sanctions, the presence of third-party partners, and plant defense. It is probably useful to place obligate pollination mutualisms within a broader framework of plant–herbivore interactions, in which the roles of plant defense and trophic control are far better understood.

13.5 Evolution of Pollinator's Host Specificity

Obligate pollination mutualisms are classic examples of species-specific plant–insect associations. Although increasing evidence suggests that strict one-to-one specificity is not met in any of the mutualisms, it is striking that the majority of fig wasp, yucca moth, and leafflower moth species are monophagous. The question of whether host specificity is reinforced among pollinators is contentious. Ecological and phylogenetic studies in the yucca moth family Prodoxidae have found that close relatives of the pollinators are also highly host-specific herbivores (Pellmyr & Thompson 1992; Pellmyr 1999; Pellmyr et al. 2006), prompting the idea that high degrees of pollinator specificity are the fortuitous result of their inherently parasitic lifestyle (Thompson 1994, 2005). Conversely, the degree of host specificity among *Epicephala* is distinctly higher than that of related leaf-feeding gracillariids, pointing to the possibility that mutualism reinforces host specificity (Kawakita et al. 2010). In the fig system, nonpollinating fig wasps that are closely related to and co-occur with pollinating wasps tend to be less host-specific than the pollinators are (Weiblen & Bush 2002; Marussich & Machado 2007), although some studies report similar levels of host specificity between the two groups (Lopez-Vaamonde et al. 2001; Jousselin et al. 2006, 2008).

How and under what circumstances mutualisms reinforce host specificity is highly relevant to our general understanding of plant–insect interactions. Because shared pollinators can result in hybridization among closely related, co-occurring plants, there is a clear benefit for plants that attract specialist pollinators and thereby achieve conspecific pollination. Pollinators, on the other hand, have no constraint on their selection of host plants, as long as the plants provide suitable food for their larvae. Diet alone is unlikely to be the driver of extreme specialization in pollinating seed parasites because the two derived nonpollinating yucca moth species both utilize 4–6 yucca hosts (Pellmyr 2003; but see Moe and Weiblen 2012 for a notable example of fig wasp failing to develop on nonnatal fig host). An exciting possibility, therefore, is that plants are able to attract specific pollinator species selectively or, alternatively, repel pollinators that have high affinity to other host species. The floral odors of co-occurring *Glochidion* species are remarkably distinct, even to the human nose, which is striking, given that these plants are the product of a relatively recent radiation and are thus very closely related (<10 my; Kawakita and Kato 2009). A similar difference in floral scent is also found among co-occurring fig species. Because the fitness of the pollinators is heavily dependent on their ability to locate host plants, their chemical perception ability may be maximally tuned to the odor of the host plants with which they are normally associated, perhaps at the expense of their ability to perceive the odor of other species. If so, these insects may not be able to detect the scents of more than one host species, provided that the host odors are sufficiently divergent. Whether the observed levels of interspecific difference in floral scent are the result of divergent selection requires further study. However, the presence of these differences is itself a strong indication that plants are selected to attract specific pollinators; otherwise, the scents of co-occurring *Glochidion* should converge to attract the whole local *Epicephala* community.

A study using *Breynia vitis-idaea* suggests that a mixture of two conventional volatile compounds is sufficient to attract its specialist pollinator, *Epicephala vitisidaea*, at the same rate as do real flowers (Svensson et al. 2010). However, many plants pollinated by *Epicephala*, including *B. vitis-idaea*, produce 20–30 volatile compounds whose functions are not yet fully understood (Okamoto et al. 2013). Studies aimed at clarifying the roles of seemingly redundant compounds should lead to improved understanding of plant–pollinator specificity in obligate pollination mutualisms. Plants and pollinators may also be coadapted phenologically and/or morphologically; thus, pollinators are more constrained in their selection of host plants than are their close relatives that feed on vegetative tissues.

Although the evolutionary processes leading to high plant–pollinator specificity remain unknown, the resulting specificity has important ecological consequences. In the southern islands of Japan, four *Glochidion* species regularly co-occur and flower at the same time during May–June. Although the species differ slightly in their microhabitat preferences, any pair of species can grow side by side (see Fig. 7.4 in Chapter 7), apparently without producing hybrids (note that hybrids do occur at extremely low frequencies). Because fertile hybrid seeds can be easily produced by artificial heterospecific pollination, pollinator isolation is likely the primary mechanism by which co-occurring plants maintain reproductive isolation. The importance of pollinator specificity in maintaining species integrity is well documented in figs (Moe and Weiblen 2012; Wang et al. 2016) and is also true for yuccas. In a lowland tropical rainforest of Malaysia, as many as nine *Glochidion* species co-occur without sharing pollinators, and dozens of *Ficus* species can be found in a single forest. Pollinator specialization thus allows multiple closely related plants to co-occur in a single area, and contributes to the maintenance of overall species richness of plants pollinated by seed parasites.

13.6 Mutualism and Reciprocal Diversification

Perhaps the most outstanding feature of the fig–wasp, yucca–moth, and leafflower–moth mutualisms is their high species richness (Table 13.1). High reciprocal dependence and specialization in these mutualisms have fueled predictions that plants and pollinators may undergo accelerated reciprocal diversification through cospeciation (Kiester et al. 1984; Sanderson and Donoghue 1996; Schluter 2000). Specifically, because pollinating seed parasites are highly dependent on host flowers for their own reproduction, changes in key floral traits, such as floral volatiles, are likely to be tracked by adaptation in the pollinators; similarly, changes in pollinator traits may also cause modifications in floral traits. Such a process may accelerate the divergence between populations if the plants and pollinators in each population evolve unique sets of coadapted traits (Thompson 1994, 2005). In addition, because pollinators are solely responsible for the movement of gametes among plants, reproductive isolation in plants can arise as a direct result of population differentiation in seed-parasitic pollination mutualisms.

Although the idea of coevolution-driven diversification has received theoretical support (Kiester et al. 1984; Thompson 1994, 2005), empirical demonstration of the hypothesis has been difficult. The best documented example is that of *Yucca brevifolia* and *Y. jaegeriana*, which are pollinated by *Tegeticula synthetica* and *T. antithetica*, respectively (Pellmyr and Seraves 2003; Lenz 2007; Smith et al. 2008b, 2009; Godsoe et al. 2008, 2009). These two plant–pollinator pairs are distributed on different sides of the Mojave Desert in the Western United States, and members of each pair evolved coadapted morphological traits (stylar morphology and ovipositor length) that effectively maintain species specificity and reproductive isolation in a narrow contact zone in Southern Nevada. Although other studies suggest that coevolution may not have been important in shaping the current diversity of yuccas and yucca moths (Smith et al. 2008a; Althoff et al. 2012), the above example illustrates the power of coevolution in generating and maintaining new plant and pollinator species. Demonstration of cospeciation in figs is difficult because fig species typically have broad geographic ranges, making it necessary to sample across a vast geographic area to find any pattern. A notable example is the study of three fig–fig wasp pairs that diverged in situ on the oceanic Ogasawara Islands of Japan (Yokoyama 2003). The three fig species (*Ficus nishimurae*, *F. boninshimae*, and “Higashidaira type”) are distributed allopatrically on Chichijima Island, and each is associated with a genetically distinct fig wasp with an olfactory preference for the scent produced by their host figs. This suggests that chemical coadaptation, in addition to geographic isolation, may have promoted species divergence. These findings in yuccas and figs provide strong evidence that cospeciation driven by coevolution is a powerful driver of diversification in obligate pollination mutualisms.

Population-level analysis of cospeciation is still lacking in the leafflower–moth mutualism, but observations indicate that cospeciation may also be common in this system. Many species of New Caledonian *Phyllanthus* are locally restricted and have very narrow distributions that may span less than 50 km. Pairs of allopatric, closely related species are usually adapted to different edaphic and/or climatic conditions, thus abiotic factors are the primary driver of species divergence. However, there are cases in which plants in such pairs are each associated with distinct moth species that are closely related to one another, indicating that cospeciation, whether or not accompanied by coevolution, is occurring (Fig. 13.5). There are many spectacular plant radiations in New Caledonia, but notably, *Phyllanthus* is the largest plant genus on the islands, with >110 species (Morat 1993). Although there are numerous explanations for variation in species

Fig. 13.5 Closely related, allopatric species pair of *Phyllanthus* on Grand Terre Island, New Caledonia. Blue and green dots indicate populations of *P. cf. koniamboensis* and *P. poumensis*, respectively. They are associated with species-specific *Epicephala* species that are also closely related

richness among plant clades, it is tempting to speculate that coevolution has driven the remarkable diversification of *Phyllanthus* and *Epicephala* in New Caledonia.

The above view of cospeciation-driven diversification, however, is not well supported by broad phylogenetic patterns. Tests of cospeciation in figs, yuccas, and leafflowers have all found clear cases of host switch and species specificity breakdown, and there are many cases in which plant and pollinator distribution ranges do not match. However, inferring the mode of speciation from current distributions or association data can be misleading because range expansions and subsequent contact with other species may obscure the pattern at the time of speciation. This may explain why some of the clearest examples of cospeciation come from recently diverged species. Estimates of divergence dates in *Yucca* and *Tegeticula* suggest that the derived *Tegeticula* clade currently associated with capsular- and fleshy-fruited yuccas diversified much later than did their host plants, which is interpreted as an indication that the plants diverged first, followed by displacement of the original pollinators (perhaps *Parategeticula*) as *Tegeticula* radiated on these yuccas (Althoff et al. 2012). Similarly, across Southeast Asia, *Phyllanthus reticulatus* is pollinated by >10 *Epicephala* species that are well diverged genetically and distributed allopatrically. This suggests that, in this case, the plant *P. reticulatus* may have displaced other closely related plants that were once associated with these *Epicephala* species. In any case, the original process that generated species diversification may be cospeciation, with subsequent processes overriding the original pattern. Such a view is consistent with the results of phylogenetic studies showing that, although they are not strictly identical, there is a significant level of similarity between plant and pollinator phylogenies, suggesting that the association is characterized by background cospeciation.

Finally, intimate plant–pollinator associations may contribute to overall species richness, not only because they promote speciation but also because they allow recently diverged species to coexist stably in a given area. In many parts of the tropics, figs and leafflowers both flower throughout the year, and heterospecific crosses easily result in fertile hybrid seeds (e.g., Moe and Weiblen 2012). Thus, in the absence of species-specific pollinators, closely related fig or leafflower species would not coexist stably regardless of whether they have different habitat preferences. Assuming a situation in which all leafflower species have the same range size, overall species richness would be much higher when multiple leafflower species can co-occur in one area than when only one species can occur and all species are allopatrically distributed. Thus, specialized pollinators allow more species to be packed within an entire range of the clade, thereby contributing to species richness independent of the speciation process. This may explain, at least in part, why figs and leafflowers have attained diversity an order of magnitude greater than that in yuccas; up to 10 leafflower species or dozens of fig species can co-occur locally, whereas the number of yucca species that coexist in any location is usually only two.

13.7 Conclusions

Obligate pollination mutualisms are important model systems in the study of ecology and evolution. Over the years, significant findings have greatly enhanced our understanding of mutualisms and the coevolutionary process. Whereas we once considered obligate pollination mutualisms to be extremely rare associations resulting from evolutionary contingency, they now appear to be more widely occurring relationships that are logical outcomes of evolution. Rather than being stable associations locked by sanction mechanisms, they seem to be more evolutionarily flexible and prone to mutualism reversals. Moreover, we now view these mutualisms as dynamic associations mixed with host shift, partner displacement, and species-specificity breakdown, rather than strictly cospeciating associations. Undoubtedly, further study of obligate pollination mutualisms will bring major improvements to our understanding of ecology and evolution and further revisions to our views of these remarkable associations. The Phyllanthaceae–*Epicephala* association is distributed globally, and unexpected natural history findings are still being made every year. Obligate pollination mutualism in Phyllanthaceae thus holds promise as a prime model in the next generation of research on obligate pollination mutualisms.

Literature Cited

Aanen DK, Eggleton P, Lefèvre CR, Frøslev TG, Rosendahl S, Boomsma JJ (2002) The evolution of fungus-growing termites and their mutualistic fungal symbionts. Proc Natl Acad Sci U S A 99:14887–14892

Ackerman JD (1989) Geographic and seasonal variation in fragrance choices and preferences of male euglossine bees. Biotropica 21:340–347

Addicott JF (1996) Cheaters in yucca/moth mutualism. Nature 380:114–115

Addicott JF, Bao T (1999) Limiting the costs of mutualism: multiple modes of interaction between yuccas and yucca moths. Proc R Soc Lond B 266:197–202

Airy Shaw HK (1978) Notes on Malesian and other Asiatic Euphorbiaceae. Kew Bull 33:25–77

Als TD, Vila R, Kandul NP, Nash DR, Yen S-H, Hsu Y-F, Mignault AA, Boomsma JJ, Pierce NE (2004) The evolution of alternative parasitic life histories in large blue butterflies. Nature 432:386–390

Althoff DM, Groman JD, Segraves KA, Pellmyr O (2001) Phylogeographic structure of the bogus yucca moth *Prodoxus quinquepunctellus*: comparisons with coexisting pollinator moths. Mol Phylogenet Evol 21:117–127

Althoff DM, Segraves KA, Pellmyr O (2005) Community context of an obligate mutualism: pollinator and florivore effects on *Yucca filamentosa*. Ecology 86:905–913

Althoff DM, Segraves KA, Smith CI, Leebens-Mack J, Pellmyr O (2012) Geographic isolation trumps coevolution as a driver of yucca and yucca moth diversification. Mol Phylogenet Evol 62:898–906

Althoff DM, Xiao W, Sumoski S, Segraves KA (2013) Florivore impacts on plant reproductive success and pollinator mortality in an obligate pollination mutualism. Oecologia 173:1345–1354

Angiosperm Phylogeny Group (2003) An update of the Angiosperm Phylogeny Group classification for the orders and families of flowering plants: APG II. Bot J Linn Soc 141:300–436

Angiosperm Phylogeny Group (2009) An update of the Angiosperm Phylogeny Group classification for the orders and families of flowering plants: APG III. Bot J Linn Soc 161:105–121

Antal JS, Prasad M (1996) Some more leaf-impressions from the Himalayan foot-hills of Darjeeling District, West Bengal, India. Palaeobotanist 43:1–9

Askew RR (1980) The diversity of insect communities in leaf-mines and plant galls. J Anim Ecol 49:817–829

Axelrod R, Hamilton WD (1981) The evolution of cooperation. Science 211:1390–1396

Ayasse M, Schiestl FP, Paulus HF, Ibarra F, Francke W (2003) Pollinator attraction in a sexually deceptive orchid by means of unconventional chemicals. Proc R Soc Lond B 270:517–522

M. Kato, A. Kawakita (eds.), *Obligate Pollination Mutualism*, Ecological Research Monographs, DOI 10.1007/978-4-431-56532-1

Bailey R, Schönrogge K, Cook JM, Melika G, Csóka G, Thuróczy C, Stone GN (2009) Host niches and defensive extended phenotypes structure parasitoid wasp communities. PLoS Biol 7: e1000179

Baldwin BG, Wagner WL (2010) Hawaiian angiosperm radiations of North American origin. Ann Bot 105:849–879

Bao T, Addicott JF (1998) Cheating in mutualism: defection of *Yucca baccata* against its yucca moths. Ecol Lett 1:155–159

Barriault I, Barabé D, Cloutier L, Gibernau M (2010) Pollination ecology and reproductive success in Jack-in-the-pulpit (*Arisaema triphyllum*) in Québec (Canada). Plant Biol 12:161–171

Bashford R (2002) The insect fauna inhabiting *Uromycladium* (Uredinales) rust galls on silver wattle (*Acacia dealbata*) in Tasmania. Aust Entomol 29:81–95

Bawa KS (1990) Plant–pollinator interactions in tropical rain forests. Annu Rev Ecol Syst 21:399–422

Beardsley PM, Yen A, Olmstead RG (2003) AFLP phylogeny of *Mimulus* section *Erythranthe* and the evolution of hummingbird pollination. Evolution 57:1397–1410

Beattie AJ, Hughes L (2002) Ant–plant interactions. In: Herrera CM, Pellmyr O (eds) Plant–animal interactions: an evolutionary approach. Blackwell Science, Oxford, pp 211–235

Beattie AJ, Turnbull CL, Knox RB, Williams EG (1984) Ant inhibition of pollen function: a possible reason why ant pollination is rare. Am J Bot 71:421–426

Bennett GM, O'Grady PM (2013) Historical biogeography and ecological opportunity in the adaptive radiation of native Hawaiian leafhoppers (Cicadellidae: *Nesophrosyne*). J Biogeogr 40:1512–1523

Bentley BL (1977) Extrafloral nectaries and protection by pugnacious bodyguards. Annu Rev Ecol Syst 8:407–427

Beraldi-Campesi H (2013) Early life on land and the first terrestrial ecosystems. Ecol Process 2:1

Bidartondo MI (2005) The evolutionary ecology of myco-heterotrophy. New Phytol 167:335–352

Bidartondo MI, Bruns TD (2001) Extreme specificity in epiparasitic Monotropoideae (Ericaceae): widespread phylogenetic and geographical structure. Mol Ecol 10:2285–2295

Bidartondo MI, Read DJ, Trappe JM, Merckx V, Ligrone R, Duckett JG (2011) The dawn of symbiosis between plants and fungi. Biol Lett 7:574–577

Blanco MA, Barboza G (2005) Pseudocopulatory pollination in *Lepanthes* (Orchidaceae: Pleurothallidinae) by fungus gnats. Ann Bot 95:763–772

Bland KP (1980) Nigerian Gracillariidae. J Lepidop Soc 34:25–35

Bogler DJ, Neff JL, Simpson BB (1995) Multiple origins of the yucca–yucca moth association. Proc Natl Acad Sci U S A 92:6864–6867

Brantjes N (1976a) Riddles around pollination of *Melandrium album* (Mill) Garcke (Caryophyllaceae) during oviposition by *Hadena bicruris* Hufn (Noctuidae Lepidoptera), 1. Proc K Ned Akad Wet C 79:1–12

Brantjes N (1976b) Riddles around pollination of *Melandrium album* (Mill) Garcke (Caryophyllaceae) during oviposition by *Hadena bicruris* Hufn (Noctuidae Lepidoptera), 2. Proc K Ned Akad Wet C 79:127–141

Braun AF (1909) Notes on Chamber's species of Tineina. Entomol News 20:428–434

Bray JR, Curtis JT (1957) An ordination of the upland forest communities of southern Wisconsin. Ecol Monogr 27:325–349

Brian WM, Trautwein MD, Winkler IS, Barr NB, Kim JW, Lambkin C, Bertone MA, Cassel BK, Bayless KM, Heimberg AM, Wheeler BM, Peterson KJ, Pape T, Sinclair BJ, Skevington JH, Blagoderov V, Caravas J, Kutty SN, Schmidt-Ott U, Kampmeier GE, Thompson FC, Grimaldi DA, Beckenbach AT, Courtney GW, Friedrich M, Meier R, Yeates DK (2011) Episodic radiations in the fly tree of life. Proc Natl Acad Sci 108:5690–5695

Brito R, Gonçalves GL, Vargas HA, Moreira GR (2013) A new Brazilian *Passiflora* leafminer: *Spinivalva gaucha*, gen. n., sp. n. (Lepidoptera, Gracillariidae, Gracillariinae), the first gracillariid without a sap-feeding instar. Zookeys 17:1–26

Brooke M de L, Jones PJ (1995) The diet of the Henderson fruit dove *Ptilinopus insularis*. I. Field observations of fruit choice. Biol J Linn Soc 56:149–165

Brouat C, Garcia N, Andary C, McKey D (2001) Plant lock and key: pairwise coevolution of an exclusion filter in an ant–plant mutualism. Proc R Soc Lond B 268:2131–2141

Brower AVZ (1994) Rapid morphological radiation and convergence among races of the butterfly *Heliconius erato* inferred from patterns of mitochondrial DNA evolution. Proc Natl Acad Sci U S A 91:6491–6495

Brown JM, Pellmyr O, Thompson JN, Harrison RG (1994) Mitochondrial DNA phylogeny of the Prodoxidae (Lepidoptera: Incurvarioidea) indicates a rapid ecological diversification of the yucca moths. Ann Entomol Soc Am 87:795–802

Bshary R, Grutter AS (2006) Image scoring and cooperation in a cleaner fish mutualism. Nature 441:975–978

Buchmann SL (1987) The ecology of oil flowers and their bees. Annu Rev Ecol Syst 18:343–369

Bull JJ, Rice WR (1991) Distinguishing mechanisms for the evolution of co-operation. J Theor Biol 149:63–74

Burkill IH, Crosby CS (1901) The flora of Vavau, one of the Tonga islands. J Linn Soc Lond Bot 35:20–65

Busck A (1934) Microlepidoptera of Cuba. Entomol Am 13:151–217

Butaud J-F (2009) Les Gambier, un archipel à la végétation naturelle relictuelle et à la flore patrimoniale menacée. Bull Soc Étud Océan 315–316:99–140

Butaud J-F, Jacq F (2009) Atolls soulevés des Tuamotu: guide floristique. Direction de l'Environnement, Government of French Polynesia, Papeete. http://www.environnement.pf/spip.php?article125. Accessed 28 June 2015

Butaud J-F, Gérard J, Guibal D (2008) Guide des arbres de Polynésie française: bois et utilisations. Éditions Au vent des îles. Pirae, French Polynesia

Byers KJ, Bradshaw HD Jr, Riffell JA (2014) Three floral volatiles contribute to differential pollinator attraction in monkeyflowers (*Mimulus*). J Exp Biol 217:614–623

Cantley JT, Swenson NG, Markey A, Keeley SC (2014) Biogeographic insights on Pacific *Coprosma* (Rubiaceae) indicate two colonizations to the Hawaiian Islands. Bot J Linn Soc 174:412–424

Carlquist S (1974) Island biology. Columbia University Press, New York

Carroll SP, Bord C (1992) Host race radiation in the oapberry bug: natural history with the history. Evolution 46:1052–1069

Casquet J, Bourgeois YXC, Cruaud C, Gavory F, Gillespie RG, Thébaud C (2015) Community assembly on remote islands: a comparison of Hawaiian and Mascarene spiders. J Biogeogr 42:39–50

Chakrabarty T, Balakrishnan NP (2009) Transfer of Indian species of *Glochidion*, *Breynia*, and *Sauropus* to *Phyllanthus* (Phyllanthaceae): new combinations and new names. J Econ Taxon Bot 33:712–716

Chakrabarty T, Gangopadhyay M (1995) The genus *Glochidion* (Euphorbiaceae) in the Indian subcontinent. J Econ Taxon Bot 19:173–234

Chakrabarty T, Gangopadhyay M (1996) The genus *Breynia* (Euphorbiaceae) in the Indian subcontinent. J Econ Taxon Bot 20:501–512

Chantaranothai P (2005) Taxonomic notes on the genus *Phyllanthus* L. (Euphorbiaceae) in Thailand. Thai For Bull 33:16–20

Chen C, Song Q, Proffit M, Bessière JM, Li Z, Hossaert-McKey M (2009) Private channel: a single unusual compound assures specific pollinator attraction in *Ficus semicordata*. Funct Ecol 23:941–950

Chiba S (2004) Ecological and morphological patterns in communities of land snails of the genus *Mandarina* from the Bonin Islands. J Evol Biol 17:131–143

Cibois A, Beadell JS, Graves GR, Pasquet E, Slikas B, Sonsthagen SA, Thibault J-C, Fleischer RC (2011) Charting the course of reed-warblers across the Pacific islands. J Biogeogr 38:1963–1975

Clark JR, Wagner WL, Roalson EH (2009) Patterns of diversification and ancestral range reconstruction in the southeast Asian-Pacific angiosperm lineage *Cyrtandra* (Gesneriaceae). Mol Phylogenet Evol 53:982–994

Clarke JFG (1971) The Lepidoptera of Rapa Island. Smithson Contr Zool 56:1–282

Clarke JFG (1986) Pyralidae and microlepidoptera of the Marquesas Archipelago. Smithson Contr Zool 416:1–485

Clarke D, Whitney H, Gregory G, Robert D (2013) Detection and learning of floral electric fields by bumblebees. Science 340:66–69

Clayton DH, Bush SE, Goates BM, Johnson KP (2003) Host defense reinforces host–parasite cospeciation. Proc Natl Acad Sci U S A 100:15694–15699

Clouard V, Bonneville A (2005) Ages of seamounts, islands, and plateaus on the Pacific plate. In: Foulger GR, Natland JH, Presnall DC, Anderson DL (eds) Plates, plumes, and paradigms, vol 388. Geological Society of America Special Paper, Boulder, pp 71–90

Colley H (2009) Fiji, geology. In: Gillespie RG, Clague DA (eds) Encyclopedia of islands. University of California Press, Berkeley, pp 305–309

Compton SG (1990) A collapse of host specificity in some African fig wasps. S Afr J Sci 86:39–40

Compton SG, Holton KC, Rashbrook VK, van Noort S, Vincent SL, Ware AB (1991) Studies of *Ceratosolen galili*, a non-pollinating agaonid fig wasp. Biotropica 23:188–194

Condit IJ (1947) The fig. Chronica Botanica Co., Waltham

Conow C, Fielder D, Ovadia Y, Libeskind-Hadas R (2010) Jane: a new tool for the cophylogeny reconstruction problem. Algorithm Mol Biol 5:16

Cook JM, Rasplus JY (2003) Mutualists with attitude: coevolving fig wasps and figs. Trends Ecol Evol 18:241–248

Cornille A, Underhill JG, Cruaud A, Hossaert-McKey M, Johnson SD, Tolley KA, Kjellberg F, van Noort S, Proffit M (2012) Floral volatiles, pollinator sharing and diversification in the fig–wasp mutualism: insights from *Ficus natalensis*, and its two wasp pollinators (South Africa). Proc R Soc B 279:1731–1739

Coyne JA, Orr HA (2004) Speciation. Sinauer Associates, Sunderland

Crabb BA, Pellmyr O (2006) Impact of the third trophic level in an obligate mutualism: do yucca plants benefit from parasitoids of yucca moths? Int J Plant Sci 167:119–124

Cronk Q, Ojeda I (2008) Bird-pollinated flowers in an evolutionary and molecular context. J Exp Bot 59:715–727

Csotonyi and Addicott (2001)

Culley TM, Weller SG, Sakai AK (2002) The evolution of wind pollination in angiosperms. Trends Ecol Evol 17:361–369

Cunningham DD (1888) On the phenomena of fertilization in *Ficus roxburghii* Wall. Ann R Bot Gard Calcutta 1:11–51

Currie CR, Wong B, Stuart AE, Schultz TR, Rehner SA, Mueller UG, Sung GH, Spatafora JW, Straus NA (2003) Ancient tripartite coevolution in the attine ant–microbe symbiosis. Science 299:386–388

Dafni A, Kevan PG, Husband BC (2005) Practical pollination biology. Enviroquest, Cambridge

Darwin C (1859) On the origin of species by means of natural selection. John Murray, London

Davidson DW, McKey D (1993) The evolutionary ecology of symbiotic ant–plant relationships. J Hymenopt Res 2:13–83

Davis DR (1967) A revision of the moths of the subfamily Prodoxinae (Lepidoptera: Incurvariidae). US Nat Mus Bull 255:1–170

Davis CC, Webb CO, Wurdack KJ, Jaramillo CA, Donoghue MJ (2005) Explosive radiation of Malpighiales supports a mid-cretaceous origin of modern tropical rain forests. Am Nat 165: E36–E65

Davis CC, Latvis M, Nickrent DL, Wurdack KJ, Baum DA (2007) Floral gigantism in Rafflesiaceae. Science 315:1812

De Prins J, De Prins W (2005) World catalogue of insects, volume 6: Gracillariidae (Lepidoptera). Apollo Books, Svendborg

De Prins J, De Prins W (2016) Global taxonomic database of Gracillariidae (Lepidoptera). Available at: http://www.gracillariidae.net

Després L, Ibanez S, Hemborg ÅM, Godelle B (2007) Geographic and within population variation in the globeflower–globeflower fly interaction: the costs and benefits of rearing pollinators' larvae. Oecologia 151:240–250

Diakonoff A (1955) Microlepidoptera of New Guinea. Results of the third Archbold Expedition (American-Netherlands Indian Expedition 1938–1939). Part V. Verhandelingen der Koninklijke Nederlandse Akademie van Wetenschappen, Afdeling Natuurkunde 50:1–210

Diamond J (2002) Evolution, consequences and future of plant and animal domestication. Nature 418:700–707

Dobson HEM (2006) Relationship between floral fragrance composition and type of pollinator. In: Pichersky E, Dudareva N (eds) Biology of floral scent. CRC Press, Boca Raton, pp 147–198

Dötterl S, Füssel U, Jürgens A, Aas G (2005) 1, 4-Dimethoxybenzene, a floral scent compound in willows that attracts an oligolectic bee. J Chem Ecol 31:2993–2998

Dötterl S, Burkhardt D, Weißbecker B, Jürgens A, Schütz S, Mosandl A (2006) Linalool and lilac aldehyde/alcohol in flower scents: electrophysiological detection of lilac aldehyde stereoisomers by a moth. J Chromatogr A 1113:231–238

Drake DR, Motley TJ, Whistler WA, Imada CT (1996) Rain forest vegetation of 'Eua Island, Kingdom of Tonga. N Z J Bot 34:65–77

Dunn DW, Segar ST, Ridley J, Chan R, Crozier RH, Yu DW, Cook JM (2008) A role for parasites in stabilising the fig–pollinator mutualism. PLoS Biol 6:e59

Dupont YL, Hansen DM, Valido A, Olesen JM (2004) Impact of introduced honey bees on native pollination interactions of the endemic *Echium wildpretii* (Boraginaceae) on Tenerife, Canary Islands. Biol Conserv 118:301–311

Economo EP, Sarnat EM (2012) Revisiting the ants of Melanesia and the taxon cycle: historical and human-mediated invasions of a tropical archipelago. Am Nat 180:E1–E16

Edwards DP, Hassall M, Sutherland WJ, Yu DW (2006) Assembling a mutualism: Ant symbionts locate their host plants by detecting volatile compounds. Insect Soc 53:172–176

Ehrlich PR, Raven PH (1964) Butterflies and plants: a study in coevolution. Evolution 18:586–608

Engelmann G (1872) The flower of *Yucca* and its fertilization. Bull Torrey Bot Club 3:33

Espíndola A, Carstens BC, Alvarez N (2014) Comparative phylogeography of mutualists and the effect of the host on the genetic structure of its partners. Biol J Linn Soc 113:1021–1035

Etienne M, Michea G, Díaz E (1982) Flora, vegetacion y potencial pastoral de Isla de Pascua. Boletín Técnico 47:1–29. Facultad de Ciencias Agrarias, Veterinarias y Forestales, Universidad de Chile, Santiago

Farrell BD (1998) 'Inordinate fondness' explained: why are there so many beetles? Science 281:555–559

Farrell BD (2001) Evolutionary assembly of the milkweed fauna: cytochrome oxidase I and the age of *Tetraopes* beetles. Mol Phylogenet Evol 18:467–478

Federle W, Maschwitz U, Fiala B, Riederer M, Hölldobler B (1997) Slippery ant-plants and skillful climbers: selection and protection of specific ant partners by epicuticular wax blooms in *Macaranga* (Euphorbiaceae). Oecologia 112:217–224

Fenster C, Armbruster W, Wilson P, Dudash M, Thomson J (2004) Pollination syndromes and floral specialization. Annu Rev Ecol Evol Syst 35:375–403

Field KJ, Pressel S, Duckett JG, Rimington WR, Bidartondo MI (2015) Symbiotic options for the conquest of land. Trends Ecol Evol 30:477–486

Fleming TH, Holland JN (1998) The evolution of obligate pollination mutualisms: senita cactus and senita moth. Oecologia 114:368–375

Fleming TH, Tuttle MD, Horner MA (1996) Pollination biology and the relative importance of nocturnal and diurnal pollinators in three species of Sonoran Desert columnar cacti. Southwest Nat 41:257–269

Fleming TH, Sahley CT, Holland JN, Nason JD, Hamrick JL (2001) Sonoran Desert columnar cacti and the evolution of generalized pollination systems. Ecol Monogr 71:511–530

Fleming TH, Geiselman C, John Kress W (2009) The evolution of bat pollination: a phylogenetic perspective. Ann Bot 104:1017–1043

Flenley JR, King ASM, Teller IT, Prentice ME, Jackson J, Chew C (1991) The Late Quaternary vegetational and climatic history of Easter Island. J Quat Sci 6:85–115

Fletcher TB (1921) Life-histories of Indian Insects. Microlepidoptera. VI. Gracillariadae [sic]. Mem Dept Agr ic India Entomol Ser 6:1–217

Fletcher TB (1933) Life histories of Indian Microlepidoptera (Second Series). Cosmopterygidae to Neopseustidae. Scientific monograph, vol 4. Imperial Council of Agricultural Research, New Delhi, pp 1–85

Fletcher QE, Boutin S, Lane JE, LaMontagne JM, McAdam AG, Krebs CJ, Humphries MM (2010) The functional response of a hoarding seed predator to mast seeding. Ecology 91:2673–2683

Florence J (1997a) Flore de la Polynésie française, vol 1. Éditions de l'ORSTOM, Paris

Florence J (1997b) New species of *Alyxia* Banks ex R. Brown (Apocynaceae) and *Glochidion* J. R. & J. G. Forster (Euphorbiaceae) from the Pitcairn Islands (South East Pacific). Novon 7:27–31

Florence J, Lorence DH (1997) Introduction to the flora and vegetation of the Marquesas Islands. Allertonia 7:226–237

Florence J, Waldren S, Chepstow-Lusty AJ (1995) The flora of the Pitcairn Islands: a review. Biol J Linn Soc 56:79–119

Fordyce JA (2010) Host shifts and evolutionary radiations of butterflies. Proc R Soc B 277:3735–3743

Fosberg FR, Falanruw MVC, Sachet M-H (1975) Vascular flora of the Northern Marianas Islands. Smithson Contrib Bot 22:1–45

Franke S, Ibarra F, Schulz C, Twele R, Poldy J, Barrow R, Peakall R, Schiestl FP, Francke W (2009) The discovery of 2, 5-dialkylcyclohexan-1, 3-diones as a new class of natural products. Proc Natl Acad Sci U S A 106:8877–8882

Friberg M, Schwind C, Roark LC, Raguso RA, Thompson JN (2014) Floral scent contributes to interaction specificity in coevolving plants and their insect pollinators. J Chem Ecol 40:955–965

Galil J, Eisikowitch D (1969) Further studies on the pollination ecology of *Ficus sycomorus* L. (Hymenoptera, Chalcidoidea, Agaonidae). Tijdschr Entomol 112:1–13

Gaskett AC (2011) Orchid pollination by sexual deception: pollinator perspectives. Biol Rev Camb Philos Soc 86:33–75

Gaunt MW, Miles MA (2002) An insect molecular clock dates the origin of the insects and accords with palaeontological and biogeographical landmarks. Mol Biol Evol 19:748–761

Gemici Y (1993) A new species of *Flueggea* (Euphorbiaceae) from Anatolia. Edinb J Bot 50:75–77

Gilbert MG, Thulin M (2008) A new species of *Phyllanthus* (Euphorbiaceae) from Somalia. Nord J Bot 13:171–173

Gillespie RG (2004) Community assembly through adaptive radiation in Hawaiian spiders. Science 303:356–359

Godsoe W, Yoder JB, Smith CI, Pellmyr O (2008) Coevolution and divergence in the Joshua tree/yucca moth mutualism. Am Nat 171:816–823

Godsoe W, Strand E, Smith CI, Yoder JB, Esque TC, Pellmyr O (2009) Divergence in an obligate mutualism is not explained by divergent climatic requirements. New Phytol 183:589–599

Gómez JM, Zamora R (2006) Ecological factors that promote the evolution of generalization in pollination systems. In: Waser N, Ollerton J (eds) Plant–pollinator interactions: from generalization to specialization. University of Chicago Press, Chicago, pp 145–166

Goodman KR, Welter SC, Roderick GK (2012) Genetic divergence is decoupled from ecological diversification in the Hawaiian *Nesosydne* planthoppers. Evolution 66:2798–2814

Goor A (1965) The history of the fig in the Holy Land from ancient times to the present. Econ Bot 19:124–135

Goto R, Okamoto T, Kiers ET, Kawakita A, Kato M (2010) Selective flower abortion maintains moth cooperation in a newly discovered pollination mutualism. Ecol Lett 13:321–329

Govaerts R, Frodin RG, Radcliffe-Smith A (2000) World checklist and bibliography of Euphorbiaceae. Royal Botanic Gardens, Kew

Graham LK, Wilcox LW (2000) The origin of alternation of generations in land plants: a focus on matrotrophy and hexose transport. Philos Trans R Soc Lond B 355:757–767

Grandcolas P, Murienne J, Robillard T, Desutter-Grandcolas L, Jourdan H, Guilbert E, Deharveng L (2008) New Caledonia: a very old Darwinian island? Philos Trans R Soc B 363:3309–3317

Grangier J, Dejean A, Malé PJG, Solano PJ, Orivel J (2009) Mechanisms driving the specificity of a myrmecophyte–ant association. Biol J Linn Soc 97:90–97

Grant PR, Grant BR (2008) How and why species multiply: the adaptive radiation of Darwin's finches. Princeton University Press, Princeton

Gressitt JL (1956) Some distribution patterns of Pacific island faunae. Syst Zool 5:11–32

Grison-Pigé L, Bassière J, Hossaert-McKey M (2002) Specific attraction of fig-pollinating wasps: role of volatile compounds released by tropical figs. J Chem Ecol 28:283–295

Grison-Pigé L, Hossaert-McKey M, Greeff JM, Bassière J (2003) Fig volatile compounds—a first comparative study. Phytochemistry 61:61–71

Grodzinski U, Clayton NS (2010) Problems faced by food-caching corvids and the evolution of cognitive solutions. Philos Trans R Soc Lond B 365:977–987

Gruas-Cavagnetto CE, Köhler E (1992) Pollens fossils d'Euphorbiacées de l'Eocène français. Grana 31:291–304

Guimarães PR, Rico-Gray V, Oliveira PS, Izzo TJ, dos Reis SF, Thompson JN (2007) Interaction intimacy affects structure and coevolutionary dynamics in mutualistic networks. Curr Biol 17:1797–1803

Haine ER, Martin J, Cook JM (2006) Deep mtDNA divergences indicate cryptic species in a fig-pollinating wasp. BMC Evol Biol 6:83

Hansen DM, Olesen JM, Jones CG (2002) Trees, birds and bees in Mauritius: exploitative competition between introduced honey bees and endemic nectarivorous birds? J Biogeog 29:721–734

Hardoim PR, van Overbeek LS, Berg G, Pirttilä AM, Compant S, Campisano A, Döring M, Sessitsch A (2015) The hidden world within plants: ecological and evolutionary considerations for defining functioning of microbial endophytes. Microbiol Mol Biol Rev 79:293–320

Hartmann S, Nason JD, Bhattacharya D (2002) Phylogenetic origins of *Lophocereus* (Cactaceae) and the senita cactus–senita moth pollination mutualism. Am J Bot 89:1085–1092

Hata H, Kato M (2006) A novel obligate cultivation mutualism between damselfish and *Polysiphonia* algae. Biol Lett 2:593–596

Hata H, Watanabe K, Kato M (2010) Geographic variation in the damselfish-red alga cultivation mutualism in the Indo-West Pacific. BMC Evol Biol 10:185

Hayden WJ (1987) The identity of the genus *Neowawraea*. Brittonia 39:268–277

Heil M, McKey D (2003) Protective ant–plant interactions as model systems in ecological and evolutionary research. Annu Rev Ecol Evol Syst 34:425–453

Heil M, Rattke J, Boland W (2005) Postsecretory hydrolysis of nectar sucrose and specialization in ant/plant mutualism. Science 308:560–563

Heinrich B (1979) Bumblebee economics. Harvard University Press, Cambridge

Hembry DH, Balukjian B (2016) Molecular phylogeography of the Society Islands (Tahiti; South Pacific) reveals departures from hotspot archipelago models. J Biogeogr 43:1372–1387

Hembry DH (2013) Herbarium specimens reveal putative insect extinction on the deforested island of Mangareva (Gambier Archipelago, French Polynesia). Pac Sci 67:553–560

Hembry DH, Okamoto T, Gillespie RG (2012) Repeated colonization of remote islands by specialized mutualists. Biol Lett 8:258–261

Hembry DH, Kawakita A, Gurr NE, Schmaedick MA, Baldwin BG, Gillespie RG (2013a) Non-congruent colonizations and diversification in a coevolving pollination mutualism on oceanic islands. Proc R Soc B 280:20130361

Hembry DH, Okamoto T, McCormack G, Gillespie RG (2013b) Phytophagous insect community assembly through niche conservatism on oceanic islands. J Biogeogr 40:225–235

Hendrix SD (1980) An evolutionary and ecological perspective of the insect fauna of ferns. Am Nat 115:171–196

Heraty JM, Burks RA, Cruaud A, Gibson GAP, Liljeblad J, Munro J, Rasplus J-Y, Delvare G, Janšta P, Gumovsky A, Huber J, Woolley JB, Krogmann L, Heydon S, Polaszek A, Schmidt S, Darling DC, Gates MW, Mottern J, Murray E, Dal Molin A, Triapitsyn S, Baur H, Pinto JD, van Noort S, George J, Yoder M (2013) A phylogenetic analysis of the megadiverse Chalcidoidea (Hymenoptera). Cladistics 29:466–542

Herre EA (1989) Coevolution of reproductive characteristics in 12 species of New World figs and their pollinator wasps. Experientia 45:367–347

Herre EA, Machado CA, Bermingham E, Nason JD, Windsor DM, McCafferty S, Van Houten W, Bachmann K (1996) Molecular phylogenies of figs and their pollinator wasps. J Biogeogr 23:521–530

Herre EA, Knowlton N, Mueller UG, Rehner SA (1999) The evolution of mutualisms: exploring the paths between conflict and coorperation. Trends Ecol Evol 14:49–53

Herre EA, Jandér KC, Machado CA (2008) Evolutionary ecology of figs and their associates: recent progress and outstanding puzzles. Annu Rev Ecol Evol Syst 39:439–458

Hibbett DS, Gilbert LB, Donoghue MJ (2000) Evolutionary instability of ectomycorrhizal symbioses in basidiomycetes. Nature 407:506–508

Hoffmann P (2008) Revision of *Heterosavia*, stat. nov., with notes on *Gonatogyne* and *Savia* (Phyllanthaceae). Brittonia 60:136–166

Hoffmann P, McPherson G (2003) Transfer of Madagascan *Glochidion* to *Phyllanthus* (Euphorbiaceae s.l. or Phyllanthaceae). Novon 13:307–310

Hoffmann P, Kathriarachchi H, Wurdack KJ (2006) A phylogenetic classification of Phyllantheae (Malpighiales; Euphorbiaceae sensu lato). Kew Bull 61:37–53

Holland JN (2000) Ecological and evolutionary conditions for fruit abortion to regulate pollinating seed-eaters and increase plant reproduction. Theor Pop Biol 61:251–263

Holland BS, Cowie RH (2009) Land snail models in island biogeography: a tale of two snails. Am Malacol Bull 27:59–68

Holland JN, Fleming TH (1999) Mutualistic interactions between *Upiga virescens* (Pyralidae), a pollinating seed-consumer, and *Lophocereus schottii* (Cactaceae). Ecology 80:2074–2084

Holland JN, Fleming TH (2002) Co-pollinators and specialization in the pollinating seed-consumer mutualism between senita cacti and senita moths. Oecologia 133:534–540

Hölldobler B, Wilson EO (1990) The ants. Harvard University Press, Cambridge

Hollis S, Brummitt RK (1992) World geographic scheme for recording plant distributions. Hunt Institute for Botanical Documentation, Pittsburgh

Holm-Nielsen LB (1979) Comments on the distribution and evolution of the genus *Phyllanthus*. In: Larsen K, Holm-Nielsen LB (eds) Tropical Botany. Academic Press, London, pp 277–290

Hosokawa T (1935) Materials of the botanical research towards flora of Micronesia III. T Nat Hist Soc Formosa 25:17–39

Hosokawa T, Kikuchi Y, Nikoh N, Shimada M, Fukatsu T (2006) Strict host–symbiont cospeciation and reductive genome evolution in insect gut bacteria. PLoS Biol 4:1841–1851

Hossaert-McKey M, Gibernau M, Frey JE (1994) Chemosensory attraction of fig wasps to substances produced by receptive figs. Entomol Exp Appl 70:185–191

Howe HF, Smallwood J (1982) Ecology of seed dispersal. Annu Rev Ecol Syst 13:201–228

Hu BB, Wang SX, Zhang J, Li HH (2011) Taxonomy and biology of two seed-parasitic gracillariid moths (Lepidoptera, Gracillariidae), with description of one new species. ZooKeys 83:43–56

Humphreys CP, Franks PJ, Rees M, Bidartondo MI, Leake JR, Beerling DJ (2010) Mutualistic mycorrhiza-like symbiosis in the most ancient group of land plants. Nat Commun 1:103

Huth CJ, Pellmyr O (1999) Yucca moth oviposition and pollination behavior is affected by past flower visitors: evidence for a host-marking pheromone. Oecologia 119:593–599

Huth CJ, Pellmyr O (2000) Pollen-mediated selective abortion in yuccas and its consequences for the plant–pollinator mutualism. Ecology 81:1100–1107

Ibanez S, Gallet C, Dommanget F, Després L (2009) Plant chemical defense: a partner control mechanism stabilising plant–seed-eating pollinator mutualisms. BMC Evol Biol 9:261

Imada Y, Kato M (2016a) Bryophyte-feeding of *Litoleptis* (Diptera: Rhagionidae) with descriptions of new species from Japan. Zootaxa 4097:41–58

Imada Y, Kato M (2016b) Bryophyte-feeders in a basal brachyceran lineage (Diptera: Rhagionidae: Spaniinae): adult oviposition behavior and changes in the larval mouthpart morphology accompanied with the diet shifts. PLoS ONE 11:e0165808

Imada Y, Kawakita A, Kato M (2011) Allopatric distribution and diversification without niche shift in a bryophyte-feeding basal moth lineage (Lepidoptera: Micropterigidae). Proc R Soc B 176:3026–3033

Itioka T (2005) Diversity of anti-herbivore defenses in *Macaranga*. In: Roubik DW, Sakai S, Karim AAH (eds) Pollination ecology and the rain forest: Sarawak studies. Springer, New York

Itioka T, Nomura N, Inui Y, Itino T, Inoue T (2000) Difference in intensity of ant defense among three species of *Macaranga* myrmecophyte in a Southeast Asian dipterocarp forest. Biotropica 32:318–326

Jaeger N, Després L (1998) Obligate mutualism between *Trollius europaeus* and its seed-parasite pollinators *Chiastocheta* flies in the Alps. C R Acad Sci III 321:789–796

James TY, Kauff F, Schoch C, Matheny PB, Hofstetter V, Cox C, Celio G, The AFTOL Working Group (2006) Reconstructing the early evolution of the fungi using a six gene phylogeny. Nature 443:818–822

Jandér KC, Herre EA (2010) Host sanctions and pollinator cheating in the fig tree–fig wasp mutualism. Proc R Soc B 277:1481–1488

Jandér KC, Herre EA, Simms EL (2012) Precision of host sanctions in the fig tree–fig wasp mutualism: consequences for uncooperative symbionts. Ecol Lett 15:1362–1369

Janzen DH (1971) Euglossine bees as long-distance pollinators of tropical plants. Science 171:203–205

Janzen DH (1979) How to be a fig. Annu Rev Ecol Syst 10:13–51

Jersáková J, Johnson SD, Kindlmann P (2006) Mechanisms and evolution of deceptive pollination in orchids. Biol Rev 81:219–235

Jobb G (2011) TREEFINDER version of March 2011. Munich, Germany. Distributed by the author at http://www.treefinder.de/

Jousselin E, Hossaert-McKey M, Herre EA, Kjellberg F (2003) Why do fig wasps actively pollinate monoecious figs? Oecologia 134:381–387

Jousselin E, van Noort S, Rasplus JY, Greeff JM (2006) Patterns of diversification of Afrotropical Otiteselline fig wasps: phylogenetic study reveals a double radiation across host figs and conservatism of host association. J Evol Biol 19:253–266

Jousselin E, van Noort S, Berry V, Rasplus JY, Rønsted N, Erasmus JC, Greeff JM (2008) One fig to bind them all: host conservatism in a fig wasp community unravelled by cospeciation analyses among pollinating and nonpollinating fig wasps. Evolution 62:1777–1797

Jürgens A, Wee SÄ, Shuttleworth A, Johnson SD (2013) Chemical mimicry of insect oviposition sites: a global analysis of convergence in angiosperms. Ecol Lett 16:1157–1167

Kahn JG, Nickelsen C, Stevenson J, Porch N, Dotte-Sarout E, Christensen CC, May L, Athens JS, Kirch PV (2015) Mid- to late Holocene landscape change and anthropogenic transformations on Mo'orea, Society Islands: a multi-proxy approach. The Holocene 25:333–347

Kandul NP, Lukhtanov VA, Dantchenko AV, Coleman JW, Sekercioglu CH, Haig D, Pierce NE (2004) Phylogeny of *Agrodiaetus* Hübner 1822 (Lepidoptera: Lycaenidae) inferred from mtDNA sequences of COI and COII and nuclear sequences of EF1-a: karyotype diversification and species radiation. Syst Biol 53:278–298

Kathriarachchi H, Hoffmann P, Samuel R, Wurdack KJ, Chase MW (2005) Molecular phylogenetics of Phyllanthaceae inferred from five genes (plastid *atpB*, *matK*, 3'*ndhF*, *rbcL*, and nuclear *PHYC*). Mol Phylogenet Evol 36:112–134

Kathriarachchi H, Samuel R, Hoffmann P, Mlinarec J, Wurdack KJ, Ralimanana H, Stuessy TF, Chase MW (2006) Phylogenetics of the tribe Phyllantheae (Phyllanthaceae; Euphorbiaceae sensu lato) based on nrITS and plastid *matK* DNA sequence data. Am J Bot 93:637–655

Kato M (1988) Bumblebee visits to *Impatiens* spp.: Pattern and efficiency. Oecologia 76:364–370

Kato M (1996a) Plant–pollinator interactions in the understory of a lowland mixed dipterocarp forest in Sarawak. Am J Bot 83:732–743

Kato M (1996b) Effects of parasitoid community structure upon the population dynamics of the honeysuckle leafminer, *Chromatomyia suikazurae* (Diptera: Agromyzidae). Res Popul Ecol 38:27–40

Kato M (2002) First records of herbivory on Lycopodiaceae (Lycopsidales) by a dipteran (Pallopteridae) leaf/stem miner. Can Entomol 134:699–702

Kato M, Inoue T (1994) Origin of insect pollination. Nature 368:195

Kato M, Kawakita A (2004) Plant–pollinator interactions in New Caledonia influenced by introduced honey bees. Am J Bot 91:1814–1827

Kato M, Nagamasu H (1995) Dioecy in the endemic genus *Dendrocacalia* (Compositae) on the Bonin (Ogasawara) Islands. J Plant Res 108:443–450

Kato M, Shibata A, Yasui T, Nagamasu H (1999) Impact of introduced honeybees, *Apis mellifera*, upon native bee communities in the Bonin (Ogasawara) Islands. Res Popul Ecol 41:217–228

Kato M, Takimura A, Kawakita A (2003) An obligate pollination mutualism and reciprocal diversification in the tree genus *Glochidion* (Euphorbiaceae). Proc Natl Acad Sci U S A 100:5264–5267

Kato M, Kosaka Y, Kawakita A, Okuyama Y, Kobayashi C, Phimminith T, Thongphan D (2008) Plant–pollinator interactions in tropical monsoon forests in Southeast Asia. Am J Bot 95:1375–1394

Kawahara AY, Sohn J-C, De Prins J, Cho S (2010) Five species of Gracillariidae (Lepidoptera) new to Korea. Entomol Res 40:131–135

Kawahara AY, Plotkin D, Ohshima I, Lopez-Vaamonde C, Houlihan P, Breinholt JW, Kawakita A, Xiao L, Regier JC, Davis DR, Kumata T, Sohn J-C, De Prins J, Mitter C (2017) A molecular phylogeny and revised higher-level classification for the leaf-mining moth family Gracillariidae and its implications for larval host use evolution. Syst Entomol 42:60–81

Kawakita A (2010) Evolution of obligate pollination mutualism in the tribe Phyllantheae (Phyllanthaceae). Plant Spec Biol 25:3–19

Kawakita A (2012) Zettai sōfun-kyōsei wa ika ni umi o watatta ka – Komikansō-ka/Hanahosoga-zoku kyōseikei no tōsho seibutsuchiri (How did obligate pollination mutualisms cross the ocean? Island biogeography of the Phyllanthaceae-*Epicephala* pollination mutualism). Jpn J Ecol 62:317–327. [In Japanese]

Kawakita A, Kato M (2004a) Obligate pollination mutualism in *Breynia* (Phyllanthaceae): further documentation of pollination mutualism involving *Epicephala* moths (Gracillariidae). Am J Bot 91:1319–1325

Kawakita A, Kato M (2004b) Evolution of obligate pollination mutualism in New Caledonian *Phyllanthus* (Euphorbiaceae). Am J Bot 91:410–415

Kawakita A, Kato M (2006) Assessment of the diversity and species specificity of the mutualistic association between *Epicephala* moths and *Glochidion* trees. Mol Ecol 15:3567–3581

Kawakita A, Kato M (2009) Repeated independent evolution of obligate pollination mutualism in the Phyllantheae–*Epicephala* association. Proc R Soc B 276:417–426

Kawakita A, Kato M (2016) Revision of the Japanese species of *Epicephala* Meyrick with descriptions of seven new species (Lepidoptera, Gracillariidae). ZooKeys 568:87–118

Kawakita A, Takimura A, Terachi T, Sota T, Kato M (2004) Cospeciation analysis of an obligate pollination mutualism: Have *Glochidion* trees (Euphorbiaceae) and pollinating *Epicephala* moths (Gracillariidae) diversified in parallel? Evolution 58:2201–2214

Kawakita A, Okamoto T, Goto R, Kato M (2010) Mutualism favours higher host specificity than does antagonism in plant–herbivore interaction. Proc R Soc B 277:2765–2774

Kawakita A, Mochizuki K, Kato M (2015) Reversal of mutualism in a leafflower–leafflower moth association: the possible driving role of a third-party partner. Biol J Linn Soc 116:507–518
Kennett D, Anderson A, Prebble M, Conte E, Southon J (2006) Prehistoric human impacts on Rapa, French Polynesia. Antiquity 80:340–354
Kephart S, Reynolds RJ, Rutter MT, Fenster CB, Dudash MR (2006) Pollination and seed predation by moths on *Silene* and allied Caryophyllaceae: evaluating a model system to study the evolution of mutualisms. New Phytol 169:667–680
Kerdelhue C, Le Clainche I, Rasplus JY (1999) Molecular phylogeny of the *Ceratosolen* species pollinating *Ficus* of the subgenus *Sycomorus* sensu stricto: biogeographical history and origins of the species-specificity breakdown cases. Mol Phylogenet Evol 3:401–414
Kiers ET, Denison RF (2008) Sanctions, cooperation, and the stability of plant–rhizosphere mutualisms. Annu Rev Ecol Evol Syst 39:215–236
Kiers ET, Rousseau RA, West SA, Denison RF (2003) Host sanctions and the legume–rhizobium mutualism. Nature 425:78–81
Kiers ET, Duhamel M, Beesetty Y, Mensah JA, Franken O, Verbruggen E, Fellbaum CR, Kowalchuk GA, Hart MM, Bago A, Palmer TM, West SA, Vandenkoornhuyse P, Jansa J, Bucking H (2011) Reciprocal rewards stabilize cooperation in the mycorrhizal symbiosis. Science 333:880–882
Kiester AR, Lande R, Schemske DW (1984) Models of coevolution and speciation in plants and their pollinators. Am Nat 124:220–243
Kirch PV (2004) Environmental and ethnographic background. In: Conte E, Kirch PV (eds) Archaeological investigations in the Mangareva Islands (Gambier Archipelago), French Polynesia, vol 62. Archeological Research Facility, University of California, pp 16–32
Kjellberg F, Jousselin E, Bronstein JL, Patel A, Yokoyama J, Rasplus JY (2001) Pollination mode in fig wasps: the predictive power of correlated traits. Proc R Soc London, Ser B 268:1113–1121
Knudsen JT, Olesen JM (1993) Buzz-pollination and patterns in sexual traits in North European Pyrolaceae. Am J Bot 80:900–913
Knudsen JT, Tollsten L (2004) Trends in floral scent chemistry in pollination syndromes: floral scent composition in moth-pollinated taxa. Bot J Linn Soc 113:263–284
Knudsen JT, Eriksson R, Gershenzon J, Ståhl B (2006) Diversity and distribution of floral scent. Bot Rev 72:1–120
Kolle M, Lethbridge A, Kreysing M, Baumberg JJ, Aizenberg J, Vukusic P (2013) Bio-inspired band-gap tunable elastic optical multilayer fibers. Adv Mater 25:2239–2245
Koponen A (1990) Entomophily in the Splachnaceae. Bot J Linn Soc 104:115–127
Kristensen NP (1984) Studies on the morphology and systematics of primitive Lepidoptera (Insecta). Steenstrupia 10:141–191
Kumata T (1966) Descriptions of twenty new species of the genus *Caloptilia* Hübner from Japan including the Ryukyu Islands (Lepidoptera: Gracillariidae). Insecta Matsumurana 29:1–21
Kumata T (1982) A taxonomic revision of the *Gracillaria* group occurring in Japan (Lepidoptera: Gracillariidae). Insecta Matsumurana 26:1–186
Kumata T (1998) Japanese species of the subfamily Oecophyllembiinae Real et Balachowsky. (Lepidoptera: Gracillariidae), with descriptions of a new genus and eight new species. Insecta Matsumurana 54:77–131
Kumata T, Kuroko H, Ermolaev VP (1988a) Japanese species of the *Acrocercops*-group (Lepidoptera: Gracillariidae). Part II. Insecta Matsumurana 40:1–133
Kumata T, Kuroko H, Ermolaev VP (1988b) Japanese species of the *Acrocercops*-group (Lepidoptera: Gracillariidae). Part I. Insecta Matsumurana 38:1–111
Kuznetzov VI (1979) A review of the genera of Gracillariidae (Lepidoptera) of the Palaearctic fauna. Entomol Obozr 58:835–856
Labandeira CC (2010) The pollination of mid Mesozoic seed plants and the early history of long-proboscid insects. Ann Mo Bot Gard 97:469–513

Labouche AM, Bernasconi G (2013) Cost limitation through constrained oviposition site in a plant–pollinator/seed predator mutualism. Funct Ecol 27:509–521

Lack DL (1947) Darwin's finches. Cambridge University Press, Cambridge

Larson G, Piperno D, Allaby RG, Purugganan MD, Andersson L, Arroyo-Kalin M, Barton L, Climer VC, Denham T, Dobney K, Doust AN, Gepts P, Gilbert MTP, Gremillion K, Lucas L, Lukens L, Marshall FB, Olsen KM, Pires JC, Richerson PJ, Rubio de Casas R, Sanjur OI, Thomas MG, West-Eberhard MJ, Fuller D (2014) Current perspectives and the future of domestication studies. Proc Natl Acad Sci U S A 111:6139–6146

Lavin M, Herendeen PS, Wojciechowski MF (2005) Evolutionary rates analysis of Leguminosae implicates a rapid diversification of lineages during the Tertiary. Syst Biol 54:575–594

Lebrun J-P, Stork AL (2006) Tropical African flowering plants: ecology and distribution, vol 2: Euphorbiaceae–Dichapetalaceae. Conservatoire Botanique de Genève, Geneva

Legendre P, Desdevises Y, Bazin E (2002) A statistical test for host–parasite coevolution. Syst Biol 51:217–234

Lengyela S, Govec AD, Latimerd AM, Majerc JD, Dunna RR (2010) Convergent evolution of seed dispersal by ants, and phylogeny and biogeography in flowering plants: a global survey. Perspect Plant Ecol Evol Syst 12:43–55

Lenz LW (2007) Reassessment of *Yucca brevifolia* and recognition of *Y. jaegeriana* as a distinct species. Aliso 24:97–104

Lerner HR, Meyer M, James HF, Hofreiter M, Fleischer RC (2011) Multilocus resolution of phylogeny and timescale in the extant adaptive radiation of Hawaiian honeycreepers. Curr Biol 21:1838–1844

Li H, Yang X (2015) Three new species of *Epicephala* Meyrick (Lepidoptera, Gracillariidae) associated with *Phyllanthus microcarpus* (Benth.) (Phyllanthaceae). ZooKeys 484:71–81

Li H, Zhang Z (2016) Five species of the genus *Epicephala* Meyrick, 1880 (Lepidoptera: Gracillariidae) from China. Zootaxa 4084:391–405

Li B, Qiu H, Ma J, Zhu H, Gilbert MG, Esser H-J, Dressler S, Hoffmann P, Gillespie LJ, Vorontsova M, McPherson GD (2008) Euphorbiaceae. In: Wu ZY, Raven PH, Hong DY (eds) Flora of China, vol 11 (Oxalidaceae through Aceraceae). Science Press/Missouri Botanical Garden Press, Beijing/St. Louis

Li H, Wang Z, Hu B (2015) Four new species of *Epicehala* Meyrick, 1880 (Lepidoptera, Gracillariidae) associated with two species of *Glochidion* (Phyllanthaceae) from Hainan Island in China. ZooKeys 508:53–67

Lloyd DG, Webb C (1977) Secondary sex characters in plants. Bot Rev 43:177–216

Lopez-Vaamonde C, Rasplus JY, Weiblen GD, Cook JM (2001) Molecular phylogenies of fig wasps: partial cocladogenesis of pollinators and parasites. Mol Phylogenet Evol 21:55–71

Lopez-Vaamonde C, Dixon DJ, Cook JM, Rasplus JY (2002) Revision of the Australian species of *Pleistodontes* (Hymenoptera: Agaonidae) fig-pollinating wasps and their host–plant associations. Zool J Linnean Soc 136:637–683

Lopez-Vaamonde C, Wikström N, Labandeira C, Godfray HCJ, Goodman SJ, Cook JM (2006) Fossil-calibrated molecular phylogenies reveal that leafmining moths radiated millions of years after their host plants. J Evol Biol 19:1314–1326

Luo Z, Zhang D, Renner SS (2008) Why two kinds of stamens in buzz-pollinated flowers? Experimental support for Darwin's division-of-labour hypothesis. Funct Ecol 22:794–800

Luo S-X, Esser HJ, Zhang D, Renner SS (2011a) Nuclear ITS sequences help disentangle *Phyllanthus reticulatus* (Phyllanthaceae), an Asian species not occurring in Africa, but introduced to Jamaica. Syst Bot 36:99–104

Luo S, Li Y, Chen S, Zhang D, Renner SS (2011b) Gelechiidae moths are capable of chemically dissolving the pollen of their host plants: first documented sporopollenin breakdown by an animal. PLoS ONE 6:e19219

Lutzoni F, Pagel M, Reeb V (2001) Major fungal lineages are derived from lichen symbiotic ancestors. Nature 411:937–940

MacArthur RH, Wilson EO (1967) The theory of island biogeography. Princeton University Press, Princeton

Macbride JF (1951) Flora of Peru: Euphorbiaceae, spurge family. Field Museum of Natural History, Chicago

Machado CA, Jousselin E, Kjellberg F, Compton SG, Herre EA (2001) Phylogenetic relationships, historical biogeography and character evolution of fig-pollinating wasps. Proc R Soc B 268:685–694

Machado CA, Robbins N, Gilbert MTP, Herre EA (2005) Critical review of host specificity and its coevolutionary implications in the fig/fig-wasp mutualism. Proc Natl Acad Sci U S A 102:6558–6565

Mann D, Edwards J, Chase J, Beck W, Reanier R, Mass M, Finney B, Loret J (2008) Drought, vegetation change, and human history on Rapa Nui (Isla de Pascua, Easter Island). Quat Res 69:16–28

Martin-Ordas G, Haun D, Colmenares F, Call J (2010) Keeping track of time: evidence for episodic memory in great apes. Anim Cogn 13:331–340

Marussich WA, Machado CA (2007) Host-specificity and coevolution among pollinating and nonpollinating New World fig wasps. Mol Ecol 16:1925–1946

Maury RC, Legendre C, Chauvel C, Guille G, Blais S, Guillou H, Rossi P (2014) An atypical hotspot chain. In: Meyer J-Y, Claridge E (eds) Terrestrial biodiversity of the Austral Islands. French Polynesia. Muséum national d'histoire naturelle, Paris, pp 21–37

McClatchey W, Thaman R, Vodonaivalu S (2000) A preliminary checklist of the flora of Rotuma with Rotuman names. Pac Sci 54:345–363

McCormack G (2007) Cook Islands biodiversity database. Cook Islands Natural Heritage Trust, Rarotonga. http://cookislands.bishopmuseum.org (Version 2007.2). Accessed 29 June 2015

McKain MR, McNeal JR, Kellar PR, Eguiarte LE, Pires JC, Leebens-Mack J (2016) Timing of rapid diversification and convergent origins of active pollination within Agavoideae (Asparagaceae). Am J Bot 103:1717–1729

McNutt M, Menard HW (1978) Lithospheric flexure and uplifted atolls. J Geophys Res 83:1206–1212

Meier-Kolthoff JP, Auch AF, Huson DH, Göker M (2007) COPYCAT: cophylogenetic analysis tool. Bioinformatics 23:898–900

Meyer J-Y (2004) Threat of invasive alien plants to native flora and forest vegetation of Eastern Polynesia. Pac Sci 58:357–375

Meyer J-Y (2007a) Rapport de mission dans l'île de Mai'ao (Société) du 24 au 29 mai 2007: inventaire préliminaire de la flore et des différents types de végétation et observations sur la faune terrestre. Délégation à la Recherche, Government of French Polynesia, Papeete http://www.li-an.fr/jyves/Meyer_2007_Rapport_mission_Maiao.pdf. Accessed 29 June 2015

Meyer J-Y (2007b) Rapport de mission sur l'île d'Uvea (Wallis & Futuna) du 6 au 17 novembre 2007: inventaire préliminaire de la flore vasculaire secondaire. Délégation à la Recherche, Government of French Polynesia, Papeete http://www.li-an.fr/jyves/Meyer_2007_Rapport_Plantes_Introduites_Wallis.pdf. Accessed 29 June 2015

Meyer J-Y (2010) Montane cloud forests on remote islands of Oceania: the example of French Polynesia (south Pacific Ocean). In: Bruijnzeel LA, Scatena FN, Hamilton LS (eds) Tropical montane cloud forests: science for conservation and management. Cambridge University Press, Cambridge, pp 121–129

Meyer J-Y (2013) Rapport de mission sur l'atoll de Maria (archipel des Australes) du 05 au 07 mai 2013: inventaire floristique, observations ornithologiques et recommandations de gestion et de conservation. Délégation à la Recherche, Government of French Polynesia, Papeete. http://www.li-an.fr/jyves/Meyer_2013_Rapport_Mission_Atoll_Maria.pdf. Accessed 29 June 2015

Meyer J-Y, Fourdrigniez M, Taputuarai R (2009) Evaluation de la biodiversité terrestre de l'île de Me'eti'a (archipel de la Société). *Contribution à la Biodiversité de Polynésie française 18. Sites Naturels d'Intérêt Ecologique VII*. Délégation à la Recherche, Government of French Polynesia, Papeete. http://www.jymeyer.com/article-22996318.html. Accessed 29 June 2015

Meyer J-Y, Chevillotte H, Motley TJ (2014) Vascular flora, general traits, and main threats. In: Meyer J-Y, Claridge E (eds) Terrestrial biodiversity of the Austral Islands, French Polynesia. Muséum national d'histoire naturelle, Paris, pp 117–131

Meyrick E (1881) Descriptions of Australian micro-Lepidoptera. III Tineina Proc Linnean Soc NSW 5(132–182):204–271

Meyrick E (1908a) Descriptions of Indian Micro-Lepidoptera. VIII. J Bombay Nat Hist Soc 18:806–832

Meyrick E (1908b) Descriptions of African Micro-Lepidoptera. Proc Zool Soc London 47:716–756

Meyrick E (1910) Notes and descriptions of Indian Micro-Lepidoptera. Rec Indian Mus 5:217–232

Meyrick E (1918) Exotic Microlepidoptera. Exotic Microlepidoptera (Marlborough) 2:161–192

Meyrick E (1922) Exotic Microlepidoptera. Exotic Microlepidoptera (Marlborough) 2:545–576

Meyrick E (1927) Micro-Lepidoptera. Insects of Samoa 3:65–116. British Museum of Natural History, London

Meyrick E (1928) The micro-lepidoptera of the "St. George" expedition. T Ent Soc London 76:489–521

Meyrick E (1930) Exotic microlepidoptera. Exotic Microlepidoptera (Marlborough) 3:577–608

Meyrick E (1931) Exotic microlepidoptera. Exotic Microlepidoptera (Marlborough) 4:161–192

Meyrick E (1935) Exotic microlepidoptera. Exotic Microlepidoptera (Marlborough) 4:577–608

Meyrick E (1936) Exotic microlepidoptera. Exotic Microlepidoptera (Marlborough) 5:1–64

Michaloud G, Carriere S, Kobbi M (1996) Exceptions to the one:one relationship between African fig trees and their fig wasp pollinators: possible evolutionary scenarios. J Biogeogr 23:513–520

Michener CD (2007) The bees of the world. Johns Hopkins Univ Press, Baltimore

Mikheyev AS, Mueller UG, Abbot P (2006) Cryptic sex and many-to-one coevolution in the fungus-growing ant symbiosis. Proc Natl Acad Sci U S A 103:10702–10706

Mitter C, Farrell B, Wiegmann B (1988) The phylogenetic study of adaptive zones: has phytophagy promoted insect diversification? Am Nat 132:107–128

Moe AM, Weiblen GD (2012) Pollinator-mediated reproductive isolation among dioecious fig species (*Ficus*, Moraceae). Evolution 66:3710–3721

Molbo D, Machado CA, Sevenster JG, Keller L, Herre EA (2003) Cryptic species of fig-pollinating wasps: implications for the evolution of the fig–wasp mutualism, sex allocation, and precision of adaptation. Proc Natl Acad Sci U S A 100:5867–5872

Momose K, Yumoto T, Nagamitsu T, Kato M, Nagamasu M, Sakai S, Harrison RD, Itioka T, Inoue T (1998) Pollination biology in a lowland dipterocarp forest in Sarawak, Malaysia. I. Characteristics of the plant–pollinator community in a lowland dipterocarp forest. Am J Bot 85:1477–1501

Moran NA, Telang A (1998) The evolution of bacteriocyte-associated endosymbionts in insects. Bioscience 48:295–304

Morat P (1993) Our knowledge of the flora of New Caledonia: endemism and diversity in relation to vegetation types and substrates. Biodivers Lett 1:72–81

Morat P, Veillon J-M (1985) Contribution à la connaissance de la végétation et de la flore de Wallis & Futuna. Bull. Mus. Natn. Hist. Nat., Paris, 4e sér., 7, section B. Adansonia 3:259–329

Morishita M (1959) Measuring of the dispersion and analysis of the distribution patterns. Mem Fac Sci Kyushu Univ Series E 2:215–235

Mueller UG, Rehner SA, Schultz TR (1998) The evolution of agriculture in ants. Science 281:2034–2038

Mueller UG, Gerado NM, Aanen DK, Six DL, Schultz TR (2005) The evolution of agriculture in insects. Annu Rev Ecol Evol Syst 36:563–595

Nguyen NT (2007) Taxonomy of Euphorbiaceae in Vietnam. Vietnam National University Publishers, Hanoi

Nylander JAA (2004) MrModeltest v2. Program distributed by the author. Evolutionary Biology Centre, Uppsala University, Sweden

O'Connor TK, Humphrey PT, Lapoint RL, Whiteman NK, O'Grady PM (2014) Microbial interactions in the ecology and evolution of Hawaiian Drosophilidae. Front Microbiol 5:1–8

Oelschlägel B, Nuss M, Tschirnhaus M, Pätzold C, Neinhuis C, Dötterl S, Wanke S (2014) The betrayed thief—the extraordinary strategy of *Aristolochia rotunda* to deceive its pollinators. New Phytol 206:342–351

Ohshima I (2008) Host race formation in the leaf-mining moth *Acrocercops transecta* (Lepidoptera: Gracillariidae). Biol J Linn Soc 93:135–145

Okamoto T, Kawakita A, Kato M (2007) Interspecific variation of floral scent composition in *Glochidion* and its association with host-specific pollinating seed parasite (*Epicephala*). J Chem Ecol 33:1065–1081

Okamoto T, Kawakita A, Kato M (2008) Floral adaptations to nocturnal moth pollination in *Diplomorpha* (Thymelaeaceae). Plant Species Biol 23:192–201

Okamoto T, Kawakita A, Goto R, Svensson GP, Kato M (2013) Active pollination favours sexual dimorphism in floral scent. Proc R Soc B 280:20132280

Okamoto T, Okuyama Y, Goto R, Tokoro M, Kato M (2015) Parallel chemical switches underlying pollinator isolation in Asian *Mitella*. J Evol Biol 28:590–600

Okuyama Y, Kato M, Murakami M (2004) Pollination by fungus gnats in four species of the genus *Mitella* (Saxifragaceae). Bot J Linn Soc 144:449–460

Oliver TH, Leather SR, Cook JM (2009) Tolerance traits and the stability of mutualism. Oikos 118:346–352

Ollerton J, Winfree R, Tarrant S (2011) How many flowering plants are pollinated by animals? Oikos 120:321–326

Ort BS, Bantay RM, Pantoja NA, O'Grady PM (2012) Fungal diversity associated with Hawaiian *Drosophila* host plants. PLoS ONE 7:e40550

Patel A, Hossaert-McKey M (2000) Components of reproductive success in two dioecious fig species, *Ficus exasperata* and *Ficus hispida*. Ecology 81:2850–2866

Patt JM, Rhoades DF, Corkill JA (1988) Analysis of the floral fragrance of *Platanthera stricta*. Phytochemistry 27:91–95

Peakall R, Whitehead MR (2014) Floral odour chemistry defines species boundaries and underpins strong reproductive isolation in sexually deceptive orchids. Ann Bot 113:341–355

Pellissier L, Litsios G, Fiedler K, Pottier J, Dubuis A, Pradervand JN, Salamin N, Guisan A (2012) Loss of interactions with ants under cold climate in a regional myrmecophilous butterfly fauna. J Biogeogr 39:1782–1790

Pellmyr O (1989) The cost of mutualism: interactions between *Trollius europaeus* and its pollinating parasites. Oecologia 78:53–59

Pellmyr O (1992) The phylogeny of a mutualism: evolution and coadaptation between *Trollius* and its seed-parasitic pollinators. Biol J Linn Soc 47:337–365

Pellmyr O (1997) Pollinating seed eaters: why is active pollination so rare? Ecology 78:1655–1660

Pellmyr O (1999) A systematic revision of the yucca moths in the *Tegeticula yuccasella* complex north of Mexico. Syst Entomol 24:243–271

Pellmyr O (2002) Pollination by animals. In: Herrera CM, Pellmyr O (eds) Plant–animal interactions. Blackwell Publishing, Oxford, pp 157–184

Pellmyr O (2003) Yuccas, yucca moths, and coevolution: a review. Ann Mo Bot Gard 90:35–55

Pellmyr O (2012) Pollen load in an active pollinator, the yucca moth *Tegeticula yuccasella* (Prodoxidae). J Lepid Soc 66:50–51

Pellmyr O, Augenstein EJ (1997) Pollination biology of *Hesperaloe parviflora* (Agavaceae). Southwest Nat 42:182–187

Pellmyr O, Huth CJ (1994) Evolutionary stability of mutualism between yuccas and yucca moths. Nature 372:257–260

Pellmyr O, Krenn HW (2002) Origin of a complex key innovation in an obligate insect–plant mutualism. Proc Natl Acad Sci U S A 99:5498–5502

Pellmyr O, Leebens-Mack J (1999) Forty million years of mutualism: evidence for Eocene origin of the yucca–yucca moth association. Proc Natl Acad Sci U S A 96:9178–9183

Pellmyr O, Seraves KA (2003) Pollinator divergence within an obligate mutualism: two yucca moth species (Lepidoptera; Prodoxidae: *Tegeticula*) on the Joshua tree (*Yucca brevifolia*; Agavaceae). Ann Entomol Soc Am 96:716–722

Pellmyr O, Thompson JN (1992) Multiple occurrences of mutualism in the yucca moth lineage. Proc Natl Acad Sci U S A 89:2927–2929

Pellmyr O, Thompson JN, Brown JM, Harrison RG (1996a) Evolution of pollination and mutualism in the yucca moth lineage. Am Nat 148:827–847

Pellmyr O, Leebens-Mack J, Huth CJ (1996b) Non-mutualistic yucca moths and their evolutionary consequences. Nature 380:155–156

Pellmyr O, Balcázar-Lara M, Althoff DM, Segraves KA, Leebens-Mack J (2006) Phylogeny and life history evolution of *Prodoxus* yucca moths (Lepidoptera: Prodoxidae). Syst Entomol 31:1–20

Pellmyr O, Segraves KA, Althoff DM, Balcázar-Lara M, Leebens-Mack J (2007) The phylogeny of yuccas. Mol Phylogenet Evol 43:493–501

Peng YQ, Duan ZB, Yang DR, Rasplus JY (2008) Co-occurrence of two *Eupristina* species on *Ficus altissima* in Xishuangbanna, SW China. Symbiosis 45:9–14

Pettersson MW (1991a) Flower herbivory and seed predation in *Silene vulgaris* (Caryophyllaceae). Effects of pollination and phenology. Holarct Ecol 14:45–50

Pettersson MW (1991b) Pollination by a guild of fluctuating moth populations: option for unspecialization in *Silene vulgaris*. J Ecol 79:591–604

Poulsen M, Boomsma JJ (2005) Mutualistic fungi control crop diversity in fungus-growing ants. Science 307:741–744

Powell JA (1992) Interrelationships of yuccas and yucca moths. Trends Ecol Evol 7:10–15

Prasad M (1994) Siwalik (Middle Miocene) leaf impressions from the foothills of the Himalayas, India. Tertiary Res 15:53–90

Prebble M (2014) The paleobotanical record of Rapa: implications for the phytogeography. In: Meyer J-Y, Claridge E (eds) Terrestrial biodiversity of the Austral Islands, French Polynesia. Muséum national d'histoire naturelle, Paris, pp 149–169

Prebble M, Wilmshurst JM (2009) Detecting the initial impact of humans and introduced species on island environments in Remote Oceania using palaeoecology. Biol Invasions 11:1529–1556

Price P (1980) Evolutionary biology of parasites. Princeton University Press, Princeton

Proctor M, Yeo P, Lack A (1996) The natural history of pollination. Timber Press, Portland

Proffit M, Schatz B, Bessière JM, Chen C, Soler C, Hossaert-McKey M (2008) Signalling receptivity: comparison of the emission of volatile compounds by figs of *Ficus hispida* before, during and after the phase of receptivity to pollinators. Symbiosis 45:15–24

Proffit M, Chen C, Soler C, Bessière JM, Schatz B, Hossaert-McKey M (2009) Can chemical signals, responsible for mutualistic partner encounter, promote the specific exploitation of nursery pollination mutualisms?—the case of figs and fig wasps. Entomol Exp Appl 131:46–57

Pruesapan K, Telford IRH, Bruhl JJ, Draisma SGA, van Welzen PC (2008) Delimitation of *Sauropus* (Phyllanthaceae) based on plastid *matK* and nuclear ribosomal ITS DNA sequence data. Ann Bot 102:1007–1018

Pruesapan K, Telford IRH, Bruhl JJ, van Welzen PC (2012) Phylogeny and proposed circumscription of *'Breynia'*, *'Sauropus'* and *'Synostemon'* (Phyllanthaceae), based on chloroplast and nuclear DNA sequences. Aust Syst Bot 25:313–330

Quek SP, Davies SJ, Itino T, Pierce NE (2004) Codiversification in an ant–plant mutualism: stem texture and the evolution of host use in *Crematogaster* (Formicidae: Myrmicinae) inhabitants of *Macaranga* (Euphorbiaceae). Evolution 58:554–570

Quek SP, Davies SJ, Ashton PS, Itino T, Pierce NE (2007) The geography of diversification in mutualistic ants: a gene's-eye view into the Neogene history of Sundaland rain forests. Mol Ecol 16:2045–2062

Radcliffe-Smith A (1996) Notes on African Euphorbiaceae XXX: *Phyllanthus* (V) & c. Kew Bull 51:301–331

Radcliffe-Smith A (2001) Genera Euphorbiacearum. Royal Botanic Gardens, Kew

Raguso RA, Willis MA (2005) Synergy between visual and olfactory cues in nectar feeding by wild hawkmoths, *Manduca sexta*. Anim Behav 69:407–418

Raguso RA, Henzel C, Buchmann SL, Nabhan GP (2003) Trumpet flowers of the Sonoran Desert: floral biology of *Peniocereus* cacti and sacred *Datura*. Int J Plant Sci 164:877–892

Ralimanana H, Hoffmann P (2011) Taxonomic revision of *Phyllanthus* (Phyllanthaceae) in Madagascar and the Comoro Islands I: synopsis and subgenera *Isocladus*, *Betsileani*, *Kirganelia* and *Tenellanthus*. Kew Bull 66:331–335

Ralimanana H, Hoffmann P (2014) Taxonomic revision of *Phyllanthus* L. (Phyllanthaceae) in Madagascar and the Comoro Islands II: subgenera *Anisonemoides* (Jean F. Brunel) Ralim. & Petra Hoffm., stat. nov. and *Menarda* (Müll. Arg.) Ralim. & Petra Hoffm., stat. nov. Adansonia 36:265–301

Ralimanana H, Hoffmann P, Rajeriarison C (2013) Taxonomic revision of *Phyllanthus* (Phyllanthaceae) in Madagascar and the Comoro Islands III: subgenera *Swartziani*, *Afroswartziani* and *Emblica*. Kew Bull 68:535–558

Ramírez W (1969) Fig wasps: mechanisms of pollen transfer. Science 163:580–581

Ramírez WB (1974) Coevolution of *Ficus* and Agaonidae. Ann Mo Bot Gard 61:770–780

Redecker D, Kodner R, Graham LE (2000) Glomalean fungi from the Ordovician. Science 289:1920–1921

Regier JC, Shultz JW, Zwick A, Hussey A, Ball B, Wetzer R, Martin JW, Cunningham CW (2010) Arthropod relationships revealed by phylogenomic analysis of nuclear protein-coding sequences. Nature 463:1079–1083

Ren R, Labandeira CC, Santiago-Blay JA, Rasnitsyn A, Shih CK, Bashkuev A, Amelia M, Logan V, Hotton CL, Dilcher D (2009) A probable pollination mode before angiosperms: Eurasian, long-proboscid scorpionflies. Science 326:840–847

Richter KS, Weis AE (1995) Differential abortion in yucca. Nature 376:557–558

Riley CV (1872) The fertilization of the yucca plant by *Pronuba yuccasella*. Can Entomol 4:182

Riley CV (1873) On a new genus in the lepidopterous family Tineidae, with remarks on the fertilization of yucca. Trans Acad Sci St Louis 3:55–64

Riley CV (1880) The true and bogus yucca moth, with remarks on the pollination of *Yucca*. Am Entomol 3:141–145

Riley CV (1881) Further notes on the pollination of *Yucca* and on *Pronuba* and *Prodoxus*. Proc Am Assoc Adv Sci 29:617–639

Riley CV (1892) The yucca moth and yucca pollination. Annu Rep MO Bot Gard 3:99–158

Robinson GS, Tuck KR, Shaffer M (1994) A field guide to the smaller moths of South-East Asia. Malaysian Nature Society, Kuala Lumpur, pp 1–309

Robinson GS, Ackery PR, Kitching IJ, Beccaloni GW, Hernández LM (2001) Hostplants of the moth and butterfly caterpillars of the Oriental Region. The Natural History Museum, London, pp 1–744

Rodda P (1994) Geology of Fiji. In: Stevenson AJ, Herzer RH, Balance PF (eds) Geology and submarine resources of the Tonga-Lau-Fiji region. SOPAC Technical Bulletin 8:131–151. South Pacific Applied Geoscience Commission (SOPAC) Secretariat, Suva, Fiji

Roderick GK (1997) Herbivorous insects and the Hawaiian silversword alliance: coevolution or cospeciation? Pac Sci 51:440–449

Romero GQ, Benson WW (2005) Biotic interactions of mites, plants and leaf domatia. Curr Opin Plant Biol 8:436–440

Ronquist F, Huelsenbeck JP (2003) MrBayes 3: Bayesian phylogenetic inference under mixed models. Bioinformatics 19:1572–1574

Rønsted N, Weiblen GD, Cook JM, Salamin N, Machado CA, Savolainen V (2005) 60 million years of co-divergence in the fig–wasp symbiosis. Proc R Soc B 272:2593–2599

Roubik DW (1992) Ecology and natural history of tropical bees. Cambridge University Press, Cambridge

Sachs JL, Simms EL (2006) Pathways to mutualism breakdown. Trends Ecol Evol 21:585–592

Sachs JL, Mueller UG, Wilcox TP, Bull JJ (2004) The evolution of cooperation. Q Rev Biol 79:135–160

Sakagami SF, Maeta Y (1989) Compatibility and incompatibility of solitary life with eusociality in two normally solitary bees *Ceratina japonica* and *Ceratina okinawana* (Hymenoptera: Apoidea) with notes on the incipient phase of eusociality. Jpn J Ent 57:417–439

Sakai S (2002) A review of brood-site pollination mutualism: plants providing breeding sites for their pollinators. J Plant Res 115:161–168

Sakai S, Kato M, Nagamasu H (2000) *Artocarpus* (Moraceae)–gall midge pollination mutualism mediated by a male-flower parasitic fungus. Am J Bot 87:440–445

Sakai S, Harrison RD, Momose K, Kuraji K, Nagamasu H, Yasunari T, Chong L, Nakashizuka T (2006) Irregular droughts trigger mass flowering in aseasonal tropical forests in Asia. Am J Bot 93:1134–1139

Sanderson MJ, Donoghue MJ (1996) Reconstructing shifts in diversification rates on phylogenetic trees. Trends Ecol Evol 11:15–20

Sawamura M, Kawakita A, Kato M (2009) Fern–spore-feeder interaction in temperate forests in Japan: sporing phenology and spore-feeding insect community. Am J Bot 96:594–604

Schiestl F, Peakall R (2005) Two orchids attract different pollinators with the same floral odour compound: ecological and evolutionary implications. Funct Ecol 19:674–680

Schiestl FP, Ayasse M, Paulus HF, Löfstedt C, Hansson BS, Ibarra F, Francke W (1999) Orchid pollination by sexual swindle. Nature 399:421

Schluter D (2000) The ecology of adaptive radiation. Oxford University Press, New York

Schmid M (1991) *Phyllanthus*. In: Morat P, Mackee HS (eds) Flore de la Nouvelle-Calédonie et Dépendances, vol 17. Muséum national d'histoire naturalle, Paris, pp 31–320

Segraves KA (2003) Understanding stability in mutualisms: can extrinsic factors balance the yucca–yucca moth interaction? Ecology 84:2943–2951

Segraves KA (2008) Florivores limit cost of mutualism in the yucca–yucca moth association. Ecology 89:3215–3221

Shapiro JM, Addicott JF (2003) Regulation of moth–yucca mutualisms: mortality of eggs in oviposition-induced 'damage zones'. Ecol Lett 6:440–447

Shuttleworth A, Johnson SD (2009) The importance of scent and nectar filters in a specialized wasp-pollination system. Funct Ecol 23:931–940

Simon L, Bousquet J, Levesque C, Lalonde M (1993) Origin and diversification of endomycorrhizal fungi and coincidence with vascular land plants. Nature 363:67–69

Simon R, Holderied MW, Koch CU, von Helversen O (2011) Floral acoustics: conspicuous echoes of a dish-shaped leaf attract bat pollinators. Science 333:631–633

Singer RB (2002) The pollination biology of *Sauroglossum elatum* Lindl. (Orchidaceae: Spiranthinae): moth-pollination and protandry in neotropical Spiranthinae. Bot J Linn Soc 138:9–16

Smith AC (1981) Flora Vitiensis nova (A new flora of Fiji), vol 2. National Tropical Botanical Garden, Lawai

Smith CI, Pellmyr O, Althoff DM, Balcazar-Lara M, Leebens-Mack JH, Segraves KA (2008a) Pattern and timing of diversification in *Yucca* (Agavaceae): specialized pollination does not escalate rates of diversification. Proc R Soc B 275:249–258

Smith CI, Godsoe WKW, Tank S, Yoder JB, Pellmyr O (2008b) Distinguishing coevolution from covicariance in an obligate pollination mutualism: asynchronous divergence in Joshua tree and its pollinators. Evolution 62:2676–2687

Smith CI, Yoder JB, Godsoe W, Pellmyr O (2009) Host specificity and reproductive success of yucca moths (*Tegeticula* spp., Lepidoptera: Prodoxidae) mirror patterns of gene flow between host plant varieties of Joshua tree (*Yucca brevifolia*: Agavaceae). Mol Ecol 18:5218–5229

Soler C, Proffit M, Chen C, Hossaert-McKey M (2010) Private channels in plant–pollinator mutualisms. Plant Signal Behav 5:893–895

Soler CC, Proffit M, Bessiere JM, Hossaert-McKey M, Schatz B (2012) Evidence for intersexual chemical mimicry in a dioecious plant. Ecol Lett 15:978–985

Song Q, Yang D, Zhang G, Yang C (2001) Volatiles from *Ficus hispida* and their attractiveness to fig wasps. J Chem Ecol 27:1929–1942

Song B, Chen G, Stöcklin J, Peng DL, Niu Y, Li ZM, Sun H (2014) A new pollinating seed-consuming mutualism between *Rheum nobile* and a fly fungus gnat, *Bradysia* sp., involving pollinator attraction by a specific floral compound. New Phytol 203:1109–1018

Song B, Stöcklin J, Gao Y-Q, Peng D-L, Song M-S, Sun H (2016) Oviposition by mutualistic seed-consuming pollinators reduces fruit abortion in a recently discovered pollination mutualism. Sci Rep 6:29886

Sorensen AE (1986) Seed dispersal by adhesion. Annu Rev Ecol Syst 17:443–463

Sota T, Hayashi M (2007) Comparative historical biogeography of *Plateumaris* leaf beetles (Coleoptera: Chrysomelidae) in Japan: interplay between fossil and molecular data. J Biogeogr 34:977–993

St. John H (1977) The flora of Niuatoputapu Island, Tonga. Phytologia 36:374–390

St. John H, Smith AC (1971) The vascular plants of the Horne and Wallis Islands. Pac Sci 25:313–348

Stainton HT (1856) Descriptions of three species of Indian Micro-Lepidoptera. Trans Entomol Soc Lond NS 3:301–304

Stainton HT (1859) Descriptions of twenty-five species of Indian Micro-Lepidoptera. Trans Entomol Soc Lond NS 5:111–126

Steadman DW (2006) Extinction and biogeography of tropical Pacific birds. Chicago University Press, Chicago

Steinebrunner F, Twele R, Francke W, Leuchtmann A, Schiestl FP (2008) Role of odour compounds in the attraction of gamete vectors in endophytic *Epichloë* fungi. New Phytol 178:401–411

Stensmyr MC, Urru I, Collu I, Celander M, Hansson BS, Angioy AM (2002) Rotting smell of dead-horse arum florets. Nature 420:625–626

Stoddart DR (1992) Biogeography of the tropical Pacific. Pac Sci 46:276–293

Stone BC (1970) The flora of Guam: a manual for the identification of the vascular plants of the island. Micronesica 6:1–659

Stone GN, Schonrogge K (2003) The adaptive significance of insect gall morphology. Trends Ecol Evol 18:512–522

Stout JC, Goulson D, Allen JA (1998) Repellent scent-marking of flowers by a guild of foraging bumblebees (*Bombus* spp.) Behav Ecol Sociobiol 43:317–326

Strauss SY, Zangerl AR (2002) Plant–insect interactions in terrestrial ecosystems. In: Herrera CM, Pellmyr O (eds) Plant–animal interactions. Blackwell Publishing, Oxford, pp 77–106

Su Z-H, Iino H, Nakamura K, Serrato A, Oyama K (2008) Breakdown of the one-to-one rule in Mexican fig–wasp associations inferred by molecular phylogenetic analysis. Symbiosis 45:73–82

Suchan T, Beauverd M, Trim N, Alvarez N (2015) Asymmetrical nature of the *Trollius–Chiastocheta* interaction: insights into the evolution of nursery pollination systems. Ecol Evol 5:4766–4777

Suetsugu K, Kawakita A, Kato M (2015) Avian seed dispersal in a mycoheterotrophic orchid *Cyrtosia septentrionalis*. Nat Plants 1:15052

Sugisima K, Kumata T, Tominaga S (2005) Discovery of *Acrocercops tricuneatella* (Gracillariidae, Gracillariinae) from Ryukyu, southern Japan, and its appropriate generic placement. Trans Lepid Soc Jap 56:257–265

Sugiura S, Yamazaki K (2005) Moth pollination of *Metaplexis japonica* (Apocynaceae): pollinaria transfer on the tip of the proboscis. J Plant Res 118:257–262

Sun G, Ji Q, Dilcher DL, Zheng S, Nixon KC, Wang X (2002) Archaefructaceae, a new basal angiosperm family. Science 296:899–904

Sun X-J, Xiao J-H, Cook JM, Feng G, Huang D-W (2011) Comparisons of host mitochondrial, nuclear and endosymbiont bacterial genes reveal cryptic fig wasp species and the effects of *Wolbachia* on host mtDNA evolution and diversity. BMC Evol Biol 11:86

Svensson GP, Okamoto T, Kawakita A, Goto R, Kato M (2010) Chemical ecology of obligate pollination mutualisms: testing the 'private channel' hypothesis in the *Breynia–Epicephala* association. New Phytol 186:995–1004

Sykes WR (1970) Contributions to the flora of Niue. Bull N Z Dept Scient Ind Res 200:1–321

Sykes WR, Campbell EO (1977) Kermadec Islands flora: an annotated check list. Bull N Z Dep Scient Ind Res 219:1–216

Sykes WR, West CJ (1996) New records and other information on the vascular flora of the Kermadec Islands. New Zeal J Bot 34:447–462

Takeuchi W (1999) New plants from Crater Mt., Papua New Guinea, and an annotated checklist of the species. Sida 18:941–986

Takeuchi W (2003) Two new species from the Bomberai Peninsula of Indonesian Papua, New Guinea. Harv Pap Bot 7:131–136

Telford IRH, Naaykens J (2015) *Synostemon hamersleyensis* (Phyllanthaceae), a new species endemic to the Pilbara, Western Australia. Nuytsia 25:31–37

Terry I, Walter GH, Moore C, Roemer R, Hull C (2007) Odor-mediated push-pull pollination in cycads. Science 318:70

The Plant List (2013) Version 1.1. Available at: http://www.theplantlist.org/

Thien LB, Bernhardt P, Gibbs GW, Pellmyr O, Bergström G, Groth I, McPherson G (1985) The pollination of *Zygogynum* (Winteraceae) by a moth, *Sabatinca* (Micropterigidae): an ancient association? Science 227:540–543

Thien LB, Sage TL, Jaffré T, Bernhardt P, Pontieri V, Weston PH, Malloch D, Azuma H, Graham SW, McPherson MA, Rai HS, Sage RF, Dupre J-L (2003) The population structure and floral biology of *Amborella trichopoda* (Amborellaceae). Ann Mo Bot Gard 90:466–490

Thin NN (2007) Taxonomy of Euphorbiaceae in Vietnam. Vietnam National University Publishers, Hanoi

Thompson JN (1994) The coevolutionary process. The University of Chicago Press, Chicago

Thompson JN (2005) The geographic mosaic of coevolution. The University of Chicago Press, Chicago

Thompson JN, Cunningham BM (2002) Geographic structure and dynamics of coevolutionary selection. Nature 417:735–738

Thompson JN, Pellmyr O (1992) Mutualism with pollinating seed parasites amid co-pollinators: constraints on specialization. Ecology 73:1780–1791

Thompson JN, Laine A-L, Thompson JF (2010) Retention of mutualism in a geographic diverging interaction. Ecol Lett 13:1368–1377

Thompson JN, Schwind C, Guimarães PR, Friberg M (2013) Divergence through multitrait evolution in coevolving interactions. Proc Natl Acad Sci U S A 110:11487–11492

Thulin M (2003) *Phyllanthus xylorrhizus* (Phyllanthaceae), a new species from Somalia. Nord J Bot 23:385–387

Tollsten L, Bergström LG (1993) Fragrance chemotypes of *Platanthera* (Orchidaceae)—the result of adaptation to pollinating moths? Nord J Bot 13:607–613

Trueblood DD, Gallagher ED, Gould DM (1994) Three stages of seasonal succession on the Savin Hill Cove mudflat, Boston Harbor. Limnol Oceanogr 39:1440–1454

Tsai HF, Liu JS, Staben C, Christensen MJ, Latch GC, Siegel MR, Schardl CL (1994) Evolutionary diversification of fungal endophytes of tall fescue grass by hybridization with *Epichloë* species. Proc Natl Acad Sci U S A 91:2542–2546

Turner AJ (1894) Descriptions of Micro-Lepidoptera from Moreton Bay. Trans Proc R Soc S Aust 18:120–138

Turner AJ (1896) Descriptions of Micro-Lepidoptera from Queensland. Trans Proc R Soc S Aust 20:1–34

Turner AJ (1900) New Micro-Lepidoptera, mostly from Queensland. Trans Proc R Soc S Aust 24:6–23

Turner AJ (1913) Studies in Australian Microlepidoptera. Proc Linnean Soc NSW 38:174–228

Turner AJ (1940) A revision of the Australian Gracilariidae [sic] (Lepidoptera). Trans Proc R Soc S Aust 64:50–69

Turner AJ (1947) Contributions to our knowledge of Australian Microlepidoptera. Proc R Soc Queensl 57:65–74

Ueda S, Quek SP, Itioka T, Inamori K, Sato Y, Murase K, Itino T (2008) An ancient tripartite symbiosis of plants, ants and scale insects. Proc R Soc B 275:2319–2326

Uhe G (1974) The composition of the plant communities inhabiting the recent volcanic ejecta of Niuafo'ou, Tonga. Trop Ecol 15:126–139

van Noort S, Compton SG (1996) Convergent evolution of agaonine and sycoecine (Agaonidae, Chalcidoidea) head shape in response to the constraints of host fig morphology. J Biogeogr 23:415–424

van Oudtshoorn KVR, van Rooyen MW (2013) Dispersal biology of desert plants. Springer, Berlin

van Welzen PC (2003) Revision of the Malesian and Thai species of *Sauropus* (Euphorbiaceae: Phyllanthoideae). Blumea 48:319–391

van Welzen PC, Chayamarit K (2015) Flora of Thailand: Euphorbiaceae. Naturalis Biodiversity Center, Leiden; Forest Herbarium, National Park, Wildlife and Plant Conservation Department, Bangkok. www.nationaalherbariumnl/thaieuph. Accessed 20 May 2015

van Welzen PC, Pruesapan K, Telford IRH, Esser H-J, Bruhl JJ (2014) Phylogenetic reconstruction prompts taxonomic changes in *Sauropus*, *Synostemon* and *Breynia* (Phyllanthaceae tribe Phyllantheae). Blumea 59:77–94

Vander Wall SB (2010) How plants manipulate the scatter-hoarding behaviour of seed-dispersing animals. Philos Trans R Soc Lond B 365:989–997

Vargas HA, Landry B (2005) A new genus and species of Gracillariidae (Lepidoptera) feeding on flowers of *Acacia macracantha* Willd. (Mimosaceae) in Chile. Acta Entomol Chil 29:47–57

Vári L (1961) South African Lepidoptera. volume 1: Lithocolletidae. Transvaal Museum, Pretoria

Visser AA, Ros VID, de Beer ZW, Debets AJM, Hartog E, Kuyper TW, Laessøe T, Slippers B, Aanen DK (2009) Levels of specificity of *Xylaria* species associated with fungus-growing termites: a phylogenetic approach. Mol Ecol 18:553–567

von Helversen D, von Helversen O (1999) Acoustic guide in bat-pollinated flower. Nature 398:759–760

Wagner WL, Lorence DH (2011) A nomenclator of Pacific oceanic island *Phyllanthus* (Phyllanthaceae), including *Glochidion*. PhytoKeys 4:67–94

Wagner WL, Herbst DR, Sohmer SH (1990) Manual of the flowering plants of Hawai'i, vol 1. Bishop Museum Press/University of Hawaii Press, Honolulu

Wagner WL, Herbst DR, Tornabene MW, Weitzman A, Lorence DH (2012) Flora of Micronesia. Smithsonian Institution/National Tropical Botanical Garden, Washington/Lāwa'i. http://botany.si.edu/pacificislandbiodiversity/micronesia/index.htm. Accessed 19 May 2015

Waldren S, Florence J, Chepstow-Lusty AJ (1995) A comparison of the vegetation communities from the islands of the Pitcairn Group. Biol J Linn Soc 56:121–144

Wang RW, Dunn DW, Sun BF (2014) Discriminative host sanctions in a fig–wasp mutualism. Ecol 95(5):1384–1393

Wang G, Cannon CH, Chen J (2016) Pollinator sharing and gene flow among closely related sympatric dioecious fig taxa. Proc R Soc B 283:20152963

Waser NM (2006) Specialization and generalization in plant–pollinator interactions: a historical perspective. In: Waser NM, Ollerton J (eds) Plant–pollinator interactions: from specialization to generalization. The University of Chicago Press, Chicago, pp 3–17

Webster GL (1956) A monographic study of the West Indian species of *Phyllanthus*. J Arnold Arbor 37:91–122, 217–268, 340–359

Webster GL (1957) A monographic study of the West Indian species of *Phyllanthus*. J Arnold Arbor 38:51–80, 170–198, 295–373

Webster GL (1958) A monographic study of the West Indian species of *Phyllanthus*. J Arnold Arbor 39(49–100):111–212

Webster GL (1970) A revision of *Phyllanthus* (Euphorbiaceae) in the continental United States. Brittonia 22:44–76

Webster GL (1984) A revision of *Flueggea* (Euphorbiaceae). Allertonia 3:259–312

Webster GL (1986) A revision of *Phyllanthus* (Euphorbiaceae) in Eastern Melanesia. Pac Sci 40:88–105

Webster GL (1994) Synopsis of the genera and suprageneric taxa of Euphorbiaceae. Ann Mo Bot Gard 81:33–144

Webster GL (2002) Three new sections and a new subgenus of *Phyllanthus*. Novon 12:290–298

Webster GL (2003) A synopsis of *Phyllanthus* section *Nothoclema* (Euphorbiaceae). Lundellia 6:19–36

Webster GL (2007) Taxonomic and nomenclatural changes in American Euphorbiaceae sensu lato. Contrib Univ Mich Herb 25:235–239

Webster GL, Carpenter KJ (2002) Pollen morphology and phylogenetic relationships in neotropical *Phyllanthus* (Euphorbiaceae). Bot J Linn Soc 138:325–338

Webster GL, Carpenter KJ (2008) Pollen morphology and systematics of palaeotropical *Phyllanthus* and related genera of Phyllanthinae (Euphorbiaceae). Bot J Linn Soc 157:591–608

Webster GL, Miller KI (1963) The genus *Reverchonia* (Euphorbiaceae). Rhodora 65:193–207

Weiblen GD (2000) Phylogenetic relationships of functionally dioecious *Ficus* (Moraceae) based on ribosomal DNA sequences and morphology. Am J Bot 87:1342–1357

Weiblen GD (2001) Phylogenetic relationships of fig wasps pollinating functionally dioecious *Ficus* based on mitochondrial DNA sequences and morphology. Syst Biol 50:243–267

Weiblen GD (2002) How to be a fig wasp. Annu Rev Entomol 47:299–330

Weiblen GD (2004) Correlated evolution in fig pollination. Syst Biol 53:128–139

Weiblen GD, Bush GL (2002) Speciation in fig pollinators and parasites. Mol Ecol 11:1573–1578

Weisler MI (1995) Henderson Island prehistory: colonization and extinction on a remote Polynesian island. Biol J Linn Soc 56:377–404

Weng J-K, Chapple C (2010) The origin and evolution of lignin biosynthesis. New Phytol 187:273–285

Westerbergh A (2004) An interaction between a specialized seed predator moth and its dioecious host plant shifting from parasitism to mutualism. Oikos 105:564–574

Weston PH, Crisp MD (1996) Trans-Pacific biogeographic patterns in the Proteaceae. In: Keast A, Miller SE (eds) The origin and evolution of Pacific island biotas, New Guinea to Eastern Polynesia: patterns and processes. SPB Academic Publishing bv, Amsterdam, pp 215–232

Whistler WA (2002) The Samoan rainforest: a guide to the vegetation of the Samoan archipelago. Isle Botanica, Honolulu

Whistler WA (2004) Rainforest trees of Samoa: a guide to the common lowland and foothill forest trees of the Samoan archipelago. Isle Botanica, Honolulu

Wiebes JT (1979) Co-evolution of figs and their insect pollinators. Annu Rev Ecol Syst 10:1–12

Willson MF, Ågren J (1989) Differential floral rewards and pollination by deceit in unisexual flowers. Oikos 55:23–29

Wilson RD, Addicott JF (1998) Regulation of mutualism between yuccas and yucca moths: is oviposition behavior responsive to selective abscission of flowers? Oikos 81:109–118

Wink M (1997) Compartmetation of secondary metabolites and xenobiotics in plant vacuoles. Adv Bot Res 25:141–169

Wise KAJ (1962) *Parectopa leucocyma* (Meyrick) (Lepidoptera: Gracillariidae) rediscovered as a leaf-miner of kauri (*Agathis australis* Salisb.) Trans R Soc NZ Zool 1:373–375

Yokoyama J (2003) Cospeciation of figs and fig-wasps: a case study of endemic species pairs in the Ogasawara Islands. Popul Ecol 45:249–256

Yumoto T, Itino T, Nagamasu H (1997) Pollination of hemiparasites (Loranthaceae) by spider hunters (Nectariniidae) in the canopy of a Bornean tropical rainforest. Selbyana 18:51–60

Zeng L, Zhang Q, Sun R, Kong H, Zhang N, Ma H (2014) Resolution of deep angiosperm phylogeny using conserved nuclear genes and estimates of early divergence times. Nat Commun 5:4956

Zhang J, Wang SX, Li HH, Hu BB, Yang XF, Wang ZB (2012a) Diffuse coevolution between two *Epicephala* species (Gracillariidae) and two *Breynia* species (Phyllanthaceae). PLoS ONE 7: e41657

Zhang J, Hu B, Wang S, Li H (2012b) Six new species of *Epicephala* Meyrick, 1880 (Lepidoptera: Gracillariidae) associated with Phyllanthaceae plants. Zootaxa 3275:43–54

Zhao J-B, Peng Y-Q, Quinnell RJ, Compton SG, Yang D-R (2014) A switch from mutualist to exploiter is reflected in smaller egg loads and increased larval mortalities in a 'cheater' fig wasp. Acta Oecol 57:51–57

Index

M. Kato, A. Kawakita (eds.), *Obligate Pollination Mutualism*, Ecological Research Monographs, DOI 10.1007/978-4-431-56532-1

H

MIX
Papier aus verantwortungsvollen Quellen
Paper from responsible sources
FSC® C105338

If you have any concerns about our products,
you can contact us on
ProductSafety@springernature.com

In case Publisher is established outside the EU,
the EU authorized representative is:
Springer Nature Customer Service Center GmbH
Europaplatz 3, 69115 Heidelberg, Germany

Printed by Libri Plureos GmbH
in Hamburg, Germany